Jay Gould
Product Marketing Manager
Embedded Solutions Marketing

13813 280th Ave NE
Duvall, WA 98019
Telephone: 425-844-9905
E-Mail: jay.gould@xilinx.com

Practical
FPGA Programming
in C

Prentice Hall Modern Semiconductor Design Series

James R. Armstrong and F. Gail Gray
VHDL Design Representation and Synthesis

Jayaram Bhasker
A VHDL Primer, Third Edition

Mark D. Birnbaum
Essential Electronic Design Automation (EDA)

Eric Bogatin
Signal Integrity: Simplified

Douglas Brooks
Signal Integrity Issues and Printed Circuit Board Design

Alfred Crouch
Design-for-Test for Digital IC's and Embedded Core Systems

Tom Granberg
Handbook of Digital Techniques for High-Speed Design

Howard Johnson and Martin Graham
High-Speed Digital Design: A Handbook of Black Magic

Howard Johnson and Martin Graham
High-Speed Signal Propagation: Advanced Black Magic

William K. Lam
Hardware Design Verification: Simulation and Formal Method-Based Approaches

Farzad Nekoogar and Faranak Nekoogar
From ASICs to SOCs: A Practical Approach

Samir Palnitkar
Design Verification with **e**

David Pellerin and Scott Thibault
Practical FPGA Programming in C

Christopher T. Robertson
Printed Circuit Board Designer's Reference: Basics

Chris Rowen
Engineering the Complex SOC

Wayne Wolf
FPGA-Based System Design

Wayne Wolf
Modern VLSI Design: System-on-Chip Design, Third Edition

Brian Young
Digital Signal Integrity: Modeling and Simulation with Interconnects and Packages

Practical
FPGA Programming
in C

David Pellerin
Scott Thibault

Prentice Hall Professional Technical Reference

Upper Saddle River, NJ • Boston• Indianapolis • San Francisco
New York • Toronto • Montreal • London • Munich • Paris • Madrid
Capetown • Sydney • Tokyo • Singapore • Mexico City

PRENTICE
HALL
PTR

The publisher offers excellent discounts on this book when ordered in quantity for bulk purchases or special sales, which may include electronic versions and/or custom covers and content particular to your business, training goals, marketing focus, and branding interests. For more information, please contact:

U. S. Corporate and Government Sales
(800) 382-3419
corpsales@pearsontechgroup.com

For sales outside the U. S., please contact:

International Sales
international@pearsoned.com

Visit us on the Web: www.phptr.com

Library of Congress Catalog Number: 2005920042

ISBN 0-13-154318-0

Text printed in the United States on recycled paper at Courier in Westford, Massachusetts.

First printing, April 2005

For Satomi and Lisa,
our patient and talented wives.

Contents

Foreword by Clive "Max" Maxfield   **xiii**

   Why is this book of interest to the hardware folks?   xiii
   And what about the software guys and gals?   xv
   So what's the catch?   xvi

Preface   **xvii**

   C Language for FPGA-Based Hardware Design?   xviii
   Compelling Platforms for Software Acceleration   xix
   The Power to Experiment   xxi
   How This Book Is Organized   xxi
   Where This Book Came From   xxii

Acknowledgments   **xxiii**

CHAPTER 1
The FPGA as a Computing Platform   **1**

   1.1   A Quick Introduction to FPGAs   2
   1.2   FPGA-Based Programmable Hardware Platforms   4
   1.3   Increasing Performance While Lowering Costs   6
   1.4   The Role of Tools   8
   1.5   The FPGA as an Embedded Software Platform   10
   1.6   The Importance of a Programming Abstraction   12
   1.7   When Is C Language Appropriate for FPGA Design?   13
   1.8   How to Use This Book   15

CHAPTER 2
A Brief History of Programmable Platforms   **17**

2.1   The Origins of Programmable Logic   18
2.2   Reprogrammability, HDLs, and the Rise of the FPGA   23
2.3   Systems on a Programmable Chip   25
2.4   FPGAs for Parallel Computing   27
2.5   Summary   29

CHAPTER 3
A Programming Model for FPGA-Based Applications   **31**

3.1   Parallel Processing Models   32
3.2   FPGAs as Parallel Computing Machines   35
3.3   Programming for Parallelism   38
3.4   Communicating Process Programming Models   39
3.5   The Impulse C Programming Model   41
3.6   Summary   43

CHAPTER 4
An Introduction to Impulse C   **45**

4.1   The Motivation Behind Impulse C   47
4.2   The Impulse C Programming Model   48
4.3   A Minimal Impulse C Program   50
4.4   Processes, Streams, Signals, and Memory   58
4.5   Impulse C Signed and Unsigned Datatypes   59
4.6   Understanding Processes   59
4.7   Understanding Streams   63
4.8   Using Output Streams   66
4.9   Using Input Streams   67
4.10 Avoiding Stream Deadlocks   69
4.11 Creating and Using Signals   73
4.12 Understanding Registers   74
4.13 Using Shared Memories   76
4.14 Memory and Stream Performance Considerations   81
4.15 Summary   86

CHAPTER 5
Describing a FIR Filter   **87**

5.1   Design Overview   87
5.2   The FIR Filter Hardware Process   88
5.3   The Software Test Bench   90
5.4   Desktop Simulation   97
5.5   Application Monitoring   98
5.6   Summary   100

CHAPTER 6
Generating FPGA Hardware   103

6.1   The Hardware Generation Flow   104
6.2   Understanding the Generated Structure   108
6.3   Stream and Signal Interfaces   112
6.4   Using HDL Simulation to Understand Stream Protocols   116
6.5   Debugging the Generated Hardware   119
6.6   Hardware Generation Notes   125
6.7   Making Efficient Use of the Optimizers   127
6.8   Language Constraints for Hardware Processes   129
6.9   Summary   131

CHAPTER 7
Increasing Statement-Level Parallelism   133

7.1   A Model of FPGA Computation   133
7.2   C Language Semantics and Parallelism   135
7.3   Exploiting Instruction-Level Parallelism   135
7.4   Limiting Instruction Stages   139
7.5   Unrolling Loops   141
7.6   Pipelining Explained   142
7.7   Summary   145

CHAPTER 8
Porting a Legacy Application to Impulse C   147

8.1   The Triple-DES Algorithm   148
8.2   Converting the Algorithm to a Streaming Model   150
8.3   Performing Software Simulation   155
8.4   Compiling to Hardware   156
8.5   Preliminary Hardware Analysis   159
8.6   Summary   160

CHAPTER 9
Creating an Embedded Test Bench   163

9.1   A Mixed Hardware and Software Approach   164
9.2   The Embedded Processor as a Test Generator   165
9.3   The Role of Hardware Simulators   168
9.4   Testing the Triple-DES Algorithm in Hardware   168
9.5   Software Stream Macro Interfaces   174
9.6   Building the Test System   175
9.7   Summary   194

CHAPTER 10
Optimizing C for FPGA Performance   **195**

   10.1 Rethinking an Algorithm for Performance   196
   10.2 Refinement 1: Reducing Size by Introducing a Loop   199
   10.3 Refinement 2: Array Splitting   199
   10.4 Refinement 3: Improving Streaming Performance   201
   10.5 Refinement 4: Loop Unrolling   203
   10.6 Refinement 5: Pipelining the Main Loop   204
   10.7 Summary   208

CHAPTER 11
Describing System-Level Parallelism   **209**

   11.1 Design Overview   210
   11.2 Performing Desktop Simulation   213
   11.3 Refinement 1: Creating Parallel 8-Bit Filters   214
   11.4 Refinement 2: Creating a System-Level Pipeline   219
   11.5 Moving the Application to Hardware   231
   11.6 Summary   256

CHAPTER 12
Combining Impulse C with an Embedded Operating System   **257**

   12.1 The uClinux Operating System   257
   12.2 A uClinux Demonstration Project   259
   12.3 Summary   277

CHAPTER 13
Mandelbrot Image Generation   **279**

   13.1 Design Overview   280
   13.2 Expressing the Algorithm in C   282
   13.3 Creating a Fixed-Point Equivalent   285
   13.4 Creating a Streaming Version   286
   13.5 Parallelizing the Algorithm   290
   13.6 Future Refinements   297
   13.7 Summary   299

CHAPTER 14
The Future of FPGA Computing   **301**

   14.1 The FPGA as a High-Performance Computer   302
   14.2 The Future of FPGA Computing   305
   14.3 Summary   307

APPENDIX A
Getting the Most Out of Embedded FPGA Processors   **309**

A.1   FPGA Embedded Processor Overview   310
A.2   Peripherals and Memory Controllers   312
A.3   Increasing Processor Performance   313
A.4   Optimization Techniques That Are Not FPGA-Specific   314
A.5   FPGA-Specific Optimization Techniques   319
A.6   Summary   322

APPENDIX B
Creating a Custom Stream Interface   **325**

B.1   Application Overview   326
B.2   The DS92LV16 Serial Link for Data Streaming   327
B.3   Stream Interface State Machine Description   329
B.4   Data Transmission   331
B.5   Summary   332

APPENDIX C
Impulse C Function Reference   **341**

APPENDIX D
Triple-DES Source Listings   **375**

APPENDIX E
Image Filter Listings   **405**

APPENDIX F
Selected References   **417**

Index   **419**

Foreword

How cool is this? A "programming" book written for both software developers and hardware design engineers.

Where I came from, the software folk and the hardware folk got along pretty much okay, but they sometimes seemed to be from different planets. The hardware engineers would roll up their sleeves and work for months without a break, while the software programmers would sit back and relax, or play ping-pong, or whatever software programmers do, until the hardware was stable and software development could begin in earnest. At this point the hardware team would be off planning their vacations, repairing their marriages or trying to remember where they lived. All to no avail, because the software team would soon be coming after them for some perceived deficiency or new feature request, and the hardware design would quickly become a hardware redesign.

It's amazing that the software and hardware teams I worked with didn't break into full-on hand-to-hand combat. I suppose that in hardware/software codesign, as in life, "good fences make good neighbors."

Why is this book of interest to the hardware folks?

As "I R A Hardware Engineer" (maybe you figured that out already), let's start by considering things from this side of the fence. Electronic design automation (EDA) refers to the software tools used to design electronic components and systems. In the early days of digital logic design, we captured our circuits as gate-level schematics.

This approach began to run out of steam in the mid- to late-1980s, but we were saved by the introduction of sophisticated hardware description languages (HDLs). These allowed us to describe the functionality of a design at a reasonably high level of abstraction called the register transfer level (RTL). In turn, these descriptions allowed alternative design scenarios to be more quickly and easily represented and simulated. And, of course, the real boost to productivity came in the form of logic synthesis technology, which could automatically compile these high-level descriptions into equivalent gate-level netlists.

Another consideration is the way in which a design is physically realized. Initially, the only options were to use lots of off-the-shelf "jelly-bean" devices and simple programmable logic devices, or to create your own application-specific device. Then, in 1984, a new component appeared on the scene: the field-programmable gate array (FPGA). The first incarnations of these devices were relatively simple and contained only a few thousand equivalent gates, but my, how things have changed. Today's offerings can boast tens of millions of equivalent gates coupled with large numbers of embedded functions like RAM blocks and multipliers.

And now we are running into a design productivity crunch again. Although languages like Verilog and VHDL are great for precisely describing a design's functional intent, they are somewhat cumbersome when it comes to representing today's multimillion-gate systems. Actually capturing and verifying (via simulation and/or formal verification) such a design in HDL is very time-consuming.

A bigger problem, however, is that describing a design using a traditional HDL requires us to make a lot of implementation-level (microarchitecture) decisions. In turn, this makes make exploring alternative design scenarios and architectures time-consuming and complex. For example, if you already have a design captured in HDL and someone says, "Let's try doubling the number of pipeline stages to see what happens," you know that you can cancel your plans for the coming weekend.

The solution is, once again, to move to a higher level of abstraction. The most natural level for hardware design engineers is the standard ANSI C programming language (don't talk to me about C++, because trying to wrap my brain around that makes my eyes water). This book shows you how to express designs using standard ANSI C extended with a small number of C-compatible library functions. Once the design has been captured and verified (and alternative solutions and architectures explored), the book also discusses the tools used to automatically convert the untimed C description into its synthesizable HDL equivalent.

The end result is that this methodology allows hardware design engineers to achieve a working prototype much faster than using an HDL-only approach. "But what of the software folks?" you cry...

And what about the software guys and gals?

Actually, the software fraternity is the primary target of this book, because the techniques described allow software-centric engineers to take advantage of the massive amount of parallelism inherent in a dedicated hardware implementation while still using a software development methodology.

Perhaps not surprisingly, the majority of software development engineers writing C code for digital signal processing (DSP) and embedded applications tend to regard "raw hardware" in the form of an FPGA with anything ranging from trepidation (on a good day) to outright loathing. But fear not, my braves, because "there's nothing to fear but fear itself," as my dear old dad used to say.

In the case of someone tasked with creating a DSP application, for example, the majority of DSP code first sees the light of day in C. As horrendous as it seems, once the algorithms have been verified at this level of abstraction, a large proportion of the code gets rewritten and "tweaked" in assembly code in a desperate attempt to achieve the required performance. This manual translation is, of course, both painful and time-consuming.

The bigger problem is that the DSP code will eventually be run on a general-purpose microprocessor or a dedicated DSP device. Both of these realizations are inherently slow, because they are based on a classical von Neuman architecture, which requires them to

- Fetch an instruction
- Fetch a piece of data
- Fetch another piece of data
- Perform an operation
- Store the result
  $\vdots$
- Do the same thing all over again

Now, consider a typical DSP-like function along these lines:

$$y = (a * b) + (c * d) + (e * f) + (g * h);$$

If you run this through a DSP chip, it will take a substantial number of operations and clock cycles to execute. Now consider an equivalent dedicated hardware implementation in an FPGA, in which all of the multiplications are performed in parallel without the need to fetch and decode the instructions. This results in orders-of-magnitude speed improvement.

Thus, one of the purposes of this book is to show you how to take your C code representation and use it to "program" an FPGA to implement your algorithms directly in hardware.

These techniques are not limited to DSP applications; they are also of interest to embedded applications developers in general. The concepts described in this book allow you to view all or part of an FPGA as being "just another programmable core." The difference is that, instead of running your code on an embedded (soft or hard) microprocessor core, you can now decide to implement appropriate portions of the code directly in hardware to achieve phenomenal increases in performance.

So what's the catch?

Actually, there is no catch. As the authors say, if you can write C for applications targeting standard processors, you can write C for programmable hardware.

There are certainly other C-based methodologies and tools available for FPGA designs, but these tend to be focused on satisfying the requirements of the hardware design engineers. By comparison, *Practical FPGA Programming in C* is unique in that it addresses the needs of both hardware and software developers. While it doesn't completely tear down the fence between hardware and software engineering teams, it at least provides another gateway.

Clive "Max" Maxfield — Christmas Eve, 2004

Preface

This is a book about software programming for FPGAs. To be more specific, this book is about using parallel programming techniques in combination with the C language to create FPGA-accelerated software applications.

We have written this book to help bridge the gap—the great chasm, in fact—that exists between software development methods and philosophies, and the methods and philosophies of FPGA-based digital design. We understand that as a result we may be writing for two quite different audiences: software application developers and digital hardware designers.

For software engineers, our goal is to present FPGAs as software-programmable computing resources. We hope to show that these devices, when programmed using appropriate methods, are not so different from other non-traditional computing platforms, such as DSPs. We will show, through example, that software development methods and software languages can be used in a practical way to create FPGA-based, high-performance computing applications, without a deep knowledge of hardware design.

For hardware designers our intent is similar, but with a caveat: we are not trying to replace your existing methods of design or suggest that the methods and tools described in this book represent a "next wave" of hardware engineering. After all, if you are an FPGA designer using VHDL or Verilog to create complex electronic systems, the title of this book, *Practical FPGA*

Programming in C, may sound like an oxymoron. How can C, a software programming language, be a practical way to describe digital electronic hardware? The truth is, sometimes the explicit and precise descriptions offered by hardware description languages (HDLs) are essential to achieve designs goals. But, as you'll see, this explicit control over hardware is not always necessary. In the same way that you might first write software in C and then re-code key portions in assembler, hardware designers can benefit from tools that allow them to mix high-level and low-level descriptions as needed to meet design goals as quickly as possible. Even when the entire hardware design will be eventually be recoded with a lower-level HDL, high-level design languages allow hardware engineers to rapidly explore the design space and create working prototypes.

So for you, the experienced hardware engineer, we'll state right up front that we agree with you. We do not believe that C and C++ (as used by legions of programmers worldwide) are practical replacements for VHDL, Verilog, or any other HDL. And we agree with you that C and C++ may not play a leading, or perhaps even significant, role as a design entry language for general-purpose ASIC and FPGA design, at least not as we know such design today. Nevertheless, we believe there is a place for C-based design in a hardware design flow. Still not convinced? Stay with us for a moment, and we'll explain.

C Language for FPGA-Based Hardware Design?

Let's think a bit more about the role of C—or lack of a role, as the case may be—in hardware design. Why is standard C not appropriate as a replacement for existing hardware design languages? Because any good programming language provides one important thing: a useful abstraction of its target. VHDL and Verilog (or, more precisely, the synthesizable subsets of these languages) succeed very well because they provide a rational, efficient abstraction of a certain class of hardware: level- and edge-sensitive registers with reset and clock logic, arbitrary logic gates, and somewhat larger clocked logic elements arranged in a highly parallel manner. All of today's FPGAs fit this pattern, and it is the pattern also found in the vast majority of today's ASIC designs, no matter how complex.

The standard C language does not provide that level of abstraction (which we call register transfer level, or RTL), so "C-based" languages for hardware could add RTL-like constructs in the form of syntax decorations, extra functions or keywords, compiler hints, and more to create some meaningful way of expressing parallelism and describing low-level hardware structures such as clocks and resets. But in this case we would just have another HDL with new syntax. On the other hand, without the benefit of RTL constructs such as these, the developers of C compilers for FPGAs and other nontraditional targets would face a nearly impossible problem: how to efficiently map algorithms and applications written for one class of processing target (the traditional microprocessor) to something entirely different (arbitrary logic gates and registers combined with somewhat higher-level logic structures). Nobody has yet figured out how to do that mapping from a pure software application with a reasonable level of efficiency, although we are getting better at it.

So why use C at all for FPGA design? There are significant advantages, including the potential for hardware-software codesign, for the creation of test benches written in C, and (if the modified C language supports it) the ability to compile and debug an FPGA application using a standard C development environment. And if a mid-level approach to hardware abstraction is taken—one that does not require that the programmer understand all the details of the hardware target, and yet is guided by the programming model toward more appropriate methods of coding—you can strike a balance between software design productivity and hardware design results, as measured in system performance and size.

COMPELLING PLATFORMS FOR SOFTWARE ACCELERATION

Here's where software developers come in. On the applications side there is an increasing trend toward using FPGAs as hardware accelerators for high-performance computing. Applications that demand such performance exist in many domains, including communications, image processing, streaming media, and other general-purpose signal processing. Many of these applications are in the embedded software space, while others represent scientific, biomedical, financial, and other larger-scale computing solutions.

When acceleration is required, these applications typically begin their lives as software models, often in C language, and are then manually rewritten and implemented in hardware using VHDL or Verilog. This manual conversion of software algorithms to hardware is a process that can be long and tedious in the extreme; hence, there is a strong demand for more rapid paths to working hardware. There is also a strong desire to avoid later redesigns of that hardware to reflect software algorithm updates. Automating the process of software-to-hardware conversion—at the very least for the purpose of creating hardware prototypes—is therefore highly compelling. This automation also brings hardware acceleration within reach of design teams and developers who do not have vast hardware design expertise, or where hardware design might be cost-prohibitive using traditional approaches. C-to-hardware compilation is an enabling technology that has the potential to open up hardware acceleration for a whole new class of applications and developers.

On the hardware side, there have been recent advances in FPGA-based programmable platforms that make FPGAs even more practical for use as application accelerators. In fact, FPGAs with embedded processor cores have now become cost-effective as replacements for entire embedded systems. The concept of building a complete "system on an FPGA" is finally a reality, thanks to the continued efforts of the major FPGA providers and third-party intellectual property (IP) suppliers. What is common about these new FPGA-based platforms is that they combine one or more general-purpose processor cores with a large amount of programmable logic. Just as significantly, they come equipped with substantial amounts of royalty-free, pretested components and an ever-growing collection of licensable third-party and open-source components that can be assembled on-chip in a matter of hours to create an amazingly diverse set of potential hardware/software processing platforms.

From a tools perspective, what is interesting about these new devices—and the new computing platforms they represent—is that they are flexible enough to support the use of methods and tools from the software development world, including traditional software debuggers. Using languages such as C and C++ in conjunction with software-to-hardware compilers makes it practical to implement software applications directly onto a mixed hardware/software platform.

It is here that FPGA designers who are questioning the use of software design languages and methods for FPGAs may be missing the point. C is not likely to replace HDLs such as VHDL and Verilog for traditional, general-purpose hardware design. C will, however, play a strong role in the kind of mixed hardware/software applications that are emerging today—applications in which the line between what is software and what is hardware is becoming increasingly blurred, and where the ability to rapidly prototype and experiment with implementation alternatives is critical.

In addition, because of the prevalence of embedded FPGA processors, the C language is already being used for FPGAs by thousands of software engineers. The problem is that the vast majority of these software engineers are only able to program for the embedded microprocessor within the FPGA. Our goal in this book is to make C-language programming applicable to the entire FPGA, not just the embedded microprocessor within it.

THE POWER TO EXPERIMENT

Using C language programming as an aid to FPGA-based hardware design gives you the power to experiment—with alternate algorithmic approaches, with alternate hardware/software partitioning, and with alternate target platforms. Using the same design entry language, you now have the ability to evaluate your applications and their constituent algorithms using different hardware and software targets.

In this book we will show you how you can, for example, set up a test in which the same C-language algorithm can, with only minor modifications, be executed in a desktop computing environment (under the control of a C debugger such as found in Visual Studio, gbd, or Eclipse) for the purpose of functional simulation, be executed in-system on an embedded processor, and be compiled directly to an FPGA as dedicated hardware. Using this approach, fundamentally different computing alternatives can be tried, and alternative hardware/software partitioning strategies can be evaluated. As you'll see, this power to quickly generate and experiment with hardware/software prototypes is the key advantage of using software-based methods and tools for FPGAs.

HOW THIS BOOK IS ORGANIZED

For software and hardware developers, we present important background information about FPGA devices, their history, and the types of tools available for them, and we also survey some of the available FPGA-based computing platforms.

After setting the stage, we then present a method of programming, using the C language, and a programming model that is appropriate for use with highly parallel programmable hardware platforms. This programming model is somewhat different from traditional, more procedural C-language programming but is easy for experienced software engineers to pick up and use.

From that foundation we then move into a series of examples, all written in C, that demonstrate how to take advantage of the massive levels of parallelism that are available in an FPGA-based platform. We will describe software coding techniques that allow better optimization of C-language statements by automated compiler tools. These techniques are not difficult to understand but can have dramatic impacts on the performance of an FPGA application.

Later chapters present examples of how a streaming programming model can be used to create even higher levels of performance. Using a multiple-process streaming programming model can result in truly astonishing levels of performance with relatively little effort, but to the traditional C programmer the methods used to achieve such levels of performance may be somewhat new and different.

WHERE THIS BOOK CAME FROM

This book, like many projects in the real world, began its life as a prototype, one that had been taken off the shelf, fiddled with, and put back repeatedly over a number of years. In its initial conception—long before the grueling, months-long process to create the book you now hold—the book was to have a different title and quite a different emphasis—on reconfigurable hardware platforms, both FPGA-based and non-FPGA-based. That book—the one that didn't get written but that survives in bits and pieces in certain chapters of this book—was conceived as a follow-on to an earlier book on programmable logic devices authored by David Pellerin and Michael Holley and titled *Practical Design Using Programmable Logic* (Prentice Hall, 1991).

When the idea for such a book was presented to Bernard Goodwin at Prentice Hall, his initial response was "But of course!" (It is the nature of acquisitions editors, we suppose, to be enthusiastic about every project prior to actually seeing the awful sludge of the first sample chapters.) As we worked on refining the proposal, Bernard suggested that we increase the emphasis on one particular area, an area of which we—due to our current roles in a technology startup—might have a greater-than-usual understanding. Hence the emphasis on C programming for FPGAs and on the design and optimization of parallel applications for FPGA targets. Bernard was, of course, eyeing the bottom line and steering us in a direction that would maximize sales (C is popular and FPGAs are popular; ergo, we will attract more readers by having both terms in the title). What Bernard did not know—what none of us know, really—is how widely C programming for FPGAs can and will be accepted. Time will tell, of course. With this book we hope to convince you that such an approach is indeed a practical one.

David Pellerin
Kirkland, Washington

Scott Thibault
Colchester, Vermont

Acknowledgments

This book was written with the help of many individuals who contributed their time and ideas to the project.

First and foremost, we give our thanks to Bernard Goodwin for championing this project at Prentice Hall. Christine Hackerd, Project Editor at Prentice Hall, made the production process easier and helped educate us in the finer points of book layout. Gayle Johnson's copy editing gave new life to the text; her thousands of red-pen decorations were greatly appreciated.

Special thanks go to Ralph Bodenner for his deep knowledge of Impulse C programming, his assistance with the image filter example in Chapter 11, and his careful technical editing of the manuscript. Thanks also to Brian Durwood, Impulse co-founder, for his support of the project.

We owe a particular debt of gratitude to Maya Gokhale and her group at Los Alamos National Laboratories. Maya's research into streams-based programming provided the basis for the methods and tools presented in this book, and was the impetus as well behind Impulse C.

Bryan Fletcher of Memec wrote the paper (first presented at the 2005 Embedded Systems Conference in March of 2005) that appears in modified form as Appendix A. His insights into embedded processor performance are timely and useful. Ross Snider and Scott Bekker of Montana State University contributed Appendix B, which provides valuable information on streaming hardware interfaces.

The uClinux example presented in Chapter 12 builds on work done by John Williams at the University of Queensland, and we thank John for his assistance and his enthusiastic support.

Some of the optimization and debugging techniques presented in this book originated from discussions with Impulse customers and early users of the Impulse C tools. Carl Ebeling at the University of Washington, along with his students George Huang and Tsz (Oscar) Ng, deserve particular credit for their ongoing work and valuable feedback. Katsumi Kurashige and Tsuyoshi Okayama of Interlink, Inc. (Japan) deserve similar credit for their probing questions and their continued focus on user education.

Bjorn Freeman-Benson provided valuable feedback on the sections relating to programming models. Bjorn's advice and friendship has helped in many ways through the years.

Devi Pellerin created many of the figures, as well as providing help in the page layout process.

To all Impulse C users we extend our thanks, and our promise to keep those bug fixes coming.

CHAPTER 1

The FPGA as a Computing Platform

As the cost per gate of FPGAs declines, embedded and high-performance systems designers are being presented with new opportunities for creating accelerated software applications using FPGA-based programmable hardware platforms. From a hardware perspective, these new platforms effectively bridge the gap between software programmable systems based on traditional microprocessors, and application-specific platforms based on custom hardware functions. From a software perspective, advances in design tools and methodology for FPGA-based platforms enable the rapid creation of hardware-accelerated algorithms.

The opportunities presented by these programmable hardware platforms include creation of custom hardware functions by software engineers, later design freeze dates, simplified field updates, and the reduction or elimination of custom chips from many categories of electronic products. Increasingly, systems designers are seeing the benefits of using FPGAs as the basis for applications that are traditionally in the domain of application-specific integrated circuits (ASICs). As FPGAs have grown in logic capacity, their ability to host high-performance software algorithms and complete applications has grown correspondingly.

In this chapter, we will present a brief overview of FPGAs and FPGA-based platforms and present the general philosophy behind using the C language for FPGA application development. Experienced FPGA users will find much of this information familiar, but nonetheless we hope you stay with us as we take the FPGA into new, perhaps unfamiliar territory: that of high-performance computing.

1.1   A QUICK INTRODUCTION TO FPGAS

A field-programmable gate array (FPGA) is a large-scale integrated circuit that can be programmed after it is manufactured rather than being limited to a predetermined, unchangeable hardware function. The term "field-programmable" refers to the ability to change the operation of the device "in the field," while "gate array" is a somewhat dated reference to the basic internal architecture that makes this after-the-fact reprogramming possible.

FPGAs come in a wide variety of sizes and with many different combinations of internal and external features. What they have in common is that they are composed of relatively small blocks of programmable logic. These blocks, each of which typically contains a few registers and a few dozen low-level, configurable logic elements, are arranged in a grid and tied together using programmable interconnections. In a typical FPGA, the logic blocks that make up the bulk of the device are based on lookup tables (of perhaps four or five binary inputs) combined with one or two single-bit registers and additional logic elements such as clock enables and multiplexers. These basic structures may be replicated many thousands of times to create a large programmable hardware *fabric*.

In more complex FPGAs these general-purpose logic blocks are combined with higher-level arithmetic and control structures, such as multipliers and counters, in support of common types of applications such as signal processing. In addition, specialized logic blocks are found at the periphery of the devices that provide programmable input and output capabilities.

Common FPGA Characteristics

FPGAs are mainly characterized by their logic size (measured either by the number of transistors or, more commonly, by the number of fundamental logic blocks that they contain), by their logic structure and processing features, and by their speed and power consumption. While they range in size from a few thousand to many millions of logic gate equivalents, all FPGAs share the same basic characteristics:

- *Logic elements.* All FPGA devices are based on arrays of relatively small digital logic elements. To use such a device, digital logic problems must be decomposed into an arrangement of smaller logic circuits that can be mapped to one or more of these logic elements, or *logic cells*, through a process of *technology mapping*. This technology mapping process may be either manual or automatic, but it involves substantial intelligence on the part of the human or (more typically) the software program performing the mapping.

- *Lookup tables.* FPGA logic elements are themselves usually composed of at least one programmable register (a flip-flop) and some input-forming logic, which is often implemented as a lookup table of n inputs, where n is generally five or less. The detailed structure of this logic element depends on the FPGA vendor (for example, Xilinx or Altera) and FPGA family (for example, Xilinx Virtex or Altera Cyclone). The lookup tables (LUTs) that make up most logic elements are very much akin to read-only memory (ROM). These LUTs are capable of implementing any combinational function of their inputs.

- *Memory resources.* Most modern FPGA devices incorporate some on-chip memory resources, such as SRAM. These memories may be accessible in a hierarchy, such as local memory within each logic element combined with globally shared memory blocks.

- *Routing resources.* Routing is the key to an FPGA's flexibility, but it also represents a compromise—made by the FPGA provider—between programming flexibility and area efficiency. Routing typically includes a hierarchy of channels that may range from high-speed, cross-chip *long lines* to more flexible block-to-block local connections. Programmable switches (which may be SRAM-based, electrically erasable, or one-time-programmable) connect the on-chip FPGA resources to each other, and to external resources as well.

- *Configurable I/O.* FPGA-based applications have a wide variety of system-level interface requirements and therefore FPGAs have many programmable I/O features. FPGA pins can be configured as TTL, CMOS, PCI, and more, allowing FPGAs to be interfaced with, and convert between, many different circuit technologies. Most FPGAs also have dedicated high-speed I/Os for clocks and global resets, and many FPGAs include PLLs and clock management schemes, allowing designs with multiple independent clock domains to be created and managed.

FPGA Programming Technologies

FPGA programming technologies range from one-time-programmable elements (such as those found in devices from Actel and Quicklogic) to electrically erasable or SRAM-based devices such as those available from Altera, Lattice Semiconductor, and Xilinx.

While most FPGAs in use today are programmed during system manufacturing to perform one specific task, it is becoming increasingly common for FPGAs to be reprogrammed while the product is in the field. For SRAM-based and electrically erasable FPGAs this field upgrading may be as simple as providing an updated Flash memory card or obtaining a new binary device image from a website or CD-ROM.

Hardware applications implemented in FPGAs are generally slower and consume more power than the same applications implemented in custom ASICs. Nonetheless, the dramatically lowered risk and cost of development for FPGAs have made them excellent alternatives to custom ICs. The reduced development times associated with FPGAs often make them compelling platforms for ASIC prototyping as well.

Many modern FPGAs have the ability to be reprogrammed in-system, in whole or in part. This has led some researchers to create dynamically reconfigurable computing applications within one or more FPGAs in order to create extremely high-performance computing systems. The technology of reconfigurable computing is still in its infancy, however, due in large part to the high cost, in terms of power and configuration time, of dynamically reprogramming an FPGA. We will examine some of this research and make predictions regarding the future of reconfigurable computing in a later chapter.

Defining the behavior of an FPGA (the hardware that it contains) has traditionally been done either using a *hardware description language* (HDL) such as VHDL or Verilog or by arranging blocks of pre-existing functions, whether gate-level logic elements or higher-level macros, using a schematic- or block diagram-oriented design tool. More recently, design tools such as those described in this book that support variants of the C, C++, and Java languages have appeared. In any event, the result of the design entry process (and of design compilation and synthesis, as appropriate) is an intermediate file, called a netlist, that can be mapped to the actual FPGA architecture using proprietary FPGA place-and-route tools. After placement and mapping the resulting binary file—the bitmap—is downloaded to the FPGA, making it ready for use.

1.2 FPGA-Based Programmable Hardware Platforms

What constitutes a programmable hardware platform, and, in particular, one appropriate for high-performance computing? The term "platform" is somewhat arbitrary but generally refers to a known, previously verified hardware and/or software configuration that may be used as the basis for one or more specific applications. For our purposes, a programmable platform may be represented by a single FPGA (or other such programmable device), a complete board with multiple FPGAs, or even a development system such as a desktop PC or workstation that is used to emulate the behavior of an FPGA.

In short, a programmable hardware platform is a platform that includes at least one programmable hardware element, such as an FPGA, and that can implement all or part of a software algorithm. As you will see, you can also extend the definition of a platform to include various "soft" components that are found within the FPGA itself, such as embedded processors and related peripheral devices, and even such things as embedded operating systems running within the FPGA. All of these elements, taken as a whole, can be used to describe a particular FPGA-based platform.

FPGA-based platforms range from individual FPGAs, with or without embedded processors, and single-board, single-FPGA prototyping platforms to high-performance FPGA computing platforms consisting of multiple FPGAs combined with other processing resources on one or more boards. New FPGA-based platforms are being announced with increasing frequency, so the sample platforms appearing in this book represent only a small fraction of the FPGA-based solutions that are available.

Today's FPGAs are capable of much higher levels of system integration than those of previous generations. In particular, the ability to combine embedded processor cores and related standard peripheral devices with custom hardware functions (intellectual property [IP] blocks as illustrated in Figure 1-1) has made it possible to custom-craft a programmable platform ideally suited to a particular task or particular application domain.

The challenges of programmable platform-based design are primarily (but not exclusively) in the domain of the system architect and the software application developer. The success or failure of a development project—as measured in terms of development time and the final performance achieved—depends to a large degree on how well a large application and its constituent algorithms are mapped onto platform resources. To a software application developer it may be less than obvious which portions of the design should go into hardware (for example, in one or more FPGA devices) and which should be implemented as software on a traditional processor, whether a discrete device or a processor core embedded within an FPGA platform. Even more fundamental is the need for software application developers to consider their applications and algorithms in entirely new ways, and to make use of parallel programming techniques that may be unfamiliar to them in order to increase algorithmic concurrency and extract the maximum possible performance.

The fundamental decisions made by an application developer in the early phases of the design, the effectiveness of the algorithm partitioning, and the mapping of the application to hardware resources through the use of automated compiler and hardware synthesis tools can impact the final system's performance by orders of magnitude.

External processor
plus FPGA

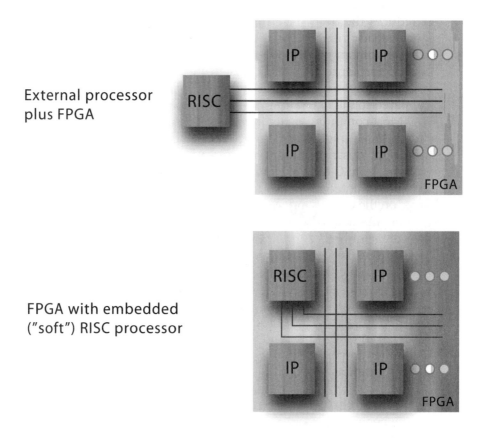

FPGA with embedded
("soft") RISC processor

Figure 1-1. FPGA-based platforms may include embedded or adjacent microprocessors combined with application-specific blocks of intellectual property.

1.3 INCREASING PERFORMANCE WHILE LOWERING COSTS

The decision to add an FPGA to a new or existing embedded system may be driven by the desire to extend the usefulness of a common, low-cost microprocessor (and avoid introducing a higher-end, specialized processor such as a DSP chip) or to eliminate the need for custom ASIC processing functions. In cases where the throughput of an existing system must increase to handle higher resolutions or larger signal bandwidths, the required performance increases may be primarily computational (requiring a scaling of computational resources) or may require completely new approaches to resolve bandwidth issues.

Digital signal processing (DSP) applications are excellent examples of the types of problems that can be effectively addressed using high-density

FPGAs. Implementing signal processing functions within an FPGA eliminates or reduces the need for an instruction-based processor. There has therefore been a steady rise in the use of FPGAs to handle functions that are traditionally the domain of DSP processors. System designers have been finding the cost/performance trade-offs tipping increasingly in favor of FPGA devices over high-performance DSP chips and—perhaps most significantly—when compared to the risks and up-front costs of a custom ASIC solution.

Most computationally intensive algorithms can be described using a relatively small amount of C source code, when compared to a hardware-level, equivalent description. The ability to quickly try out new algorithmic approaches from C-language models is therefore an important benefit of using a software-oriented approach to design. Reengineering low-level hardware designs, on the other hand, can be a tedious and error-prone process.

FPGAs help address this issue in two ways. First, they have the potential to implement high-performance applications as dedicated hardware without the up-front risk of custom ASICs. Mainstream, relatively low-cost FPGA devices now have the capacity and features necessary to support such applications. Second, and equally important, FPGAs are becoming dramatically easier to use due to advances in design tools. It is now possible to use multiple software-oriented, graphical, and language-based design methods as part of the FPGA design process.

When implementing signal processing or other computationally intensive applications, FPGAs may be used as prototyping vehicles with the goal of a later conversion to a dedicated ASIC or structured ASIC. Alternatively, FPGAs may be used as actual product platforms, in which case they offer unique benefits for field software upgrades and a compelling cost advantage for low- to medium-volume products.

High-capacity FPGAs also make sense for value engineering. In such cases, multiple devices (including processor peripherals and "glue" logic) may be consolidated into a single FPGA. While reduced size and system complexity are advantageous by-products of such consolidation, the primary benefit is lower cost. Using an FPGA as a catchall hardware platform is becoming common practice, but such efforts often ignore the benefits of using the FPGAs for primary processing as well as just using them for traditional hardware functions.

In many DSP applications, the best solution turns out to be a mixed processor design, in which the application's less-performance-critical components (including such elements as an operating system, network stack, and user interface) reside on a host microprocessor such as an ARM, PowerPC, or Pentium. More computationally intensive components reside in either a high-end DSP, in dedicated hardware in an FPGA, or in a custom ASIC as appropriate. It is not unusual, in fact, for such systems to include all three types of processing resources, allocated as needed. Developing such a system requires multiple tools and knowledge of hardware design methods and tools but provides the greatest benefit in terms of performance per unit cost.

For each processor type in the system (standard processor, DSP, or FPGA), there are different advantages, disadvantages, and levels of required design expertise to consider. For example, while DSPs are software-programmable and require a low initial investment in tools, they require some expertise in DSP-specific design techniques and often require assembly-level programming skills. FPGAs, on the other hand, require a relatively large investment in design time and tools expertise, particularly when hardware design languages are used as the primary design input method.

When compared to the expertise and tools investment required for custom ASIC design, however, FPGAs are clearly the lower-cost, lower-risk solution for developing custom hardware.

FPGAs provide additional advantages related to the design process. By using an FPGA throughout the development process, a design team can incrementally port and verify algorithms that have been previously prototyped in software. This can be done manually (by hand-translating C code to lower-level HDL), but C-based design tools such as those described in this book can speed this aspect of the design process.

1.4   THE ROLE OF TOOLS

Software development tools, whether intended for deeply embedded systems or for enterprise applications, add value and improve the results of the application development process in two fundamental ways. First, a good set of tools provides an appropriate and easily understood abstraction of a target platform (whether an embedded processor, a desktop PC, or a supercomputer). A good abstraction of the platform allows software developers to create, test, and debug relatively portable applications while encouraging them to use programming methods that will result in the highest practical performance in the resulting end product.

The second fundamental value that tools provide is in the mechanical process of converting an application from its original high-level description (whether written in C or Java, as a dataflow diagram or in some other representation) into an optimized low-level equivalent that can be implemented—loaded and executed—on the target platform.

In an ideal tool flow, the specific steps of this process would be of no concern to the programmer; the application would simply operate at its highest possible efficiency through the magic of automated tools. In practice this is rarely the case: any programmer seeking high performance must have at least a rudimentary understanding of how the optimization, code generation, and mapping process works, and must exert some level of control over the process either by adjusting the flow (specifying compiler options, for example) or by revisiting the original application and optimizing at the algorithm level, or both.

An Emphasis on Software-Based Methods

To fulfill the dual role of tools just described, emerging tools for automated hardware generation are focusing both on the automatic compilation/optimization problem and on delivering programming abstractions, or *programming models*, that make sense for the emerging classes of FPGA-based programmable platforms. All of these emerging tools focus on creating a software-oriented design experience. Software-oriented design tools are appropriate because

- Software provides a higher level of abstraction than traditional RTL design, thus helping to manage the growing complexity of platform-based systems.

- Algorithms are often specified, tested, and verified as software, so a software-oriented design environment requires fewer manual (and error-prone) translations.

- Microprocessors that inherently run software have become part of virtually every modern system. With emerging technologies enabling the ability to directly compile software into hardware implementations, software is becoming the lingua franca of system design.

Software-oriented programming, simulation, and debugging tools that provide appropriate abstractions of FPGA-based programmable platforms allow software and system designers to begin application development, experiment with alternative algorithms, and make critical design decisions without the need for specific hardware knowledge. This is of particular importance during design prototyping. As illustrated in Figures 1-2 and 1-3, the traditional hardware and software design process can be improved by introducing software-to-hardware design methods and tools. It is important to realize, however, that doing so will almost certainly not eliminate the need for hardware engineering skills. In fact, it is highly unlikely that a complete and well-optimized hardware/software application can be created using only software knowledge. On the plus side, it is certainly true that working prototypes can be more quickly generated using hardware and software design skills in combination with newly emerging tools for software-to-hardware compilation.

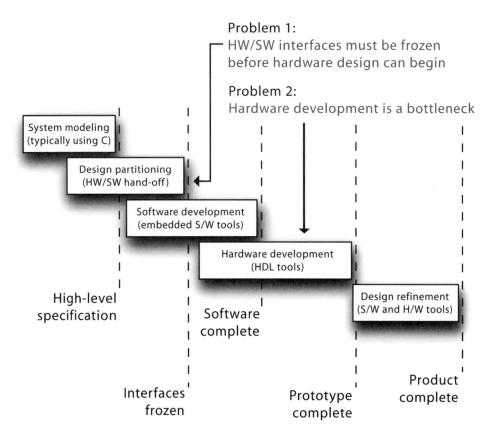

Figure 1-2. In a traditional hardware/software development process, hardware design may represent a significant bottleneck.

1.5   THE FPGA AS AN EMBEDDED SOFTWARE PLATFORM

Because of their reprogrammability, designing for FPGAs is conceptually similar to designing for common embedded processors. Simulation tools can be used to debug and verify the functionality of an application prior to actually programming a physical device, and there are tools readily available from FPGA vendors for performing in-system debugging. Although the tools are more complex and design processing times are substantially longer (it can take literally hours to process a large application through the FPGA place-and-route process), the basic design flow can be viewed as one of software rather than hardware development.

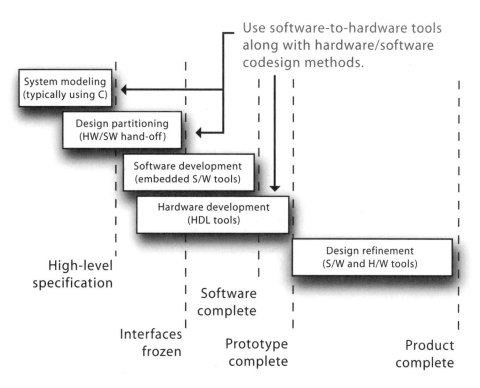

Figure 1-3. By introducing software-to-hardware compilation into the design process, it's possible to get to a working prototype faster, with more time available for later design refinement.

As any experienced FPGA application designer will tell you, however, the skills required to make the most efficient use of FPGAs, with all their low-level peculiarities and vendor-specific architectural features, are quite specialized and often daunting. To put this in the proper perspective, however, it's important to keep in mind that software development for embedded processors such as DSPs can also require specialized knowledge. DSP programmers, in fact, often resort to assembly language in order to obtain the highest possible performance and use C programming only in the application prototyping phase. The trend for both FPGA and processor application design has been to allow engineers to more quickly implement applications, or application prototypes, without the need to understand all the intricate details of the target, while at the same time providing access (through custom instructions, built-in functions/macros, assembly languages, and hardware description languages as appropriate) to low-level features for the purpose of extracting the maximum possible performance.

The Impulse C toolset used extensively in the book represents a relatively simple, C-based approach to hardware/software partitioning and hardware/software process synchronization that is coupled with efficient, FPGA-specific hardware compilers to produce a complete design path for mixed processor and FPGA implementations. This design flow has been integrated into traditional embedded and desktop programming tool flows without the need for custom behavioral simulators or other EDA-centric technologies. The combined environment enables systems-oriented development of highly parallel applications for FPGA-based platforms using the familiar C language, which may or may not be combined with existing HDL methods of design for those parts of an application that are not naturally expressed as software.

One of the key attributes of such a software-oriented system design flow is the ability to implement a design specification, captured in software, in the most appropriate platform resource. If the most appropriate resource is a microprocessor, this should be a simple matter of cross-compiling to that particular processor. If, however, the best-fitting resource is an FPGA, traditional flows would require a complete rewrite of the design into register transfer level (RTL) hardware description language. This is not only time-consuming, but also error-prone and acts as a significant barrier to the designer in exploring the entire hardware/software solution space. With a software-oriented flow, the design can simply be modified in its original language, no matter which resource is targeted.

1.6   THE IMPORTANCE OF A PROGRAMMING ABSTRACTION

As the line between what is hardware and what is software continues to blur, the skill sets of the hardware engineer and the software engineer have begun to merge. Hardware engineers have embraced language-based design (HDLs are simply a way to abstract a certain class of hardware as a written description). Software engineers—in particular, those developing embedded, multiprocessor applications—have started to introduce important hardware-oriented concepts such as concurrency and message-driven process synchronization to their design arsenals. Until recently, however, it has been difficult or impossible for software engineers to make the transition from software application programming to actual hardware implementation—to turn a conceptual application expressed in a high-level language into an efficient hardware description appropriate for the selected hardware target.

Developments in the area of high-level language compilers for hardware coupled with advances in behavioral synthesis technologies have helped ease this transition, but it remains challenging to create an efficient, mixed hardware/software application without having substantial expertise in the area of RTL hardware design and in the optimization of applications at a very low level. Early adopters of commercial tools for C-based hardware synthesis have found that they must still drop to a relatively low level of abstraction to create the hardware/software interfaces and deal with the timing-driven nature of hardware design. The availability of high-level language synthesis has not yet eliminated the need for hardware design expertise. A major goal of this book, then, is to show when software-based design methods and tools for FPGAs are appropriate, while acknowledging that not every application is well-suited to such an approach.

Emerging research and technologies are making the FPGA design process easier and placing more control in the hands of software-savvy engineers. Improved software compilers and specialized optimizers for FPGAs and other hardware targets play an important role in this change, but as we've pointed out, higher-level system and algorithm design decisions can have more dramatic impacts than can be achieved through improvements in automated tools. No optimizer, regardless of its power, can improve an application that has been poorly conceived, partitioned, and optimized at an algorithmic level by the original programmer.

1.7 WHEN IS C LANGUAGE APPROPRIATE FOR FPGA DESIGN?

Experimenting with mixed hardware/software solutions can be a time-consuming process due to the historical disconnect between software development methods and the lower-level methods required for hardware design, including design for FPGAs. For many applications, the complete hardware/software design is represented by a collection of software and hardware source files that are not easily compiled, simulated, or debugged with a single tool set. In addition, because the hardware design process is relatively inefficient, hardware and software design cycles may be out of sync, requiring system interfaces and fundamental software/hardware partitioning decisions to be prematurely locked down.

With the advent of C-based FPGA design tools, however, it is now possible to use familiar software design tools and standard C language for a much larger percentage of a given application—in particular, those parts of the design that are computationally intensive. Later performance tweaks may introduce handcrafted HDL code as a replacement for the automatically generated hardware, just as DSP users often replace higher-level C code with handcrafted assembly language. Because the design can be compiled directly from C code to an initial FPGA implementation, however, the point at which a hardware engineer needs to be brought in to make such performance tweaks is pushed further back in the design cycle, and the system as a whole can be designed using more productive software design methods.

These emerging hardware compiler tools allow C-language applications to be processed and optimized to create hardware, in the form of FPGA netlists, and also include the necessary C language extensions to allow highly parallel, multiple-process applications to be described. For target platforms that include embedded processors, these tools can be used to generate the necessary hardware/software interfaces as well as generating low-level hardware descriptions for specific processes.

One key to success with these tools, and with hardware/software approaches in general, is to partition the application appropriately between software and hardware resources. A good partitioning strategy must consider not only the computational requirements of a given algorithmic component, but also the data bandwidth requirements. This is because hardware/software interfaces may represent a significant performance bottleneck.

Making use of a programming model appropriate for highly parallel applications is also important. It is tempting to off-load specific functions to an FPGA using traditional programming methods such as remote procedure calls (whereby values are pushed onto some stack or stack equivalent, a hardware function is invoked, and the processor waits for a result) or by creating custom processor instructions that allow key calculations to be farmed out to hardware. Research has demonstrated, however, that alternate, more dataflow-oriented methods of programming are more efficient and less likely to introduce blockages or deadlocks into an application. In many cases, this means rethinking the application as a whole and finding new ways to express data movement and processing. The results of doing so, however, can be dramatic. By increasing application-level parallelism and taking advantage of programmable hardware resources, for example, it is possible to accelerate common algorithms by orders of magnitude over a software-only implementation.

During the development of such applications, design tools can be used to visualize and debug the interconnections of multiple parallel processes. Application monitoring, for example, can provide an overall view of the application and its constituent processes as it runs under the control of a standard C debugger. Such instrumentation can help quantify the results of a given partitioning strategy by identifying areas of high data throughput that may represent application bottlenecks. When used in conjunction with familiar software profiling methods, these tools allow specific areas of code to be identified for more detailed analysis or performance tweaking. The use of cycle-accurate or instruction-set simulators later in the development process can help further optimize the application.

What about SystemC?

SystemC is becoming increasingly popular as an alternative to existing hardware and system-level modeling languages—in particular, VHDL and Verilog.

SystemC provides (according to www.SystemC.org) "hardware-oriented constructs within the context of C++ as a class library implemented in standard C++." What this means is that SystemC includes features, implemented as C++ class and template libraries, that allow the definition of hardware behaviors at many levels, from the high-level representation of a complete system to the smallest details of transistor-level timing.

While SystemC does provide the features needed to describe parallel systems of interconnected processes, the complexity of the libraries coupled with a historically low acceptance of C++ as a language for embedded software development has limited its appeal to mainstream embedded systems programmers. This is not to say that SystemC is inappropriate for FPGA-based applications; on the contrary, most of the concepts and specific programming techniques covered in this book are applicable to SystemC FPGA design flows.

1.8   HOW TO USE THIS BOOK

Our goal in writing this book is to make you a more productive designer of mixed hardware/software applications, using FPGAs as a vehicle, and to provide you with an alternate method of design entry, debugging, and compilation. We have not written this book with the goal of eliminating, or even reducing the importance of, existing hardware and software design methods.

If you are an experienced software developer or embedded systems designer who is considering FPGAs for hardware acceleration of key algorithms, this book is designed primarily, but not exclusively, with you in mind. We will introduce you to FPGAs, provide some history and background on their use as general-purpose hardware devices, and help you understand some of the advantages and pitfalls of introducing them into an embedded system as an additional computing resource. We will then spend a substantial amount of time introducing you to the concepts of parallel programming and creating mixed hardware/software applications using C.

If you are an experienced hardware developer currently using VHDL or Verilog, our goal is to help you use the C language as a complement to your existing methods of design. Using C will allow you to more quickly generate prototype hardware for algorithms that have previously been expressed in C and give you the freedom to experiment with alternative methods of creating large, interconnected systems in hardware. This flexibility will in turn accelerate your development of custom hardware elements. The C language will also provide you with alternative methods of creating hardware/software test benches and related components for the purpose of unit or system testing.

Finally, both hardware and software developers can learn from this book about the demands of designing for the "other side" of the narrowing hardware/software divide. We hope that as development methods for hardware and software begin to converge, so too will the cooperation between hardware and software designers be improved.

CHAPTER 2

A Brief History of Programmable Platforms

You may think that writing software for programmable hardware is a relatively new phenomenon, but the history of programmable hardware and related design tools is nearly as long as the history of the microprocessor. In fact, if you consider that researchers at companies including IBM, Intel, Texas Instruments, Harris Semiconductor, and others were experimenting with general-purpose computing using the earliest forms of programmable logic arrays as far back as the late 1960s, it's not much of an exaggeration to claim that the use of programmable hardware for creating what are essentially software applications may actually predate the birth of the microprocessor, which is generally accepted to have occurred in 1970 with the release of the Intel 4004.

In this chapter we'll present a summary of the developments that have led to today's most prevalent programmable hardware platform, the field-programmable gate array (FPGA). We will also present a short history of the design tools that have evolved during this time and, in doing so, show that a gradual convergence of software and hardware design methods has been in progress for a great many years.

Keep in mind that FPGA-based computing platforms are evolving at a rapid pace. Our goal in this chapter, then, is not to describe the current state of the art in such platforms, but instead to show how we have arrived at where we are today.

2.1 THE ORIGINS OF PROGRAMMABLE LOGIC

Most industry watchers point to the introduction in 1975 of the field pro-grammable logic array (the precursor of today's FPGA) as the birth of pro-grammable logic, but the history of programmable logic actually goes back some years earlier. The programmable devices we know today are direct an-cestors of experimental devices produced in the late 1960s and early 1970s by Harris Semiconductor (then known as Radiation, Inc.) and General Electric.

The first of these devices were primitive diode matrices that were used as controllers for such diverse applications as television channel selectors, HAM radio tuners, and pipe organ switching systems, as well as in emerging defense and space applications. These relatively crude devices were soon re-placed by more capable logic devices based on arrays of combinatorial gates, typically AND and OR gates arranged to form a sum-of-products or product-of-sums architecture.

The key development that made programmable logic a practical alterna-tive for general-purpose logic design (and for use in performing special-purpose computations) was the ability to program the devices after manufac-turing using low-cost equipment. The basic technology involved creating a mask layer of fusible links that were selectively removed—literally melted away—using high voltages applied to device pins by specialized program-ming equipment. This reliable and low-cost programming method, which was initially developed to create one-time-programmable read-only memo-ries (or ROMs) is still used today for many simpler programmable logic devices.

The first field-programmable logic array (FPLA) devices were released in 1975 by competitors Intersil Corporation and Signetics Corporation. These first FPLAs were quite powerful for their time and provided enough capacity and I/O resources to implement a wide variety of functions including state machines, decoding logic, and basic arithmetic operations. When collected into larger multi-chip programmed systems, these devices were capable of creating high-performance computational devices including minicomputers such as the DEC VAX 11/750 and Data General Eclipse.

New Methods of Design Are Required

The Signetics 82S100 FPLA device was the first programmable logic device to achieve a modest level of commercial success. Its history provides a useful analogy for today's newly emerging programmable platforms and the shift in design methods required when moving to radically different types of logic and/or computing platforms.

When it was first introduced, the 82S100 was a thing quite alien to most embedded systems designers, and to digital logic designers as well; was it a logic device or some kind of controller? Using off-the-shelf TTL devices (which represented basic logic functions such as ANDs, ORs, exclusive ORs, and set/reset flip-flops) had become a standard method of digital design in 1975, and these new devices did not fit into the schematic-oriented, building-block design methods of the time. Even with a relatively high level of documentation and applications support, most digital designers rejected the new devices as being too difficult to use. This difficulty stemmed from the conceptual differences between how designers had expressed their applications in the past and how they were forced to express them when using programmable logic devices.

Applications that were to be implemented using FPLAs had to be described in an unfamiliar tabular format that bore little resemblance to the block diagrams and Boolean equations that had previously been used to work out a logic problem. The data from this tabular form was subsequently entered into a hardware device programmer into which the chip was inserted for programming. This was quite different from the TTL design process, in which components were selected from a data book and arranged to create larger systems of interconnected logic.

Early HDLs Increase Design Abstraction

Programmable logic devices didn't gain widespread acceptance until the late 1970s, when Monolithic Memories, Inc., introduced its Programmable Array Logic (PAL) series of devices. These devices, which were designed by John Birkner and H. T. Chua, had a number of advantages over the earlier FPLA devices, including more standard packaging, higher speeds, and more emphasis on compatibility with existing and widely popular TTL devices.

One important factor in the success of the PAL was a design language called PALASM (PAL Assembler), which was a compiler program that converted applications described using familiar Boolean equations into the necessary programming data. PALASM, in fact, was one of the first hardware description languages (HDLs) and remains in use today by many thousands of engineers and students. A sample PALASM source file is shown in Figure 2-1. As shown in this example, PALASM provides more than just a logic description. It also provides, in the function table section, a set of tests to be applied to the device after programming. This was an important language feature because device programming was not always successful. The devices could have faulty programmable fuse elements, the programming equipment could be out of spec, or the logic equations themselves could be incorrect. Providing hardware test data in the form of test vectors right in the design itself eliminated the vast majority of later in-system problems.

```
PAL16R4 PAL
COUNT4
4-bit counter with synchronous clear
Clk  Clr    NC NC NC NC NC NC NC GND
OE  NC     NC /Q3 /Q2 /Q1 /Q0 NC NC VCC

  Q3 := Clr
     + /Q3 * /Q2 * /Q1 * /Q0
     + Q3 * Q0
     + Q3 * Q1
     + Q3 * Q2

  Q2 := Clr
     + /Q2 * /Q1 * /Q0
     + Q2 * Q0
     + Q2 * Q1

  Q1 := Clr
     + /Q1 * /Q0
     + Q1 * Q0

  Q0 := Clr
     + /Q0

FUNCTION TABLE
OE CLR Clk  /Q0 /Q1 /Q2 /Q3
-----------------------------------------
  L H    C    L L L L
  L L    C    H L L L
  L L    C    L H L L
  L L    C    H H L L
  L L    C    L L H L
  L H    C    L L L L
-----------------------------------------
```

Figure 2-1. PALASM design file representing a 4-bit counter with synchronous clear.

Regardless of its original rationale, this simple idea of using the same language to describe both the application and the test conditions for that application has been a key aspect of all subsequent hardware description languages. As devices became more complex and the capabilities of programmable logic design languages and tools increased, these test vectors (and other higher-level abstractions for testing) were used not only to test the programmed devices, but to provide input to software-based device simulations as well.

Larger Devices, More-Complex Programming Tools

By the early 1980s, many companies were providing programmable logic devices, each with its own unique set of architectural features. As these devices became more complex and varied, the need grew for higher levels of design abstraction and for automated conversion of higher-design descriptions to programmable logic implementations. Simulation tools also became increasingly important; the size and complexity of programmable logic applications were now growing to the point where testing based on device inputs and outputs (by providing simple device test vectors) was not sufficient.

In response to these challenges, in March of 1983 a company called Assisted Technology released its CUPL (Common Universal tool for Programmable Logic) compiler. Shortly thereafter, Data I/O Corporation (the largest manufacturer of device programming hardware at the time) released its ABEL (Advanced Boolean Expression Language) product. Both of these software products, and the design languages they encapsulated, were evolutionary advances over the earlier PALASM and PALASM-like tools. They represented the first commercially successful attempts at creating relatively target-independent methods of programming hardware from a higher (software-like) level of abstraction.

Both ABEL and CUPL allowed logic designs to be described in formats ranging from higher-level logic equations (which included such things as relational and math operators) to large state diagrams describing complex control logic (as shown in Figure 2-2). Using such tools, it was now feasible for an embedded systems designer who had little or no knowledge of circuit design fundamentals and only the most rudimentary understanding of digital logic to describe and compile a high-performance algorithm directly into hardware.

Although the use of programmable logic devices for creating logic circuits grew dramatically in the early to mid-1980s, they did not take hold among embedded systems designers and programmers due to their relatively small size and the wide availability and low cost of standard microcontrollers. While microcontrollers were (and remain) exceptionally slow in terms of data throughput relative to what is possible using programmable logic, the tremendous flexibility provided by microcontrollers and microprocessors, along with their ability to host much larger algorithms and applications, made them the obvious choice for all but the simplest and most performance-critical functions.

```
module SequenceDetect
title '8-bit 10111101 sequence detector example'

declarations
DIN, CLK, ARST pin;        " Inputs
SEQ pin istype 'reg';      " Registered output
Q2,Q1,Q0 pin istype 'reg'; " State registers

" Define the state values using optimal encoding
SREG = [Q2,Q1,Q0];
  S0  = [ 0, 0, 0];
  S1  = [ 0, 0, 1];
  S2  = [ 0, 1, 1];
  S3  = [ 1, 1, 1];
  S4  = [ 1, 0, 1];
  S5  = [ 1, 1, 0];
  S6  = [ 0, 1, 0];
  S7  = [ 1, 0, 0];

equations
"Define the async reset and clock signals
[Q2,Q1,Q0].AR = ARST;
[Q2,Q1,Q0].CLK = CLK;

"Define the state transitions
state_diagram SREG
STATE S0:
  IF DIN == 1 THEN S1 ELSE S0;
STATE S1:
  IF DIN == 0 THEN S2 ELSE S0;
STATE S2:
  IF DIN == 1 THEN S3 ELSE S0;
STATE S3:
  IF DIN == 1 THEN S4 ELSE S0;
STATE S4:
  IF DIN == 1 THEN S5 ELSE S0;
STATE S5:
  IF DIN == 1 THEN S6 ELSE S0;
STATE S6:
  IF DIN == 0 THEN S7 ELSE S0;
STATE S7:
  IF DIN == 1 THEN S0 WITH Z:=1; ELSE S0;
end SequenceDetect
```

Figure 2-2. ABEL design file using state diagram syntax to describe a simple sequence detector (test vectors are removed for brevity).

2.2 REPROGRAMMABILITY, HDLS, AND THE RISE OF THE FPGA

Three factors combined in the mid- to late 1980s to cause a dramatic change in how programmable logic was applied, and in its applicability to hardware acceleration of software computations. These factors were reprogrammability, more flexible interconnect architectures, and tools for the automated synthesis of circuitry from standard hardware description languages.

Reprogrammability—the ability of a device to be erased and/or overwritten with new programming data—was a critical advancement over first-generation programmable logic devices, which, once programmed, were in effect fixed-function custom hardware devices. Reprogrammability had enormous implications for hardware designers creating complex systems. Not only could these systems be quickly developed and tested as prototypes (with no need to throw away experimental versions of a given hardware function), but field upgrades of hardware algorithms were now possible, just as it had become possible to provide field upgrades of software and PROM-based firmware.

The new, more flexible device architectures represented by FPGA devices (which were produced by Xilinx, Inc., Altera Corporation, and others) were also critical. These devices had dramatically higher logic capacities and more general-purpose features than had existed previously, and therefore provided a new set of opportunities for performing complex computations directly in hardware.

Finally, the emergence of more powerful logic synthesis software tools and the development and eventual standardization of two powerful hardware description languages, VHDL and Verilog, made it possible for application developers to work at a higher level of abstraction, leaving to the design tools the details of actual implementation of an algorithm as low-level logic gates, registers, and the like. A sample VHDL source file describing a cyclic redundancy check (CRC) module is shown in Figure 2-3. VHDL and Verilog descriptions can be relatively abstract (they "feel" like programming languages), but if the described function is intended for actual compilation to hardware, through the use of a logic synthesis tool, it must be assumed that the programmer has a somewhat detailed knowledge of the underlying hardware and, in particular, how clocked logic is implemented in low-level registers. In fact, this is a key characteristic of programming for FPGAs using VHDL or Verilog: the designer must understand the clock boundaries and must write logic descriptions that clearly define an application's clocked (or *timed*) behavior.

By the start of the 1990s, it was clear that programmable logic devices had matured to the point where high-performance software algorithms and even complete applications could be implemented directly in hardware. Throughout the 1990s, researchers involved in the domain of reconfigurable

```
-------------------------------------
-- 8-bit serial CRC generator
--
library ieee;
use ieee.std_logic_1164.all;

entity crc8s is
   port (Clk, Rst, Din: in std_ulogic;
         CRC: out std_ulogic_vector(15 downto 0));
end crc8s;

architecture behavior of crc8s is
   signal X: std_ulogic_vector(15 downto 0);
begin
   process(Clk,Rst)
   begin
      if Rst = '1' then
         X <= (others=> '1');
      elsif rising_edge(Clk) then
         X(0)  <= Din xor X(15);
         X(1)  <= X(0);
         X(2)  <= X(1);
         X(3)  <= X(2);
         X(4)  <= X(3);
         X(5)  <= X(4) xor Din xor X(15);
         X(6)  <= X(5);
         X(7)  <= X(6);
         X(8)  <= X(7);
         X(9)  <= X(8);
         X(10) <= X(9);
         X(11) <= X(10);
         X(12) <= X(11) xor Din xor X(15);
         X(13) <= X(12);
         X(14) <= X(13);
         X(15) <= X(14);
      end if;
   end process;

   CRC <= X;

end behavior;
```

Figure 2-3. VHDL design file describing a CRC generator.

FPGA-based computing published results that were often surprising, demonstrating the huge potential for algorithm acceleration using these devices, which were growing in capacity and power in lockstep with Moore's Law, which states that the number of transistors (or the *data density* in more recent definitions) in integrated circuits will double every 18 months. Traditional CISC and RISC processors, on the other hand, continue to suffer from lower silicon efficiencies when measured as a percentage of transistors actually capable of doing useful work at any given time.

And yet, at the turn of the century, the primary use of FPGAs was still almost overwhelmingly for traditional hardware functions: controllers, interface logic, and peripheral integration. FPGA-based computing had not yet caught hold.

In recent years, two important developments have combined with ever-increasing device densities to make FPGA-based computing practical for mainstream applications. The first of these developments is the availability of full-featured embedded FPGA processor cores, including the MicroBlaze and PowerPC cores available from Xilinx, the Nios and Nios II processor cores available from Altera, and other processor cores available from third-party IP suppliers. The second development, and the primary subject of this book, is the availability of software-to-hardware compiler tools.

2.3   SYSTEMS ON A PROGRAMMABLE CHIP

Embedded (or "soft") IP cores representing common, low-complexity microprocessors and microcontrollers have been available from third-party suppliers (typically subject to a licensing fee, although free, open-source cores are also available) for well over a decade. Using such cores in combination with other related peripheral cores makes it possible to create complete systems composed of, for example, an 8- or 16-bit controller, a UART, and other such I/O devices on a single programmable chip. It has become quite common, in fact, for embedded systems designers to forgo the use of dedicated legacy controllers (such as Zilog Z-80s, Intel 8051s, PIC processors, and others) and instead implement these processors directly within programmable hardware. This has had obvious revenue implications for the original processor developers and suppliers, who often have no IP protection apart from copyrighted—and easily replaced—assembly language specifications.

More recently, major FPGA vendors have created proprietary 32-bit processor cores optimized for their own FPGA devices and have made these processor cores available to their users at a very low cost, typically a few thousand dollars for a development license with no per-unit royalties. Both Xilinx and Altera have also created hybrid processor/FPGA platforms that combine standard 32-bit processors (a PowerPC and an ARM, respectively) with general-purpose programmable FPGA fabric to create what are often referred to as "platform FPGAs."

As FPGA densities increased, it became clear that the types of systems that could be implemented in this way were truly without bounds. One limiting factor in the general acceptance of FPGA-based systems-on-chip, however, has been on-chip interface standards. Combining IP cores from multiple vendors has proven to be a challenge, and here again the FPGA vendors have offered solutions that advance the usefulness of their proprietary platforms. Altera, for example, promotes the use of its Avalon interconnect and the related SOPC Builder software tools as a way to greatly simplify the selection and interconnection of a growing number of Altera and third-party IP cores. Xilinx, with its Platform Studio software and OPB bus, provides similar capabilities. In the open-source IP community, the Wishbone standard (originally developed by Wade Peterson of Silicore, Inc.) has gained converts.

The ability to quickly assemble a programmable hardware platform that combines one or more customized embedded processors with other third-party IP cores and general-purpose programmable logic resources is a great enabler for high-performance embedded applications. For embedded systems designers, then, all that is missing is a way to create new hardware functions and integrate them cleanly with other IP cores—without requiring hardware design skills. This is, of course, where software-to-hardware compiler tools are beginning to play a role.

What constitutes a programmable hardware platform?

When we talk about programmable hardware platforms in this book, we are referring to a system—which might be represented by a single programmable integrated circuit, a single-board computer, or a large-scale network of computing nodes—that includes both traditional software processors (such as RISC or CISC CPUs) and general-purpose programmable logic.

An FPGA-based platform may be defined by many physical factors, including the type and number of FPGAs, the arrangement of external (board-level) peripherals and related I/O, but may also be defined by the contents of the FPGA itself, such as whether an embedded processor core or other "soft" components are included.

Today's FPGA-based platforms are, of course, only the beginning. Future reconfigurable and high-performance computing platforms will certainly extend this concept to create ever larger, more powerful systems of connected processing resources.

Toward Faster System Prototypes and Higher Performance

Fundamentally, an FPGA provides the opportunity to create arbitrary hardware structures and to connect these structures using a switched fabric of programmable interconnects. The flexibility of such a platform creates enormous opportunities for increasing the performance of algorithms—in particular, algorithms that can make use of parallelism. Without an efficient and semi-automated way to compile higher-level algorithms to equivalent lower-level hardware structures, however, the rapid creation of high-performance systems is difficult to achieve.

A software-to-hardware compiler that allows applications (or application components) to be compiled directly to an FPGA is a critical component of a rapid prototyping environment. By introducing C-language compilation to an existing design flow, it becomes possible to more quickly experiment with alternative design concepts and to create working prototypes. And by using the same design entry language and software-oriented methods of debugging and verification, system designers can very quickly create and evaluate different prospective hardware and software partitioning strategies and create working product prototypes. As C-to-hardware tools mature and FPGAs become larger, using such methods of design (whether the input language is C or some other language more appropriate for parallel programming), the practicality of hand-coding large applications at the level of HDLs will diminish to the point where such methods will be used for only the most performance-critical and/or irregular types of hardware components.

2.4   FPGAs for Parallel Computing

The recent history of high-performance computing research has suggested that massively parallel custom computing engines built of FPGAs (or FPGA-like structures) will play an important role in future applications outside the realm of traditional hardware or embedded systems design. Researchers worldwide have demonstrated that supercomputer-class performance can be obtained for many critical tasks at a tiny fraction of the cost of more general-purpose supercomputing hardware. A common element of the ongoing research in this area is the use of arrays of relatively small, independently synchronized processing elements connected via data streams or messaging-oriented communications. Using this concept of independently synchronized, semi-autonomous processing elements provides great flexibility and power and fits well into an FPGA-based reconfigurable computing approach.

Reconfigurable Computing and the FPGA

Reconfigurable computing is a research area that focuses on carrying out high-performance computations using highly flexible, dynamically reconfigurable computing fabrics such as FPGAs. Reconfigurable computing differs from more traditional computing methods by allowing substantial changes to be made to the data path, in addition to allowing control flow to be altered.

The concept of reconfigurable computing has been around since the 1960s, when a paper published by Gerald Estrin (see Appendix F, "Selected References," for reference) proposed a computer architecture consisting of a standard processor coupled to an array of "restructurable" hardware elements. In such a system, the main processor acted as a controller for the reconfigurable hardware, which was dynamically changed (reconfigured) to perform specific tasks as dedicated hardware. These ideas did not, at the time, lead to a practical implementation, because the underlying capability for dynamic generation of hardware did not yet exist.

Since the introduction of FPGAs in the 1980s, however, there has been a great deal of new research into creating dynamically reconfigurable computers using these relatively low-cost, easy-to-program devices. Multiple conferences are held on this specific topic, including the *IEEE Symposium on Field-Programmable Custom Computing Machines* (FCCM), held each spring in Napa, California, and the *International Conference on Field-Programmable Logic and Applications* (FPL), held in late summer in various European locations.

When it comes to reconfigurable computing, however, FPGAs are not the only story. There have also been newer types of devices, either proposed or actually developed, that lie somewhere between FPGAs and standard processors. These *coarse-grained* programmable devices promise both the hardware flexibility of FPGAs and the software programmability of traditional processors, with less overhead in terms of the on-chip interconnects and related place-and-route issues mentioned earlier. As of this writing, however, none of these new architectures (which in the commercial sector have been offered by companies including PACT, IPFlex, Quicksilver, Morphics, Chameleon, picoChip, and others) have proven to be general-purpose enough to become true microprocessor alternatives. Perhaps equally important, no unifying programming methodology has yet emerged for highly parallel, hardware-based computing. Such a common methodology is important to allow this new technology—reconfigurable computing—to become widely popular.

2.5 SUMMARY

This chapter has traced the history of programmable hardware, from its early beginnings in the late 1960s through the "golden age" of programmable logic in the late 1970s and early 1980s and into the era of the FPGA, which began in the mid-1980s and continues to this day.

If there is one thing that we hope we've shown by presenting this brief history, it is that advances in design tools and corresponding advances in the methods of programming for these devices have been a major factor in the acceptance and broad use of each succeeding generation of programmable hardware.

Although much of this book deals with more modest applications in the domain of embedded systems, it is clear that FPGAs—and FPGA-based computing—have an enormously important role to play in future high-performance, reconfigurable computing platforms. It is also clear that FPGAs themselves will continue to evolve in parallel with (and potentially in response to) developments in parallel programming tools and methods. In the next chapter, we'll begin our investigation of software-based methods for programming the newest generation of FPGA-based platforms.

CHAPTER 3

A Programming Model for FPGA-Based Applications

At this point we have described the general nature of FPGA-based programmable platforms, and we have described how embedded microprocessors may be used to implement traditional software algorithms and complete applications within the FPGA. We have also, to some extent, discussed the role of programming tools that allow software algorithms (expressed using languages such as C) to be implemented directly in programmable hardware for high performance.

What we haven't discussed yet is how such an application, which might be composed of dozens or hundreds of independently operating parallel processes, can be expressed. How are the independent functional elements described and interconnected? How is the system synchronized? This chapter begins that discussion by presenting a view from the perspective of the programming model. Choosing an appropriate programming model is important because it allows us, as software and hardware programmers, to create applications and their constituent algorithms in an abstract manner, while still remaining congizant of—and able to program efficiently for—the underlying machine.

3.1   PARALLEL PROCESSING MODELS

The popularity and usability of today's most common processor architectures (whether they are simple embedded controllers or more advanced, workstation-class processors) owe a great deal to the fact that they all share a common machine model, which was first described by computer pioneer John von Neumann in the 1940s. In a classic von Neumann computer, a central processing unit (CPU) is connected to some form of memory that is itself capable of storing both machine instructions and data. The CPU executes a stored program as a sequence of instructions that perform various operations including read and write operations on the memory.

SISD: The Original Processor Machine Model

At its simplest, such a single-processor machine is termed (in the generally accepted taxonomy of computers) a *Single Instruction, Single Data machine* (SISD). In such a machine, only one operation on one data input can be performed at any given time, and that operation is defined by only one computer instruction (an ADD operation, for example). It is for this basic machine model that the vast majority of software programming languages (and resulting application programs) have been developed.

In a programming environment for SISD machines, a software process is expressed using a series of statements, each corresponding to one or more distinct machine operations, that are punctuated by various branches, loops, and subroutine calls. The key thing to understand is that today's most common programming methods—or *programming models*—have been developed to meet the needs of the basic SISD model. This has remained true even as various levels of parallelism have crept into the machines for which we program, in the form of instruction pipelines and other such processor architecture enhancements.

The existence of a common machine model for general-purpose processors has been extremely useful for software developers. Programming languages—and the expertise of programmers who use them—have evolved in a gradual manner as the applications and operating systems that are implemented on these machines have grown increasingly complex and powerful. As support for multitasking and multithreaded operating systems and processor architectures has emerged, existing languages have been adapted in response, and new (but not fundamentally different) languages have been developed.

Throughout this evolutionary process, the application programmer has assumed that a given program is being executed on a single, sequential processing resource. There are exceptions, of course, but for the vast majority of software applications being developed today, including embedded software applications, programmers have been trained to think in terms of a linear flow of control and to consider software programs to be fundamentally sequential in their operation. But parallelism at the machine level has in fact been with us for some time.

Early in the development of processor architectures, parallelism was introduced to increase processor performance. An example of this is an instruction prefetch operation, which allows the overlapping of instruction fetches and instruction executions. This feature later evolved into generalized instruction pipelining, which allows time multiplexing of operations. More recent advances include vector processors, which support multiplexing in both the time and space domains, whereby a given instruction can operate on several inputs simultaneously.

The SIMD Machine Model

If we move out of the realm of traditional processors and into the domain of supercomputers, we can find examples of machines that support much greater degrees of parallelism, in which a controlling processor (or control unit) directs the operation of many supporting processing elements, each of which performs some specific operation in parallel with all the others. In such a machine, a single instruction (which might perform a matrix multiply operation, for example) triggers a potentially large number of processing elements to execute the same operation simultaneously, but each on its own data. This is an example of a *Single Instruction, Multiple Data* (SIMD) machine model. A number of commercial supercomputers have been constructed using this model, including machines offered by Thinking Machines, Cray, and Digital Equipment Corporation.

MIMD Machines and the Transputer

If we take parallelism to the next logical level, we can conceive of machines that are capable of executing *Multiple Instructions on Multiple Data* (MIMD). In this type of machine, many different instruction processors operate in parallel, each accepting different data and executing different instructions. This sounds like the best of all possible worlds, but programming such a system necessarily involves coordinating all the machine's independent processing elements to solve some larger problem. This is trickier than it sounds, particularly given that the programmers who would make use of such machines are used to thinking in terms of a sequential flow of control.

There has been much research into the development and programming of such "multicomputer" systems and into methods of programming for them. One such project (which has spawned many other areas of parallel processing research efforts) has been the Transputer, first described by the English company INMOS in the mid-1980s.

The Transputer was a blueprint for creating highly modular computer systems based on arrays of low-cost, single-chip computing elements. These self-supporting, independently synchronized chips were to be connected to form a complete computer of arbitrary size and complexity. The goal of this modular architecture was to allow any number of Transputers to be connected, creating a high-performance parallel computing platform with little or no need to design complex interconnections or motherboards.

Communication between Transputer processing elements was via serial links. This meant that the primary performance bottleneck in such a system might well have been data movement rather than raw processing power. For many types of applications the Transputer nonetheless demonstrated extremely high performance for relatively low cost, and the project suggested an architecture that—two decades later—makes a great deal of sense when considering high-performance computing on FPGA-based platforms.

Because of their high degree of parallelism, Transputers were programmed using a unique programming language called Occam. Occam supported parallelism at different levels, including a thread model for multi-process programs and language features for describing parallelism at the level of individual statements and statement blocks. The language supports the explicit unrolling of loops, for example, and the explicit parallelizing of individual statements.

Shared Memory MIMD Architectures

Because serial interfaces form the communications backbone of a Transputer-based system, such a machine may be characterized as a *message-passing* architecture. Message passing in its purest form assumes that there is no shared memory or other shared system resources. Instead, the data in a message-passing application moves from process to process as small packets on buffered or unbuffered communication channels. This simple interconnection strategy makes it possible for processing elements to operate with a high degree of independence and with fewer resource conflicts.

Another category of MIMD machines includes those with shared memory resources, which may be arranged in some hierarchy to provide a combination of localized high-speed storage elements (such as local, processor-specific caches), as well as more generally accessible system memory. These memories are used in conjunction with, or as alternatives to, the message-passing data communications mechanism.

3.2 FPGAs AS PARALLEL COMPUTING MACHINES

There is no argument that FPGAs provide enormous opportunities for performing parallel computations and accelerating complex algorithms. It is common—indeed, almost trivial given the right programming experience and tools—to demonstrate speedups of a thousand times or more over software approaches for certain types of algorithms and low-level computations. This is possible because the FPGA is, in many respects, a blank slate onto which a seemingly infinite variety of computational structures may be written. An FPGA's resources are not unlimited, however, and creating structures to efficiently implement a broad set of algorithms, ranging from large array-processing routines to simpler combinatorial control functions, can be challenging. As you will see in later chapters, this suggests a two-pronged approach to application development. At the low level, compiler tools can be used to extract and generate hardware for instruction-level parallelism. At a higher level, parallelism can be expressed explicitly by modeling the application as a set of blocks operating in parallel.

The key to success with FPGAs as computing machines is to apply automated compiler tools where they are practical, but at the same time use programming techniques that are appropriate for parallel computing. Although tools have been developed that will extract parallelism from large, monolithic software applications—applications that were not written with parallelism in mind—this technique is not likely to produce an efficient implementation; the compiler does not have the same knowledge of the application that the programmer possesses. Hence, it cannot make the system-level decisions and optimizations that are needed to make good use of the available parallel structures. In addition, it should be understood that compiler tools for FPGAs (including those described in this book and all others currently in existence) are still in their infancy. This means that for maximum performance it may be necessary for a hardware designer to step in and rewrite certain parts of the application at a low level. It is therefore important to use a programming and partitioning strategy that allows an application to be represented by a collection of any number of semi-independent modular components, such that hardware-level reengineering is practical and does not represent a wholesale redesign of the application.

The approach of partitioning an application for system-level parallelism suggests the need for a different conceptual model of program execution than is common in traditional software development. In this model, functionally independent subprograms are compiled into hardware blocks rather than into the assembly language of a processor. Within these blocks there is no CPU with its fetch-and-execute cycle. Rather, whole components of the program can execute in parallel, to whatever degree the compiler (and the software programmer) can handle.

In support of such a machine model—one in which multiple program blocks are simultaneously operating on multiple data streams, and in which each program block is itself composed of parallel structures—we need a different kind of programming model, one that is both parallel and at the same time procedural. The C language libraries used in this book support such a model of programming while retaining the general look and feel of—and compatibility with—C language programming.

As you have seen from the previous descriptions of machine models, a programming model does not need to exactly mirror the underlying machine model. In fact, the more a programming model can be abstracted away from the machine model, the more reliance the program can place on the compiler tools, and the easier the machine is to program. There is a downside to this, however, which is that the program's efficiency also depends on the compiler's capabilities.

Why use C?

Why use C, rather than a language such as Occam that was specifically designed for parallelism? The primary reason for using C is that it is the language that the vast majority of embedded systems and desktop application developers are most familiar with. This is important because parallel programming—expressing the operation of multiple interacting processes—is not trivial. At the same time, coarse-grained parallel programming does not require a fundamentally different method of expressing individual blocks of functionality (a loop, for example).

The widespread understanding of the C language also makes it easier for us, in the book, to describe how programming for parallelism is different, and it provides us with a familiar way of expressing the basic statements that make up a given algorithm.

Another important benefit of C is the large body of legacy algorithms and applications. While direct translation of these algorithms into a streaming parallel model is not always practical, it is quite possible to combine legacy sequential applications and algorithms with their parallel equivalents for the purpose of testing and algorithm validation. It is also possible to create mixed software/hardware applications with relatively little effort.

The programming model described in this book takes a middle path by abstracting away details of the lower-level (instruction-level) parallelism while offering explicit control over higher, system-level parallelism.

Adding Soft Processors to the Mix

If we consider an FPGA-based computing machine to be a collection of parallel machines, each of which has its own unique computing function, there is no reason why one or more of these machines can't be a traditional microprocessor. Given the wide availability of FPGA-based "soft" processors, this is a reasonable, practical way to balance the need for legacy C and traditional programming with the need for application-specific hardware accelerators.

Such FPGA-based processors can be useful for a variety of reasons. They can run legacy code (including code that is planned for later acceleration in the FPGA fabric). They can be used during development as software test generators. They can also be used to replace most costly hardware structures for such things as embedded state machines, and for standardizing I/O. They can run operating systems and perform noncritical computations that would be too space-intensive when implemented in hardware. When arranged as a grid, multiple soft processors can even form a parallel computing platform in and of themselves—one that is more generally programmable than an equivalent platform constructed entirely of low-level FPGA gates.

The recent explosion in the use of soft processors has proven that FPGAs can provide a flexible, powerful hardware platform for complete "systems-on-programmable-chips." FPGA vendors now provide (at little or no cost) all the necessary processor and peripheral components needed to assemble highly capable single-chip computing platforms, and these platforms can include customized, highly parallel software/hardware accelerators.

Used in this way, FPGAs are excellent platforms for implementing *coarse-grained heterogeneous parallelism.* Compared to other models of machine parallelism, this approach requires less process-to-process communication overhead. If each process maintains its own local memory and has a clearly delineated task to perform, the application can easily be partitioned between different areas of the FPGA (perhaps including different clock domains) and between independent FPGA devices. Many types of calculations lend themselves quite naturally to coarse-grained parallelism, including vector array processing, pipelined image and audio processing, and other multistage signal filtering.

3.3   PROGRAMMING FOR PARALLELISM

As we implied in the preface to this book, there is a fundamental problem with attempting to program general-purpose hardware (or "non-von Neumann machines," if you will) using the C language. The problem is how to express parallelism. Parallel processing and the programming of parallel systems require support for concurrency in the language being used and an understanding of how to manage multiple quasi-independent computational elements on the part of the programmer. The standard C language does not contain any such features. VHDL and Verilog, on the other hand, which are intended for describing highly parallel systems of connected hardware components, are designed for exactly this purpose, albeit at a rather low level of abstraction.

The closest thing to a truly parallel programming model in the context of C-language programming is support for threads, which is not a standard feature of C but is popular and readily available in the form of add-on, operating system-specific libraries. Another, less common C library for this type of programming is the message-passing interface (MPI). This library is intended for the design of larger supercomputing applications implemented on clusters of standard desktop computers and other heavy-duty platforms.

So if the C language (and, by extension, any other programming language developed for von Neumann machines) is not appropriate for programming general-purpose hardware, and if the languages specifically designed for such hardware (for example, HDLs) are too low-level, what is the answer? As it turns out, a compromise solution is best. On the hardware side, we need to assemble all those undifferentiated hardware elements (the programmable gates and flip-flops that make up an FPGA) into some kind of abstract structure appropriate for higher-level programming. Fortunately, compiler tools can create this structure automatically, using knowledge of the application and of the available low-level FPGA hardware. On the software side, we need to extend the language of choice (which for the purposes of this discussion will be C) to support programming the abstract parallel processing machine that we have just assembled, as illustrated in Figure 3-1.

To summarize: to make sense of programming FPGA-based hardware (as opposed to designing it from the ground up), we need to create an abstract machine model and choose a software programming model appropriate for that abstract machine.

Parallel programming researchers have generally found that creating an abstract, multinode machine (sometimes called a multicomputer) consisting of multiple, semi-autonomous processing nodes (often called processing elements [PEs]) is a good way to express a platform for coarse-grained parallel processing. By targeting this model of an abstract machine (whether or not the underlying hardware actually implements such a system) it is relatively easy to construct a usable programming model. The communicating sequential process (CSP) programming model has been well-studied and can be used to apply formal methods for FPGA design, as well as to build actual applications.

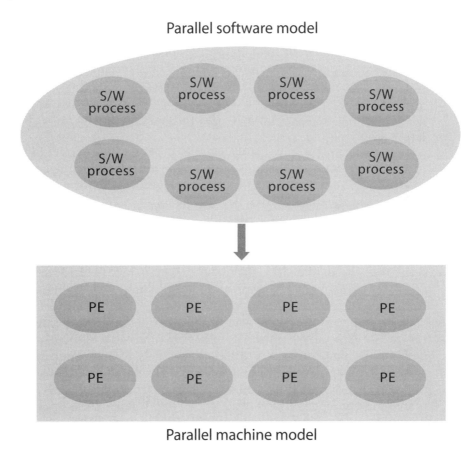

Figure 3-1. Multinode parallel machine model and multiprocess programming model.

3.4 COMMUNICATING PROCESS PROGRAMMING MODELS

The programming model we will focus on here was first described by Sir Anthony Hoare in 1978. Hoare later published his ideas in 1985 in a seminal book on the topic titled *Communicating Sequential Processes*. CSP, as described by Hoare, is both a programming model and a language for describing patterns of interactions between independently operating components called processes. Each of the processes in such a system may represent a traditional software program (which operates sequentially in terms of its internal behavior), but processes are limited to communicating with one another via well-defined channels.

Each program in a CSP application may be executing as its own hardware process, independent of all the others and acting much as a traditional von Neumann machine would operate. Rather than having a main controlling process that calls other processes as subroutines, the application as a whole is designed such that data moves through the system (and through various processes in the application) via buffered data channels. While each running process has access to local memory resources (for the purpose of performing computations and storing intermediate results), there is little or no communication between independent processes except via the data channels, which are sometimes called streams.

In an idealized CSP application, no time delay or other overhead is involved in the transfer of data (which is assumed to consist of a relatively small payload of values) between two nodes, regardless of their relative locations. In reality there may be significant bandwidth and latency considerations, particularly if the application is spread between processing elements of dramatically different types, such as FPGA hardware and embedded processors. A "pure" implementation of CSP would also not include any communication outside of the data channels/streams (such as when two processes access the same memory) because doing so would violate the notion of synchronizing the processes to the data.

An important characteristic of the CSP model is that local memory access is less expensive than access to remote (node-external) memory or to the data buffers that implement the data channels. When programming a CSP system, it is therefore important to access data locally whenever possible, as opposed to externally, to realize optimum performance. This parallel programming concept is called *locality*, and it is fundamental to understanding how to create high-performance parallel applications.

A machine model based on CSP is an ideal target for compiling coarse-grained parallel applications in C. The process components of a CSP application enclose manageable blocks of computation for which a compiler can generate hardware or software from relatively straightforward, standard C-language statements. At the system level, extensions to the C language (provided in the form of function libraries) can provide the needed interprocess communication channels to express a streams-oriented parallel programming model.

Is the resulting hybrid programming model (not quite software, not quite hardware) the right choice for all applications? Surely not, but for the types of applications that can most benefit from acceleration on FPGA-based, highly parallel platforms, it is more than adequate.

3.5   THE IMPULSE C PROGRAMMING MODEL

The Impulse C programming model used in this book is a communicating se-quential process model intended for streams-oriented, mixed hardware/soft-ware applications. These applications are becoming quite common and represent a large percentage of the computationally intensive problems re-quiring hardware acceleration. Examples of such applications are found in many domains, including image processing (for consumer products as well as security and videoconferencing), data communications (including data compression and encryption/decryption), wireless communications, geo-physics, genetic and medical analysis, and many others.

At the heart of the Impulse C programming model (shown in Figure 3-2) are processes and streams. Processes are independently synchronized, con-currently operating application components. It's useful to think of these processes as persistent subprograms that accept various data, perform spe-cific computations and generate outputs, either on data streams or by writing to memories or other system resources. Unlike traditional software subrou-tines, these processes are not "called" (repeatedly invoked during the course of the program's operation) but are instead "always on" and constantly re-sponding to data appearing on their inputs. Processes are persistent, perform-ing the work of an application by accepting data, performing computations, and generating outputs.

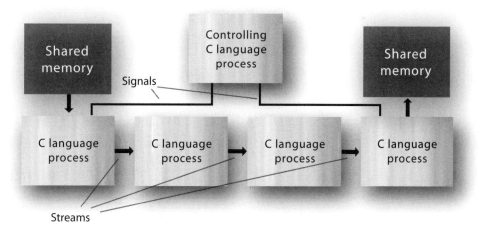

Figure 3-2. The data processed by an application flows from process to process by means of streams, or in some cases by means of signals and/or shared memories.

The CSP programming model is designed to simplify the creation of highly parallel applications using a streams-oriented approach to data movement, processing, and synchronization.

An unlimited number of applications may be expressed in this way, but applications with some or all of the following characteristics are best suited to the CSP model:

- The application features high data rates to and from data sources and between processing elements.

- Data packet sizes are fixed, with relatively small stream payloads.

- Multiple related but independent computations are required to be performed on the same data stream.

- The data consists of low- or fixed-precision values, typically fixed-width integers or fixed-point fractional values.

- There are references to local or shared memories, which may be used for storage of data arrays, coefficients, and other constants, and to deposit results of computations.

- Multiple independent processes are communicating primarily through the data being passed, with occasional synchronization being requested via messages.

Impulse C is specifically designed to address streaming applications, or other such applications in which a large amount of data must be processed through a relatively intense set of operations. (Note that other C-based hardware design languages, including the Handel-C language offered by Celoxica, provide similar capabilities.)

In Impulse C, the communicating sequential process programming model is embodied in a small set of datatypes and intrinsic library functions that can be called from a conventional C program. The C-compatible language extensions are used to specify an application as a collection of processes communicating via streams, signals, and shared memories. Intrinsic library functions defined by Impulse C are used to communicate stream and signal data between the processes and to move data in and out of shared or local memory elements. The processes are then assigned to actual resources on the target programmable platform through the use of additional Impulse C library functions.

Streams are the primary method by which multiple processes communicate with one another and synchronize their activities. An application with well-designed stream interfaces will operate efficiently when mapped to actual hardware and software. Much of the effort in creating such applications is therefore found in the design of the stream interfaces and in optimizing data movement.

3.6 SUMMARY

In this chapter we have described, in general terms, a programming model appropriate for describing highly parallel, mixed software/hardware applications. This programming model has been proven to be an effective, efficient way to describe the types of applications and algorithms for which programmable hardware platforms are best suited. In subsequent chapters you'll learn through example how to make use of this programming model within the context of the C language.

CHAPTER 4

An Introduction to Impulse C

The programming examples presented in the remainder of this book are written using Impulse C, a function library and related compiler and debugging tools provided by Impulse Accelerated Technologies. These libraries and tools are compatible with standard ANSI C and with standard C development tools such as Microsoft Visual Studio and gcc.

Impulse C supports the development of highly parallel, mixed hardware/software algorithms and applications using the communicating process programming model described in the preceding chapter. The features of Impulse C for expressing parallelism at a system level are similar to features found in other C-based languages for hardware and mixed hardware/software design, including Celoxica's Handel-C and SystemC. This means that the concepts we will describe in this and subsequent chapters are applicable to different C-based FPGA design environments and are in fact useful even if you are developing FPGA applications using some other method of design.

In the preface to this book we stated that C language programming is not a replacement for proven methods of hardware design using hardware description languages. Impulse C can be used to describe a wide variety of functions that are appropriate for compiling to FPGA hardware, but it is not intended for describing low-level hardware structures. Nor is it intended for converting large, monolithic C applications, which typically consist of many C subroutines that are invoked via remote procedure calls.

Where did Impulse C originate?

Impulse C has its philisophical roots in research carried out at Los Alamos National Laboratories under the direction of Dr. Maya Gokhale. This research, which culminated in the publicly available Streams-C compiler (www.streams-c.lanl.gov), provided a method of expressing applications for implementation on FPGA-based, board-level platforms for the purpose of high-performance, hardware-accelerated computing. Applications developed using Streams-C have been in the domains of data encryption, image processing, astrophysics, and others.

Impulse C borrows its programming model and general philoso-phy from the Streams-C programming environment but differs from Streams-C in a number of respects, the most important being its focus on maintaining compatibility with standard C programming environ-ments. By using the Impulse C libraries it's possible to describe appli-cations consisting of many (perhaps hundreds) of communicating processes and simulate their collective behavior using standard C development tools including Microsoft Visual Studio and gcc- and gdb-based environments.

To allow the compilation and simulation of highly parallel appli-cations consisting of independently synchronized processes, the Im-pulse C libraries include functions that define process interconnections (typically streams and/or signals) and emulate the behavior of multiple processes (for the purpose of desktop simula-tion) using threads.

Monitoring functions included with the Impulse C library allow specific processes in a large, parallel application to be instrumented with special debugging functions. The results of computations are dis-played in multiple windowed views, most often while the application is running under the control of a standard C debugger. This capability is introduced in this chapter and is described in more detail in subse-quent chapters.

For hardware generation, the Impulse tools include a C lan-guage compiler flow that is based in part on the publicly available SUIF (Stanford Universal Intermediate Format) tools, which are com-bined with proprietary optimization and code generation tools devel-oped at Impulse Accelerated Technologies.

The Impulse C tools include a software-to-hardware compiler that converts individual Impulse C processes to functionally equivalent hardware descriptions and that generates the necessary process-to-process interface logic. While this C-to-hardware compilation is an enormous time-saver, it is still up to you, as the application developer, to make use of the tools, the Impulse C libraries, and an appropriate multiprocess programming model to effectively develop applications appropriate for these new categories of programmable hardware platforms. In the examples and tutorials included in this and later chapters, we'll show you how.

4.1 THE MOTIVATION BEHIND IMPULSE C

The goal of Impulse C is to allow the C language to be used to describe one or more units of processing (called processes) and connect these units to form a complete parallel application that may reside entirely in hardware, as low-level logic mapped to an FPGA, or that can be spread across hardware and software resources, including embedded microprocessors and DSPs. This multiprocess, parallel approach is highly appropriate for FPGA-based embedded systems, as well as for larger platforms that consist of many (perhaps hundreds) of FPGAs interconnected with traditional processors to create a high-performance computing platform.

The Impulse C approach focuses on the mapping of algorithms to mixed FPGA/processor systems, with the goal of creating hardware implementations of processes that optionally communicate (via streams, signals and memories) with software processes either residing on an embedded microprocessor and/or implemented as software test bench functions in a desktop simulation environment.

Support for streams, signals, and memories is provided in Impulse C via C-compatible intrinsic functions and related stream, signal, and memory datatypes. For processes mapped to hardware, the C language is constrained to a subset of C, while software processes are constrained only by the limitations of the host or target C compiler.

The Impulse C programming model is that of communicating processes, as presented in the previous chapter. The Impulse C compiler generates synthesizable HDL for hardware processes as well as generating the required hardware-to-hardware and hardware-to-software interfaces implementing the specified streams, signals, and memories. The compiler can perform instruction scheduling, loop pipelining, and unrolling. It includes various pragmas (expressed using the #pragma statement in C) allowing optimization results to be tailored to meet general size/performance requirements.

The Impulse C software library supports desktop emulation/simulation of the parallel behavior of Impulse C applications when compiled using standard C development tools such as gcc and Visual Studio. The Impulse C libraries also support execution of Impulse C software processes on one or more embedded processors, providing a programming model in which system-level parallelism (expressed using multiple processes running on embedded processors and in FPGA hardware) may be expressed.

Impulse C *platform support libraries* are available for specific FPGA-based targets such as the Xilinx MicroBlaze and PowerPC-based FPGAs, as well as Altera FPGAs featuring the Nios and Nios II soft processors. Impulse C can also be used to generate hardware modules that do not interface to software processes, meaning it is not necessary to include an embedded processor to make use of Impulse C.

As you'll see in a later chapter, the ability to create mixed hardware/software applications in a common language is also helpful for creating in-system, software-driven test benches for hardware modules. In this use model, specific hardware modules, whether originally described in C or hand-crafted using HDLs, are tested using software and/or hardware test modules written in C and compiled to the embedded processor, to available FPGA resources, or to both, creating a mixed hardware/software test system.

4.2 THE IMPULSE C PROGRAMMING MODEL

Impulse C extends standard ANSI C using C-compatible predefined library functions in support of a communicating process parallel programming model. This programming model is conceptually similar to a dataflow or communicating sequential process programming model in that it simplifies the expression of highly parallel algorithms through the use of well-defined data communication, message passing, and synchronization mechanisms. The programming model supports a wide range of applications and parallel process topologies.

In Impulse C, the programming model emphasizes the use of buffered data streams as the primary method of communication between independently synchronized processes, which are implemented as persistent (rather than being repetitively called) C subroutines. This buffering of data, which is implementing using FIFOs that are specified and configured by the application programmer, makes it possible to write parallel applications at a higher level of abstraction, without the clock cycle-by-cycle synchronization that would otherwise be required.

Impulse C is designed for streams-oriented applications, but it is also flexible enough to support alternate programming models, including the use of signals and shared memory as a method of communication between parallel, independently synchronized processes (see Figure 4-1). The programming model you select depends on the requirements of your application, but also on the architectural constraints of the selected programmable platform target.

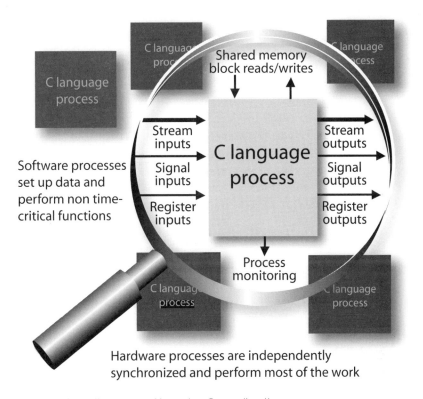

Figure 4-1. Processes form the core of Impulse C applications.

The Impulse C library consists of minimal extensions to the C language (in the form of new datatypes and predefined function calls) that allow multiple, parallel program segments to be described, interconnected, and synchronized. An Impulse C application can be compiled by standard C development tools (such as Visual Studio or gcc) for desktop simulation or can be compiled for an FPGA target using the Impulse C compiler. The Impulse C compiler translates and optimizes Impulse C programs into appropriate lower-level representations, including VHDL hardware descriptions that can be synthesized to FPGAs, and standard C (with associated library calls) that can be compiled onto supported microprocessors through the use of widely available C cross-compilers.

The complete Impulse C environment consists of a set of libraries allowing Impulse C applications to be compiled and executed in a standard desktop compiler (for simulation and debugging purposes) as well as cross-compiler and translation tools allowing Impulse C applications to be implemented on selected programmable hardware platforms. Additional tools for application profiling and co-simulation with other environments (including links to EDA tools for hardware simulation) are provided.

4.3 A MINIMAL IMPULSE C PROGRAM

In their classic book *C Programming Language*, authors Brian Kernighan and Dennis Ritchie began by presenting a simple example called "Hello World." In honor of Kernighan and Ritchie's book (which was first published in 1978 by Prentice Hall) we now present a minimal Impulse C application that we call "Hello FPGA." This will allow us to demonstrate how data moves in and out of Impulse C processes, as well as providing a quick introduction to the Impulse C library.

If you are familiar with the "Hello World" example presented by Kernighan and Ritchie, you may initially be dismayed at the complexity of this purportedly simple example. Don't worry, though; the concepts are straightforward, and you can use this basic example as a template for any number of Impulse C examples, large and small.

An important aspect of Impulse C highlighted by this simple example is the creation of a software test bench, which will provide you with your first look at how parallelism is expressed as well as demonstrating how to test streams-oriented hardware and software processes. The relationship of the software test bench to the hardware process is diagrammed in Figure 4-2. As shown in the diagram, the goal of this example is to demonstrate moving data from the software test bench to the hardware process and back. (We'll cover the concept of software test benches more fully in later chapters.)

The "Hello FPGA" example includes two source files; let's examine them in detail.

The Software Source File: HelloFPGA_sw.c

The file HelloFPGA_sw.c (see Figure 4-3) contains those portions of the application that represent software running on a traditional processor, which may be an embedded processor that is part of the target system or (as in this case) may be used only as a software test bench during desktop simulation. This part of the application is written using standard C, with few (if any) constraints being placed on the kind of C statements used.

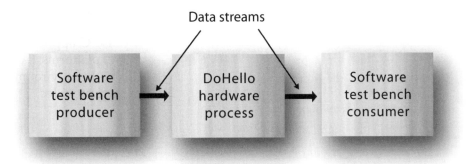

Figure 4-2. HelloFPGA hardware process and software test bench.

Examining this source file section by section and starting at the top, we find:

- A comment header, which in this example is limited to one line for the sake of brevity. As in traditional C programming, you should make extensive use of comments for internal documentation as well as for revision control. This is particularly important if your application makes certain assumptions about the target platform or relies on specific tool settings.

- #include statements referencing the Impulse C co.h file as well as the standard C library stdio.h:

```
#include "co.h"
#include <stdio.h>
```

The co.h file (which is located in the Impulse C installation directories in the /include subdirectory) contains declarations and macros representing the Impulse C function library. Functions in this library are prefixed with the co_ character combination.

- An extern declaration for the function co_initialize:

```
extern co_architecture co_initialize(int param);
```

The co_initialize function is a special function that you must include in your application, typically in the same file containing your application's configuration function (which we will describe in a moment). The extern declaration for co_initialize in HelloFPGA_sw.c is required so that co_initialize may be referenced in the application's main function.

```
// HelloWorld_sw.c: Software processes to be executed on the CPU.
#include "co.h"
#include <stdio.h>
extern co_architecture co_initialize(int param);

void Producer(co_stream output_stream) {
    int32 i;
    static char HelloWorldString[] = "Hello FPGA!";
    char *p;

    co_stream_open(output_stream, O_WRONLY, CHAR_TYPE);
    p = HelloWorldString;
    while (*p) {
        printf("Producer writing output_stream with: %c\n", *p);
        co_stream_write(output_stream, p, sizeof(char));
        p++;
    }
    co_stream_close(output_stream);
}

void Consumer(co_stream input_stream) {
    char c;

    co_stream_open(input_stream, O_RDONLY, CHAR_TYPE);
    while (co_stream_read(input_stream, &c, sizeof(char)) == co_err_none ) {
        printf("Consumer read %c from input stream\n", c);
    }
    co_stream_close(input_stream);
}

int main(int argc, char *argv[]) {
    int param = 0;
    co_architecture my_arch;

    printf("HelloFPGA starting...\n");

    my_arch = co_initialize(param);
    co_execute(my_arch);

    printf("HelloFPGA complete.\n");
    return(0);
}
```

Figure 4-3. HelloFPGA software test bench functions (HelloFPGA_sw.c).

- A **Producer** process run function, which has been described as a C function with a **void** return value:

```
void Producer(co_stream output_stream) {
```

This process run function represents the input side of a software test bench. The code in this process generates some test data (in this case, a brief stream of characters spelling out "Hello FPGA!") to test the hardware module that is described in **HelloFPGA_hw.c**. The process itself consists of some declarations and an inner code loop (a **while** loop) that iterates through the input string and writes characters to a stream declared as **output_stream**. Note the use of **co_stream_open, co_stream_write**, and **co_stream_close** in this process:

```
co_stream_open(output_stream, O_WRONLY, CHAR_TYPE);
p = HelloWorldString;
while (*p) {
    printf("Producer writing output_stream with: %c\n", *p);
    co_stream_write(output_stream, p, sizeof(char));
    p++;
}
co_stream_close(output_stream);
```

As you will see, Impulse C separates the definition of a process run function such as this one from its actual implementation, or instantiation, in hardware or software. Process run functions (which are technically procedures, not functions, because they have no return value) are described in more detail later in this chapter.

- A **Consumer** process run function, which has also been described as a C function with a **void** return value:

```
void Consumer(co_stream input_stream) {
```

This process represents the output side of our software test bench. Like the **Producer** process, this process interacts with other parts of the application (in particular, the hardware module being tested) via a data stream. In this case the data stream is incoming rather than outgoing, as indicated by the **O_RDONLY** mode specified in the call to **co_stream_open**:

```
co_stream_open(input_stream, O_RDONLY, CHAR_TYPE);
```

As in the **Producer** process, this process includes an inner code loop (which is again a **while** loop) that causes characters to be read from the input stream (which is declared here as a **co_stream** argument named input_stream) as long as characters are being generated by the module under test:

```
co_stream_open(input_stream, O_RDONLY, CHAR_TYPE);
while (co_stream_read(input_stream, &c, sizeof(char)) == co_err_none ) {
    printf("Consumer read %c from input stream\n", c);
}
co_stream_close(input_stream);
```

The consumer process uses a simple printf statement to display the test results to the console.

- A main function. The main function simply displays console messages (using printf) and, most importantly, launches the application using calls to the Impulse C functions co_initialize and co_execute. The co_initialize function is a required part of every Impulse C application, and is described in a later section. The co_execute function is an Impulse C library function that launches the application and its constituent processes:

```
int main(int argc, char *argv[]) {
    int param = 0;
    co_architecture my_arch;

    printf("HelloFPGA starting...\n");

    my_arch = co_initialize(param);
    co_execute(my_arch);

    printf("HelloFPGA complete.\n");
    return(0);
}
```

The Hardware Source File: HelloFPGA_hw.c

The file HelloFPGA_hw.c (see Figure 4-4) contains those portions of the application that represent hardware running on the FPGA. This file (and any other files containing additional hardware processes) is analyzed by the Impulse C compiler for the purpose of hardware generation. Examining this source file section by section and starting at the top, we find:

- A comment header.
- An include statement referencing the co.h include file containing Impulse C macros and function declarations:

```
#include "co.h"
```

```
// HelloWorld_hw.c: Hardware processes and configuration.

#include "co.h"

extern void Consumer(co_stream input_stream);
extern void Producer(co_stream output_stream);

// Hardware process
void DoHello(co_stream input_stream, co_stream output_stream) {
   char c;

   co_stream_open(input_stream, O_RDONLY, CHAR_TYPE);
   co_stream_open(output_stream, O_WRONLY, CHAR_TYPE);
   while (co_stream_read(input_stream, &c, sizeof(char)) == co_err_none ) {
      // Do something with the data stream here
      co_stream_write(output_stream,&c,sizeof(char));
   }
   co_stream_close(input_stream);
   co_stream_close(output_stream);
}

void config_hello(void *arg) {   // Configuration function
   co_stream s1,s2;
   co_process producer, consumer;
   co_process hello;

   s1 = co_stream_create("Stream1", CHAR_TYPE, 2);
   s2 = co_stream_create("Stream2", CHAR_TYPE, 2);
   producer = co_process_create("Producer",
                     (co_function) Producer, 1, s1);
   hello = co_process_create("DoHello",
                     (co_function) DoHello, 2, s1, s2);
   consumer = co_process_create("Consumer",
                     (co_function) Consumer, 1, s2);
   co_process_config(hello, co_loc, "PE0");  // Assign to PE0
}

co_architecture co_initialize(int param) {
   return(co_architecture_create("HelloArch","generic",
                              config_hello,(void *)param));
}
```

Figure 4-4. HelloFPGA hardware process and configuration function (HelloFPGA_hw.c).

- Extern declarations for the **Producer** and **Consumer** processes defined in HelloFPGA_sw.c:

  ```
  extern void Consumer(co_stream input_stream);
  extern void Producer(co_stream output_stream);
  ```

  These are required here because the configuration function (described in a moment) must reference these processes.

- A process run function named **DoHello**:

  ```
  void DoHello(co_stream input_stream, co_stream output_stream) {
  ```

  This process run function accepts as its arguments two stream objects. One of these streams (**input_stream**) represents an incoming stream of 8-bit character values. This stream will be connected to the single output of the **Producer** process. The other stream (**output_stream**) represents the processed data, which is also composed of 8-bit character values. This stream will be connected to the input stream of the **Consumer** process. As in the **Consumer** process, this process uses an inner code loop (represented by a while loop) that operates on the input stream for as long as data appears on the inputs. The stream operations are described using the Impulse C functions **co_stream_open**, **co_stream_read**, **co_stream_write**, and **co_stream_close**:

  ```
  co_stream_open(input_stream, O_RDONLY, CHAR_TYPE);
  co_stream_open(output_stream, O_WRONLY, CHAR_TYPE);
  while (co_stream_read(input_stream, &c, sizeof(char)) == co_err_none ) {
      // Do something with the data stream here
      co_stream_write(output_stream,&c,sizeof(char));
  }
  co_stream_close(input_stream);
  co_stream_close(output_stream);
  ```

  The **co_process_create** function call detailed later in this section creates one instance of this process and indicates the name of that process instance, as well as defining the external stream connections. Note that this process does not actually do anything with the incoming data; instead, each value appearing on the input stream is immediately written (via **co_stream_write**) to the output stream.

- A configuration subroutine named **config_hello**:

  ```
  void config_hello(void *arg) {   // Configuration function
  ```

  The configuration subroutine is a required part of every Impulse C application. The configuration subroutine defines the structure of your application in terms of the processes that are to be used and how they are to be interconnected. This configuration subroutine uses the co_process_create function to create three process instances, which are given the names producer, consumer, and hello:

```
s1 = co_stream_create("Stream1", CHAR_TYPE, 2);
s2 = co_stream_create("Stream2", CHAR_TYPE, 2);
producer = co_process_create("Producer", (co_function) Producer, 1, s1);
hello = co_process_create("DoHello", (co_function) DoHello, 2, s1, s2);
consumer = co_process_create("Consumer", (co_function) Consumer, 1, s2);
```

We have also used the function co_stream_create to create the two streams that will carry data from producer to hello and from hello to consumer, respectively. And lastly, we have used the process configuration function co_process_config to specify that one of the three processes, hello, is to be assigned to a hardware resource called PE0, which represents the target FPGA:

```
co_process_config(hello, co_loc, "PE0"); // Assign to PE0
```

This single source file line, the call to co_process_config, is the only clue we have in this application that the target of compilation is an FPGA.

- After the configuration subroutine we find a definition for the co_initialize function that was referenced from our main function (in file HelloFPGA_sw.c):

```
co_architecture co_initialize(int param) {
    return(co_architecture_create("HelloArch","generic",
                              config_hello,(void *)param));
}
```

As with the configuration subroutine, one such function (named co_initialize) is required in every Impulse C application. Within this function there is a single call to function co_architecture_create, which associates the configuration function we previously defined with the application's target architecture, which in this case is a generic hardware/software platform. (The specific programmable platform that will be the target of compilation is defined outside of the application source files, as part of the hardware compiler settings.)

You have now seen a complete Impulse C application that includes both a hardware process (DoHello) and two software test bench processes (Consumer and Producer). We have also examined how these three processes (which are more formally defined as *process run functions*) are instantiated (to form three *process instances*) and are interconnected by data channels called *streams*. The following sections describe these various Impulse C elements in more detail.

4.4 PROCESSES, STREAMS, SIGNALS, AND MEMORY

At the heart of the Impulse C programming model are *processes* and *streams*. You saw examples of both of these elements in the preceding HelloFPGA example.

Processes are independently synchronized, concurrently operating segments of your application. Processes are written using standard C (subject to the limitations of the target processing element, whether hardware or software). They perform the work of your application by accepting data, performing computations, and generating outputs.

The data processed by your application flows from process to process by means of streams, or in some cases by means of messages sent on special channels called *signals* and/or via shared memories. Streams represent one-way channels of communication between concurrent processes and are self-synchronizing with respect to the processes by virtue of buffering. The characteristics of a given stream (its width and depth) are specified at the time a stream is created in your application using co_stream_create.

As you saw in the "Hello FPGA" example, the implementation of a process is defined by a user-defined C procedure called the process run function. When compiling an application for a target platform, each process is classified as a software process or a hardware process based on its location specified in the configuration subroutine. The possible locations are defined by the target platform.

A software process is constrained only by the limitations of the target processor (whether a common RISC, a custom DSP processor, or a custom processor core), while a hardware process is typically more constrained. A process written for an FPGA, for example, must be written using a somewhat narrowly defined subset of the C language to meet the constraints of the Impulse C hardware compiler. In addition to standard C expressions, predefined functions that perform stream or signal operations may be referenced in a software or hardware process.

The Impulse C compiler generates synthesis-compatible hardware descriptions for one or more FPGAs as well as a set of communicating processes (in the form of C code compatible with the target cross-compiler) to be implemented on conventional processors. The compiler is capable of scheduling and pipelining stream operations and other computations (within loops, for example) so that the generated hardware descriptions take advantage of parallelism within the target hardware itself.

4.5 IMPULSE C SIGNED AND UNSIGNED DATATYPES

Impulse C provides predefined unsigned and signed integer datatypes for selected bit lengths ranging from 1 to 64, as shown in the following examples:

co_int1—1-bit integer type

co_int7—7-bit signed integer type

co_uint16 —16-bit unsigned type

co_uint24 —24-bit unsigned type

co_int32—32-bit signed type

co_uint64—64-bit unsigned type

A simple convention is used to name these predefined types. Signed types
have the name co_int followed by the bit length, while unsigned types have
the name co_uint followed by the bit length. Variables of these types may be
used in an Impulse C program for either software or hardware processes. A
stream may have one of these C integer types as its data element type.

During desktop simulation, types whose widths do not match one of the
standard C types (for example, a 24-bit integer) are modeled using the next
largest integer type. This can result in differences in bit-accurate behavior between the desktop simulation environment and a hardware implementation.
To prevent such differences and ensure bit-accurate modeling, you may
choose to use the bit-accurate arithmetic macro operations defined in the Impulse C library. Here are some examples of these macro operations:

UADD4(a,b)—Unsigned 4-bit addition

ISUB7(a,b)—Signed 7-bit subtraction

UMUL24(a,b)—Unsigned 24-bit multiplication

UDIV28(a,b)—Unsigned 28-bit division

Note that the bit-accurate macro operations are specified in terms of their return value bit width, and there is no enforcement or checking of bit widths of
the operands.

4.6 UNDERSTANDING PROCESSES

Processes are the fundamental units of computation in an Impulse C application. Once they have been created, assigned, and started, the processes in an
application execute as independently synchronized units of code on the target platform.

Programming with Impulse C processes is conceptually similar to programming with threads. As with thread programming, each Impulse C process has its own control flow, is independently synchronized, and has access to its own local memory resources (which will vary depending on the target platform). For this reason it is relatively easy to convert applications written in threaded C (for example, using the Posix thread library) to Impulse C. There are some key differences, however:

- In thread programming, it is assumed that globals and heap memory are shared among threads. In Impulse C, heap memory may be explicitly shared, but global variables are not supported in general.

- In thread programming, the threads are assumed to execute on the same processor. In Impulse C, the assumption is that each process is assigned to an independently synchronized processor or block of logic.

- The Impulse C programming model is specifically designed to support mixed hardware and software targets, with process communication and synchronization occurring primarily in hardware buffers (FIFOs). In thread programming, threads often communicate implicitly using shared data structures and semaphores.

- The Impulse C programming model assumes that processes are defined and created at the time the application is initialized and loaded, rather than dynamically created, invoked, and torn down as in a threaded application.

The Impulse C desktop simulation library is based on a threading model, so global variables and heap memory are shared. Also, during simulation all Impulse C processes are executed on one processing element (your desktop computer), although this may not be the case on the target platform. It is your responsibility as the programmer to avoid using global variables and heap memory allocated outside of the Impulse C library.

Note: In desktop simulation, when all processes are translated to software running on a single processor, the scheduling of instructions being executed within each process is predictable, but the scheduling of instructions across processes depends on the threading model used in the host compiler and/or the host operating system. This can result in behavioral differences between software and hardware implementations of an application. In particular, your application must not assume that one process will start and execute before another process starts and executes. If such process synchronization is required, you should use signals, which are described later in this chapter, or rethink your application's use of streams.

What is meant by "timed" and "untimed" C?

When applied to C-to-hardware programming methodologies, the terms "timed" and "untimed" refer to a requirement (or lack thereof) for the application programmer to understand and specify where the clock boundaries are within a given sequence of C statements. If the programmer must explicitly define these these boundaries, or if by definition each C statement or source file line represents a distinct clock cycle, the programming model is that of timed C. In such an environment, the programmer is given a great deal of control over how an application is mapped across clock boundaries, but this level of control also means significantly more work for the programmer, who must manually insert information into the design to indicate how otherwise-sequential statements are to be made parallel. This work is not trivial and can result in C code representing hardware that is not functionally equivalent to the original description as rendered in untimed C.

Impulse C represents an untimed method of expressing applications. In this programming model, statements within a process are optimized with the goal of minimizing the total number of instruction stages. Only more-general controls (in the form of compiler pragmas and external tool options) are provided for influencing the optimizer and meeting size/speed requirements. Impulse C provides no way to relate a specific C-language statement to a specific clock event, or to introduce any other such low-level hardware concept (such as a clock enable), except when such concepts have a direct and defined relationship to standard C coding styles.

The advantage of an untimed approach is that it more closely emulates a software programming experience. Applications can be expressed at a higher level and can then (perhaps at a much later time) be optimized through the addition of more hardware-centric compiler hints and C-level optimizations. The disadvantage, of course, is that a hardware-savvy programmer may not have the ability to precisely control the generation of hardware. As a result, he or she may be required to drop into lower-level HDL programming for some parts of the application if such a level of control is required.

Creating Processes

Processes are created and named in the configuration function of your application. In the following example, the **my_app_configuration** function declares three processes (**procHost1**, **procPe1**, and **procPe2**) and associates these processes with three corresponding process run functions named **Host1**, **Pe1**, and **Pe2**:

```
#define BUFSIZE 4
void my_app_configuration()
{
    co_process procHost1, procPe1, procPe2;
    co_stream s1, s2;

    s1 = co_stream_create("s1", INT_TYPE(16), BUFSIZE);
    s2 = co_stream_create("s2", INT_TYPE(16), BUFSIZE);

    procHost1 = co_process_create("Host1", (co_function)Host1, 1, s1);
    procPe1 = co_process_create("Pe1", (co_function)Pe1, 2, s1, s2);
    procPe2 = co_process_create("Pe2", (co_function)Pe2, 1, s2);
}
```

The **co_process_create** function is used to define both hardware and software processes. Unless otherwise assigned, all processes created using the **co_create_process** function are assumed to be software processes. The function accepts three or more arguments. The first argument must be a pointer to a character string (NULL terminated) containing a process name. This name does not have to match the variable name used within the process itself; it is only used as a label when monitoring the process externally—for example, using the application monitor.

The second argument to **co_process_create** is a function pointer of type **co_function**. This function pointer identifies the specific run function that is to be associated with (or instantiated from) the call to **co_process_create**.

The third argument to **co_process_create** indicates the number of ports (inputs and outputs) that are defined by the process. This number must match the number of actual port arguments that follow and must also match the parameters of the specified run function. For example, if the number of ports is two, there must be two ports declared as arguments four and five.

Tip: Specifying the wrong number or wrong order of ports in the co_process_create function is a common mistake and can be difficult to debug. This is particularly true when connecting streams, which may appear to compile properly, yet exhibit incorrect behavior during simulation. Check these connections carefully.

Port arguments specified for an Impulse C run process may be one of five distinct types:

- co_stream—A streaming point-to-point interface on which data is transferred via a FIFO buffer interface.

- co_signal—A buffered point-to-point interface on which messages may be posted by a sending process and waited for by a receiving process.

- co_memory—A shared memory interface supporting block reads and writes. Memory characteristics are specific to the target platform.

- co_register—A low-level, unbuffered hardware interface.

- co_parameter—A compile-time parameter.

Variables of these types must be declared in the configuration function and (in the case of co_stream, co_signal, co_register, and co_memory) the appropriate co_xxxxx_create function must be used to create an instance of the desired communication channel.

When your application creates a process using the co_process_create function, the process is configured to run, and it begins execution, when the co_execute function is called at the start of the program's execution. This is a fundamental difference between thread programming and programming in Impulse C: in Impulse C, the entire application and all its parallel processing components are set up in advance to minimize runtime overhead.

4.7   UNDERSTANDING STREAMS

Streams are unidirectional communication channels that connect multiple processes, whether hardware or software. The co_stream_create function creates a stream, defines its data width and its buffer size, and makes the stream available for use in subsequent co_process_create calls. The following is an example of a stream being created with co_stream_create:

```
strm_image_value = co_stream_create("IMG_VAL", INT_TYPE(16),BUFSIZE);
```

There are three arguments to the co_stream_create function. The first argument is an optional name that may be assigned to the stream for debugging, external monitoring, and post-processing purposes. This name has no semantic meaning in the application but may be useful for certain downstream synthesis or simulation tools.

If application monitoring will be used (as indicated by any use of cosim_ monitoring functions, which are described in a later chapter), stream names are required, and the chosen stream names must be unique across the entire application.

How are streams implemented in hardware?

The details of streams (their control signals and external protocols) are the subject of a later chapter, but for now it's useful to know that streams are implemented as first-in-first-out (FIFO) buffers. You define the characteristics of a given stream when you create the stream using the **co_stream_create** function. For example, the following stream definition:

```
co_stream pixelvalues;
pixelvalues = co_stream_create("pixval", UINT_TYPE(24), 2);
```

results in a stream of name **pixval** being generated in hardware that will carry 24-bit-wide data values on a buffer with a depth of two, for a total of 48 bits of data memory plus control signals. Note that the actual depth of the generated hardware FIFO must be a power of two, so the compiler rounds up the specified buffer size to the closest power of two.

Buffer widths (specified in the second argument to the co_stream_create function) reflect the nature of the data being transferred (whether a single character or a 32-bit word). The depth of a stream and its corresponding FIFO (specified as the third argument to co_stream_create) can be a more complex decision, however. Although there is no performance penalty for using deep FIFOs (no extra stream delay is introduced into the system), using large buffers can have a significant impact on the size of the generated hardware. Using too small a buffer, on the other hand, can result in deadlock conditions for processes that must read multiple packets of data prior to performing some computation, or for connected processes that operate at different rates (such as communicating hardware and software processes). We will cover the subject of stream optimization in later chapters.

The second argument specifies the type and size of the stream's data element. Macros are provided for defining specific stream types, including INT_TYPE, UINT_TYPE, and CHAR_TYPE.

The third and final argument to **co_stream_create** is the buffer size. This buffer size directly relates to the size of the FIFO buffer that will be created between two processes that are connected with a stream. A buffer size of 1 indicates that the stream is essentially unbuffered; the receiving process will block until the sending process has completed and moved data onto the stream. In contrast, a larger buffer size will result in additional hardware

resources (registers and corresponding interconnect resources) being generated, but may result in more efficient process synchronization. As an application designer, you will choose buffer sizes that best meet the requirements of
your particular application.

When choosing buffer sizes, keep in mind that the width and depth of a
stream (as specified in the call to co_stream_create) will have a significant
impact on the amount of hardware required to implement the process. You
should therefore choose a stream buffer size that is as small as practical for
the given process. In the following example a buffer size of four has been selected for all three streams, as indicated in the definition of BUFSIZE:

```
#define BUFSIZE 4
void my_config_function(void *arg) {
    co_stream host2controller,controller2pe,pe2host;
    co_process host1, host2;
    co_process controller;
    co_process pe;

    host2cntrlr=co_stream_create("host2cntrlr", INT_TYPE(32), BUFSIZE);
    cntrlr2pe=co_stream_create("cntrlr2pe", INT_TYPE(32), BUFSIZE);
    pe2host=co_stream_create("pe2host", INT_TYPE(32),BUFSIZE);

    host2=co_process_create("host2", (co_function)host2_run, 1, pe2host);
    pe=co_process_create("pe", (co_function)pe1_proc_run, 2, cntrlr2pe, pe2host);
    cntrlr=co_process_create("cntrlr", (co_function) cntrlr_run, 2, host2cntrlr, cntrlr2pe);
    host1=co_process_create("host1",(co_function)host1_run, 2, host2cntrlr, iterations);

    co_process_config(cntrlr, co_loc, "PE0");
    co_process_config(pe, co_loc, "PE0");
}
```

Stream I/O

Reading and writing of processes is performed within the process run functions that form an application. Each process run function in a dataflow-
oriented Impulse C application iteratively reads one or more input streams
and performs the necessary processing when data is available. If no data is
available on the stream being read, the process blocks until such time as data
is made available by the upstream process.

Other, non-dataflow processing models are supported, including
message-based process synchronization. These alternate models are described in subsequent sections.

At the start of a process run function, the co_stream_open function resets a stream's internal state, making it available for either reading or writing.
The co_stream_open function must be used to open all streams (input and

output) prior to their being read from or written to. An example of a stream being opened is as follows:

```
err = co_stream_open(input_stream, O_RDONLY, INT_TYPE(32));
```

The **co_stream_open** function accepts three arguments: the stream (which has previously been declared as a process argument of type **co_stream**), the type of stream (which may be **O_RDONLY** or **O_WRONLY**), and the data type and size, as expressed using either **INT_TYPE**, **UINT_TYPE**, or **CHAR_TYPE**.

Streams are point-to-point and unidirectional, so each stream should be opened by exactly two processes, one for reading and the other for writing. If a stream being opened with **co_stream_open** has already been opened, the **co_stream_open** function returns the error code **co_err_already_open**.

When the stream is no longer needed, it may be closed using the **co_stream_close** function:

```
err = co_stream_close(input_stream);
```

The **co_stream_close** function writes an end-of-stream (EOS) token to the output stream, which can then be detected by the downstream process when the stream is read using **co_stream_read**. The **co_stream_read** function returns an error when the EOS token is received, indicating that the stream is closed. If a stream being closed is not open (or has already been closed), the **co_stream_close** function returns the error code **co_err_not_open**.

Keep in mind that all streams must be properly closed by the reading and writing process. In particular, note that the reading process blocks when it calls **co_stream_close** until the EOS has been received, indicating that the upstream proess has also closed the stream. Once closed, a process can open a stream again for multiple sessions.

Note: Stream connections between processes must be one-to-one; broadcast patterns are not supported, and many-to-one connections are not supported. It is possible to create such data distribution patterns in an application, however, by creating intermediate stream collector and distributor processes.

4.8   USING OUTPUT STREAMS

Output streams may be written using the **co_stream_write** function as follows:

```
co_stream_open(output_stream, O_WRONLY, INT_TYPE(32));
for (i=0; i < ARRAYSIZE; i++) {
```

```
        co_stream_write(output_stream, &data[i], sizeof(int32));
    }

    co_stream_close(output_stream);
```

The stream must be a writable stream (opened with the O_WRONLY direction indicator), and the data must match the size of the stream datatype.

The co_stream_write function, when invoked, first checks to see if the specified output stream is full. If the stream is full, the function blocks until there is room in the stream buffer for a write operation to be performed. This is an important aspect of stream behavior: once a stream has been opened and is being used by a producer and consumer process, there is no need for you (as the application programmer) to manage the stream in terms of acknowledgments or other control signals.

There is, of course, a downside to the standard behavior of streams. If your producer and consumer processes are not properly designed, there can be a risk of deadlock conditions, in which neither the producer nor the consumer can proceed with its operations. We'll discuss the issue of stream synchronization and deadlocks in more detail in later chapters.

4.9 USING INPUT STREAMS

Two operations may be performed on an input stream: an end-of-stream test and a stream read. The end-of-stream test checks to see whether a "close" operation was performed on the stream by the stream writer. It does this by checking the current element at the head of the stream. If this element is determined to be an end-of-stream token, a true value is returned; otherwise, a false value is returned indicating that the stream is open. The co_stream_eos function is nonblocking, and a return value of false does not imply that there is, or will be, more data available.

The co_stream_read function attempts to read the next stream element and blocks if the stream is empty. A read operation on a closed stream returns the value true. Thus, the preferred sequence of operations is to describe a stream reading loop:

```
    err = co_stream_open(input_stream, O_RDONLY, INT_TYPE(32));
    while(co_stream_read(input_stream) == co_err_none) {
        . . .   // Process the data here
    }
    co_stream_close(input_stream);
```

When an input stream is closed using co_stream_close, all the unread data in the stream is flushed, and the EOS token is consumed. If there is no EOS in the stream (for example, the writer hasn't closed it yet), co_stream_close blocks until an EOS is detected. Note also that **co_stream_close** writes the EOS token only when called from the writer process, so it is important not to close a stream from the read side unless the EOS token has been detected.

Checking for End of Stream

The **co_stream_eos** function returns true to indicate that the writer has closed the stream. Once **co_stream_eos** returns false, all subsequent calls to **co_stream_eos** return false until the reader closes the stream. Similarly, all subsequent calls to **co_stream_read** fail with **co_err_eos** until the stream is closed and reopened for reading.

Efficient Use of Stream Reads

The efficient processing of stream-related data is a key part of programming using Impulse C. There are three possible methods of reading data from streams:

Method 1 (preferred method):

```
while(co_stream_read(input_stream) == co_err_none) {
...  // Process the data here
}
```

Method 2 (acceptable method):

```
do {
  if (co_stream_read(input_stream,&i,sizeof(i)) == co_err_none) {
    ...  // Process the data here
  }
} while(!co_stream_eos(input_stream));
```

or the following derivative:

```
while ( ! co_stream_eos(input_stream) ) {
  if ( co_stream_read(input_Stream, &i, sizeof(i)) == co_err_none ) {
    . . .
  }
}
```

Method 3 (less acceptable):

```
while(!co_stream_eos(input_stream)) {
  co_stream_read(input_stream,&i,sizeof(i));
  ...  // Process the data here
}
```

or the following derivative (which is also less acceptable):

```
do {
  co_stream_read(input_stream,&i,sizeof(i));
    ...   // Process the data here
} while(!co_stream_eos(input_stream));
```

As indicated, the first two methods are acceptable, but the third may result in problems during simulation and/or in the generated VHDL and should not be used. (The reason? It's possible that between the call to co_stream_eos and co_stream_read, the stream may be closed by the upstream process. Since method 3 does not check the return value of co_stream_read, the read buffer could contain invalid data.)

Which method you will use depends on the nature of your application, but there are significant trade-offs for processes that will be compiled to hardware. Because each control point in a hardware process will require a full cycle, method 1 is strongly preferred for efficient hardware synthesis. Also, when testing for a condition on the return value from co_stream_read, the condition must be (co_stream_read(...) == co_err_none) as shown.

The best strategy is almost always the first of these three methods. Eliminating the explicit call to co_stream_eos will result in the loop waiting (possibly forever) for the the first data element to appear on the stream and will not result in an additional wasted cycle. If, however, you need to perform conditional read operations or are operating on multiple streams simultaneously, the second method is acceptable as an alternative.

4.10 AVOIDING STREAM DEADLOCKS

Deadlocks can be one of the most difficult problems to resolve in a streaming application, and care must therefore be taken when designing complex, multiprocess applications. A stream deadlock occurs when one process is unable to proceed with its operation until another process has completed its tasks and written data to its outputs. If the two processes are mutually dependent or are dependent on some other blocked process, the system can come quickly to a halt.

The problem of deadlocks is most severe in systems having irregular data (unpredictable numbers of stream outputs for a given number of stream inputs) or in systems having variable cycle delays (such as the example presented in Chapter 13). In some cases stream deadlocks can be removed by increasing the depth of stream buffers, but in most cases this only delays finding a real solution to the problem. Many such situations, in fact, are completely independent of stream buffer sizes.

Consider, for example, the two processes shown in Figure 4-5.

```
void Supervisor_run(co_stream S1, co_stream S2, co_parameter iparam)
{
    int iterations = (int)iparam;
    int i, j;
    uint32 local_S1, local_S2;
    co_stream_open(S1, O_WRONLY, UINT_TYPE(32));
    co_stream_open(S2, O_RDONLY, UINT_TYPE(32));
    srand(iterations);   // Seed value

    // For each test iteration, send random value to the stream.
    for ( i = 0; i < iterations; i++ ) {
        // Must send 4 characters on S1 before attempting to read from S2...
        for ( j = 0; j < 4; j++ ) {
            local_S1 = rand();
            printf("S1 = %d\n", local_S1);
            co_stream_write(S1, &local_S1, sizeof(uint32));
        }
        for ( j = 0; j < 4; j++ ) {
            co_stream_read(S2, &local_S2, sizeof(uint32));
            printf("S2 = %d\n", local_S2);
        }
    }
    co_stream_close(S1);
    co_stream_close(S2);
}

// This process will reverse the order of every block of four input values.
void Worker_run(co_stream S1, co_stream S2) {
    int i;
    uint32 local_S1, local_S2;
    uint32 data[4];
    co_stream_open(S1, O_RDONLY, UINT_TYPE(32));
    co_stream_open(S2, O_WRONLY, UINT_TYPE(32));

    while (!co_stream_eos(S1)) {
        for (i = 0; i < 4; i++)
            co_stream_read(S1, &data[i], sizeof(uint32));
        for (i = 0; i < 4; i++)
            co_stream_write(S2, &data[3-i], sizeof(uint32));
    }
    co_stream_close(S1);
    co_stream_close(S2);
}
```

Figure 4-5. The producer and consumer must be designed to avoid deadlocks.

In this example, notice that the first process, called **Supervisor**, sends packets of four 32-bit unsigned values on a stream (S1), which is subsequently received by a second process called **Worker**. The first process sends the stream using a loop that generates four calls to the co_stream_write function. Similarly, the second process (after receiving the four values) writes to its output stream (S2) using the same type of loop, reversing the order of the four values written.

This is a very simple example of a process (**Worker**) that must cache some number of values locally before performing its operation (reversing the order of the values). In this example, the assumption being made by the programmer is that the worker process will accept these four values without blocking. In this case it is a valid assumption, but you can easily imagine situations in which the controlling process (the Supervisor) has not been so carefully designed and in which a deadlock is inevitable. The following code represents one such situation:

```
// Send random values to the stream, one value at a time.
for ( i = 0; i < iterations * 4; i++ ) {
    local_S1 = rand();
    printf("S1 = %d\n", local_S1);
    co_stream_write(S1, &local_S1, sizeof(uint32));
    co_stream_read(S2, &local_S2, sizeof(uint32));  // This will deadlock!
    printf("S2 = %d\n", local_S2);
}
```

In this version of the **Supervisor** processing loop, the programmer has incorrectly assumed that data will become available on stream S2 after only one data element has been placed on stream S1. Because **Worker** does not produce any output until four values have been received, **Supervisor** (and hence the rest of the system) will be deadlocked and will not produce any outputs.

Note: Be careful with the design of your streams, and always consider issues of process and stream synchronization. While debugging tools can help you find where deadlocks are occurring, it is not always trivial to resolve them in the most efficient way, or in extreme cases to resolve them at all without a substantial redesign.

Using Nonblocking Stream Reads

What if you need to create an equivalent to the **Supervisor** function that is not dependent on a particular length of the data packet? In fact, what if the **Worker** process is capable of producing reordered outputs of arbitrary and constantly changing lengths? This is a common situation, particularly for pattern matching and searching functions, and the solution is to use a non-blocking stream read function.

Figure 4-6 demonstrates one alternative way to write the Supervisor function. In this version, the co_stream_read function has been replaced with the nonblocking co_stream_read_nb function. The processing loop uses co_stream_read_nb to iteratively poll the output stream (S2). This introduces some delay into the system (polling has some cost in terms of cycle delays) but resolves the issue of deadlocking in this example.

```
void Supervisor_run(co_stream S1, co_stream S2, co_parameter iparam)
{
    int   iterations = (int)iparam;
    int   i, j;
    uint32 local_S1;
    uint32 local_S2;

    co_stream_open(S1, O_WRONLY, UINT_TYPE(32));
    co_stream_open(S2, O_RDONLY, UINT_TYPE(32));
    srand(iterations);  // Seed value

    // For each test iteration, send four random values to the stream.
    for ( i = 0; i < iterations * 4; i++ ) {
        local_S1 = rand();
        printf("S1 = %d\n", local_S1);
        co_stream_write(S1, &local_S1, sizeof(uint32));
        while (co_stream_read_nb(S2, &local_S2, sizeof(uint32)))
            printf("S2 = %d\n", local_S2);
    }
    co_stream_close(S1);
    co_stream_close(S2);
}
```

Figure 4-6. The nonblocking stream read function **co_stream_read_nb** can resolve many deadlock problems.

Deadlocks and the PIPELINE Pragma

When using the PIPELINE pragma (described in later chapters), it is possible for the generated hardware to exhibit deadlock conditions not present during desktop simulation. Pipelining is a parallelizing technique that allows multiple iterations of a loop to execute in parallel. When a loop that inputs a value and outputs a result each iteration is converted to a pipeline, the resulting hardware may not produce any output until after it has received some number of input values, because multiple iterations are being executed at the same time. Pipelining and the **PIPELINE** pragma are discussed in more detail in later chapters.

4.11 CREATING AND USING SIGNALS

The Impulse C programming model emphasizes the buffered movement of data between processes and on the management of streams. The more efficiently your application manages stream data, and the more balanced your application is (in terms of buffer loading and process utilization), the faster it will operate. Streaming makes it possible to create independently synchronized processes that have a minimum number of extraneous inputs and outputs, which simplifies system design and can lead to more reliable, tested systems.

There are times, however, when you need more direct control over the starting and stopping of processes and the synchronization of processes to external events and devices. For these times, Impulse C provides an alternate method of synchronization.

Using signals, you can communicate information from one process to another using a message passing scheme, one in which read operations (more properly called "waits") are blocking, but writes (posts) are nonblocking. The mechanism is simple. It uses a **co_signal_post** function call in the sending process, and a **co_signal_wait** function call in the receiving process, as shown in Figure 4-7. In this example, the process is synchronized to a serial input stream, which pipes that stream to an output buffer. After writing eight values to the output stream, it posts a message to a third controlling process and waits for a message back from that process before continuing with the next set of values.

Note that the call to **co_signal_post** in this example includes a data value indicating the number of stream data elements written. As this demonstrates, signals serve the dual purpose of allowing synchronization of processes and supporting the passing of data values.

Note: When using signals, keep in mind that the **co_signal_wait** function is blocking—the calling process does not continue until a message is received on the specified signal—but the **co_signal_post** function is nonblocking. This means that repeated calls to **co_signal_post** override existing message values on the output signal; the function does not wait until an existing (previously posted) message has been received by the downstream process.

```
// This process is synchronized to a serial input stream,
// and pipes that stream to an output buffer. After writing
// eight values to the output stream it posts a message
// to a third controlling process, and waits for message
// back from that process before continuing with the next
// set of values.
//
void proc1_run(co_stream stream_input,
               co_signal ack_signal_input,
               co_stream stream_output,
               co_signal ready_signal_output) {
  uint32 data, i;
  co_stream_open(stream_input, O_RDONLY, INT_TYPE(32));
  co_stream_open(stream_output, O_WRONLY, INT_TYPE(32));

  do {
    for (i = 0; i < 8; i++) {
      if (co_stream_read(stream_input, &data, sizeof(uint32)) == co_err_none)
        co_stream_write(stream_output, &data, sizeof(uint32));
      else
        printf("Unexpected end of data stream detected in proc1.\n");
    }
    co_signal_post(ready_signal_output, i);
    co_signal_wait(ack_signal_input);   // The co_wait function will block until
                                        //  a message is received
  } while (co_stream_eos(stream_input) == co_err_none);

  co_stream_close(stream_input);
  co_stream_close(stream_output);
}
```

Figure 4-7. A signal can be used to coordinate the operation of two or more processes for those applications in which stream-based synchronization is not sufficient.

4.12 UNDERSTANDING REGISTERS

The most common and convenient method of communicating between Impulse C processes is to use streams. Depending on the application requirements and the target hardware capabilities, shared memories and/or signals may also be used. What these methods of communication have in common is that they are synchronized and support buffering of the data. This makes it possible to create highly parallel systems without the need to handle low-level process synchronization, assuming the application being described lends itself to data-centric methods of process synchronization.

Many applications, however, require more direct, unsynchronized input and output of data. Some applications may interface directly to hardware devices and their corresponding control signals, while other applications may require that an unsynchronized direct connection be set up between two independent hardware processes.

For this purpose Impulse C includes the **co_register** data object, which corresponds to a wired connection (input or output) in hardware. Like streams and signals, registers are declared and created in the configuration process of your application (using **co_register_create**) and are passed as register pointers to the processes of your application in the configuration function, as in the following example:

```
void config_counter(void *arg) {
    co_register counter_dir;
    co_register counter_val;
    co_process main_process;
    co_process counter_process;

    counter_dir = co_register_create("counter_dir", UINT_TYPE(1));
    counter_val = co_register_create("counter_val", INT_TYPE(32));

    main_process = co_process_create("main_process",
                    (co_function)counter_main,
                    2, counter_dir, counter_val);

    counter_process = co_process_create("counter_process",
                    (co_function)counter_hw,
                    2, counter_dir,
                    counter_val);
}
```

In this example, two processes are declared and connected via two registers, **counter_dir** and **counter_val**. Register **counter_dir** represents an unbuffered control input to the counter process, while **counter_val** represents the output of the counter process, which is also unbuffered. Within a process, the Impulse C functions **co_register_get**, **co_register_read**, **co_register_put**, and **co_register_write** are used to access the value appearing on a register or to write a value to that register.

The following process uses **co_register_get** and **co_register_put** to describe a simple 32-bit up/down counter:

```
void counter_hw(co_register direction, co_register count) {
    int32 nValue = 0;

    while ( 1 ) {    /* Main processing loop... */
        if (co_register_get(direction) == 1)
            nValue++;
```

```
        else
          nValue--;
        co_register_put(count, nValue);
    }
  }
```

In this process, the variable **nValue** represents a local storage element (a set of clocked hardware registers), while the **direction** and **count** register parameters represent inputs and outputs that may be tied directly to device I/O pins or to other hardware elements in the system as a whole.

Controlling Registers from Software Processes

In most (perhaps all) cases you will use registers to communicate only between hardware elements of your application (hardware processes and other, external hardware interfaces). It is also possible to interface between hardware and software processes using registers—for example, to communicate status information back to the software process or to set hardware configuration registers from a software process.

You might want to create software test benches in order to test your application (including the register connections) in desktop simulation such as within Visual Studio. You can do this, but you need to keep in mind that the order in which processes (and the statements within processes) will run in a desktop operating system environment or debugger can not be guaranteed. The result is that values placed on a register in one process cannot be guaranteed to be available to the destination process unless some other synchronization method (a signal, for example) is used to pause (or yield) the originating process.

Note: The hardware implementation of registers also requires that there be only one process that writes to a given register. There is no concept of a bidirectional register, and the hardware compiler reports an error if more than one process writes to the same register, as indicated by calls to either **co_register_put** or **co_register_write**. This condition may or may not be detected during software (desktop) simulation.

4.13 USING SHARED MEMORIES

The Impulse C programming model supports the use of shared memories as an alternative to streams communication. Shared memories can be useful for initializing a process with some frequently used array values (for such things as coefficients) and, as described in the previous section, may be a more efficient, higher-performance means of transferring data between hardware and software processes for some platforms. Shared memory interfaces can also be used as generic interfaces to external memory resources not on the system bus. Using memories in this way, however, may require a certain amount of hardware-design expertise to develop the necessary memory controller interfaces.

To demonstrate how shared memory interfaces are described and used, we'll take a fresh look at the HelloFPGA example presented earlier. We'll modify this example so that it makes use of a shared memory as a replacement for one of the stream interfaces, the one that carries the text data from the Producer process to the DoHello process. To start, we'll modify the configuration subroutine such that the first stream (s1) is deleted and replaced with a resource of type co_memory and a corresponding call to the co_memory_create function:

```
void config_hello(void *arg) {   // Configuration function
    co_memory memory;
    co_stream s2;
    co_process producer, consumer, hello;

    memory = co_memory_create("Memory", "mem0", MAXLEN*sizeof(char));
    . . .
```

The co_memory_create function is similar to the co_stream_create function it replaces. The co_memory_create function allocates a specified number of bytes of memory for reading and writing and returns a handle that can be passed to any number of processes for read/write access to the memory. The actual reading and writing of the memories is accomplished using the co_memory_readblock and co_memory_writeblock functions, respectively. These functions allow a specified number of bytes of data to be read from or written to a specified offset in a previously declared memory.

Unlike co_stream_create, the co_memory_create function includes a platform-specific identifier that indicates the physical location of the memory resource. The identifier you will use in the second argument to co_memory_create depends on the platform you are targeting. In the case of the generic VHDL platform, the identifier "mem0" represents a generic dual-port synchronous RAM.

Another critical distinction between stream and memory interfaces is that stream interfaces are always synchronized by virtue of the FIFO buffers used to implement them. When passing values from one process to another on streams, the producer and consumer of data on that stream never collide, and data sent by the producer process are never in a "half baked" state from the perspective of the consumer process.

Memories, on the other hand, may require careful synchronization when used for data communication. This may mean that signals or streams (or in some cases lower-level communication lines called registers) will be required in addition to the memory interface.

In the case of HelloFPGA, if you want to replace the stream s1 with an equivalent memory interface, you need to somehow synchronize the behavior of process Producer with process DoHello such that DoHello does not attempt to read data from the shared memory until Producer has finished writing to that memory. You do this by using a signal, as follows:

```
co_memory memory;
co_signal ready;
co_stream s2;
co_process producer, consumer, hello;

memory = co_memory_create("Memory", "mem0", MAXLEN*sizeof(char));
ready = co_signal_create("Ready");
s2 = co_stream_create("Stream2", CHAR_TYPE, 2);
```

In this excerpt from the configuration subroutine, a signal called ready is created and is used as a means of communicating status from one process (Producer) to the other (DoHello). Because signals can also carry integer values, we will also use this signal to pass a character count (the number of characters in the "Hello FPGA!" string) to the DoHello process.

Figures 4-8 and 4-9 show the resulting source files. In this modified version, notice how the co_signal_post function is used by the Producer process to pass a message to the DoHello process indicating that the memory is ready. The co_signal_post function is a nonblocking operation, so posting this message does not result in the Producer pausing in its operation when the signal is posted.

On the DoHello side of the interface, notice how the co_post_wait function is used as a delaying mechanism, guaranteeing that the contents of the memory (which is subsequently accessed using the co_memory_readblock function) are ready for reading. With this kind of explicit synchronization, it is possible for two processes (such as a software process and a hardware process) to make simultaneous requests to read or write a co_memory block, and the two operations may in fact overlap.

```
// HelloWorld_sw.c: Software processes to be executed on the CPU.
//
// In this version the Producer passes the text via a shared
// memory interface instead of on a stream.
//

#include "co.h"
#include <stdio.h>

extern co_architecture co_initialize(int param);

void Producer(co_memory shared_mem, co_signal ready) {
   int32 count;
   static char HelloWorldString[] = "Hello FPGA!";

   count = strlen(HelloWorldString);
   co_memory_writeblock(shared_mem, 0, HelloWorldString, count);
   co_signal_post(ready, count);
}

void Consumer(co_stream input_stream) {
   char c;

   co_stream_open(input_stream, O_RDONLY, CHAR_TYPE);
   while (co_stream_read(input_stream, &c, sizeof(char)) == co_err_none ) {
      printf("Consumer read %c from input stream\n", c);
   }
   co_stream_close(input_stream);
}

int main(int argc, char *argv[]) {
   int param = 0;
   co_architecture my_arch;

   printf("HelloFPGA starting...\n");

   my_arch = co_initialize(param);
   co_execute(my_arch);

   printf("HelloFPGA complete.\n");
   return(0);
}
```

Figure 4-8. HelloFPGA_mem software test bench functions (HelloFPGA_mem_sw.c).

```
// HelloWorld_hw.c: Hardware processes and configuration.
//

#include "co.h"
#define MAXLEN 128

extern void Producer(co_memory shared_mem, co_signal ready);
extern void Consumer(co_stream output_stream);

// Hardware process: reads from memory, writes to stream
void DoHello(co_memory shared_mem, co_signal ready,
             co_stream output_stream) {
   int32 i, count;
   char buf[MAXLEN];
   char c;

   co_signal_wait(ready, &count);
   co_memory_readblock(shared_mem, 0, buf, count);
   co_stream_open(output_stream, O_WRONLY, CHAR_TYPE);
   for (i=0; i < count; i++) {
      c = buf[i];
      co_stream_write(output_stream,&c,sizeof(char));
   }
   co_stream_close(output_stream);
}

void config_hello(void *arg) {   // Configuration function
   co_memory memory;
   co_signal ready;
   co_process producer, consumer, hello;
   co_stream s2;

   memory = co_memory_create("Memory", "mem0", MAXLEN*sizeof(char));
   ready = co_signal_create("Ready");
   s2 = co_stream_create("Stream2", CHAR_TYPE, 2);

   producer = co_process_create("Producer", (co_function) Producer, 2, memory, ready);
   hello = co_process_create("DoHello", (co_function) DoHello, 3, memory, ready, s2);
   consumer = co_process_create("Consumer", (co_function) Consumer, 1, s2);

   co_process_config(hello, co_loc, "PE0");  // Assign to PE0
}
```

Figure 4-9. HelloFPGA_mem hardware process and configuration subroutine (HelloFPGA_mem_hw.c).

4.14 MEMORY AND STREAM PERFORMANCE CONSIDERATIONS

FPGA-based computing platforms may include many different types of memory, some of which are embedded within the FPGA, and others of which are external. Embedded memory is integrated within the FPGA fabric itself and may in fact be implemented using the same internal resources that are used for logic elements. External memory, on the other hand, is located in one or more separate chips that are connected to the FPGA through board-level connections. FPGA platforms most often include some external memory such as SRAM or Flash. All popular FPGAs today also include some configurable amount of on-chip embedded memory.

Note: There are many ways to configure a platform to use the available external or embedded memories. Platforms can be configured to have program, data, and cache storage located in either embedded or external memory or a combination of the two. Additionally, memory may be connected to the CPU and/or FPGA logic in many different ways, perhaps over multiple on-chip bus interfaces. As a result, there are many considerations in deciding how to use the available memory and, more generally, how data should move through an application. Some of these issues are discussed at greater length in Appendix A.

The Impulse C programming model supports both external and embedded memories, in many different configurations, for use as shared memory available to Impulse C processes. Using the Impulse C library, you can allocate space from a specific memory and copy blocks of data to and from that memory in any process that has access to the shared memory. This means that memories can be used as convenient alternatives to streams for moving data from process to process. So how do you know when to use shared memory for communication and when to use streams?

For communication between two hardware processes, the choice is simple. If the data is sequential, meaning that it is produced by one process in the same order it will be used by the next, a stream is by far the most efficient means of data transfer. If, however, the processes require random or irregular access to the data, a shared memory may be the only option.

For communication between a software process and a hardware process, however, the answer is more complex. The best solution depends very much on the application and the types and configuration of the memory available in the selected platform. The remainder of this section looks at some micro-benchmark results for three specific platform configurations and shows how you might use similar data from your own applications to make such memory-related decisions.

Micro-Benchmark Introduction

For the purposes of evaluating streams versus memories in a number of possible platforms, a micro-benchmark was created to test data transfer performance using the typical styles of communication found in Impulse C applications. The first three tests measure three common uses of stream communication:

- A Stream (one-way) test measures the transfer speed of using an Impulse C stream to send data from the CPU to a hardware process. This test transfers data using a local variable and represents an application in which the data is first computed on the CPU and then is passed along to hardware for further processing.

- A Stream (two-way) test is similar to the first test, but also includes the time required to return data back to the process on a second stream. This test represents a filter-type application in which data is streamed to a hardware process for processing and then back to the CPU in a loop.

- A RAM-CPU-Stream test measures the performance of the combined use of a memory and a stream with the CPU. In this case, the CPU reads data from memory and writes the data directly to a hardware process using a stream. This test represents an application in which the CPU does not need to perform any computation on the data, but the data in memory is ready for processing. The CPU is simply used to fetch data for the hardware process.

Another three tests (called Shared Mem-4B, Shared Mem-16B, and Shared Mem-1024B) measure the performance of direct transfers from memory to an Impulse C hardware process using the Impulse C shared memory support. In these three tests, the hardware process repeatedly reads blocks of data from the external memory into a local array. These tests emulate applications in which the data in memory is ready for processing and can be directly read by the hardware process without CPU intervention. The three tests differ only in the size of the blocks transferred to represent different types of applications. Applications requiring random access, for example, might need to read only four bytes at a time, whereas sequential processing algorithms might need to read much larger sequential blocks at a time.

Memory Test Results for the Altera Nios Platform

The table in Figure 4-10 displays the micro-benchmark results for our first configuration, which consists of an Altera Nios embedded soft processor implemented on an Altera Stratix S10 FPGA.

Memory Test	Transfer Rate
Stream (one-way)	1529KB
Stream (two-way)	917KB
RAM-CPU-Stream	708KB
Shared Mem (4B)	1167KB
Shared Mem (16B)	1217KB
Shared Mem (1024B)	1235KB

Figure 4-10. Memory test results for the Nios processor in an Altera Stratix FPGA.

In this particular configuration, the Nios CPU was connected to an external SRAM and to our test Impulse C hardware process via Altera's Avalon bus. These results seem to indicate that stream communications are quite efficient in this platform/configuration if the data requires some computation on the CPU and needs to be subsequently sent to a hardware process. If, however, the data in memory does not require any processing and can be directly accessed by the hardware process, the shared memory approach is more efficient. If the application is accessing data randomly, just four bytes at a time, the performance difference is not significant. In that case, stream-based communication might be used because it is easier to program, requiring less complicated process-to-process synchronization.

How do we explain these results? The Avalon bus architecture is unique in that there is not a single set of data and address signals shared by all components on the bus. Instead, the Avalon bus is customized to the set of components actually attached to the bus and is automatically generated by software. A significant result of this is that two masters can perform transfers at the same time if they are not communicating with the same slave. For an example of particular interest to us, the CPU can access Impulse C streams and signals at the same time that an Impulse C hardware process might be transferring data to and from an external RAM. This means that a software process on the CPU may be polling (waiting for) a signal, or receiving data from a hardware process, while the hardware process is simultaneously transfering data to or from external memory.

Note also that, in our test, the program running on the Nios processor was stored in the external memory. This means that the CPU may have slowed down the shared memory tests by making frequent requests for instructions from the external RAM. Another approach is to use a separate embedded memory for program storage, which would increase performance. The performance gain is due to the fact that the Avalon bus architecture permits the CPU to access the embedded memory while the hardware process simultaneously accesses the external memory. This would also increase the performance of the stream tests, because the embedded memory is much faster than external memory and program execution would be faster.

Memory Test Results for the Xilinx PowerPC Platform

Figure 4-11 displays the micro-benchmark results for our second sample configuration, which includes an embedded (but hard rather than soft) PowerPC processor as supplied in the Xilinx Virtex-II Pro FPGA.

Memory Test	Transfer Rate
Stream (one-way)	522KB
Stream (two-way)	310KB
RAM-CPU-Stream	439KB
Shared Mem (4B)	3732KB
Shared Mem (16B)	5466KB
Shared Mem (1024B)	6056KB

Figure 4-11. Memory test results for the PowerPC processor in a Xilinx Virtex-II Pro FPGA.

In this test, the FPGA was configured with both a PLB (Processor Local Bus) and an OPB (On-chip Peripheral Bus). The test program running on the PowerPC was stored in an embedded (on-chip) memory attached to the PLB, while the external memory was attached to the OPB. Although these busses are standard shared-bus architectures, using two busses allows the PowerPC to execute programs from the embedded memory on the PLB bus while a hardware process might be accessing the external memory—at the same time—on the OPB.

These results indicate that stream performance over the PLB is very poor. The reason for the low performance is currently unknown, but it might be due to the PLB-to-stream bridge components. The conclusion here is if the application does not require any computation on the PowerPC and can be directly used by the hardware process, shared memory is much faster. As a general rule, it is inefficient for external data to be accessed by a hardware process through the CPU and it is better to access that memory directly.

Memory Test Results for the Xilinx MicroBlaze Platform

Figure 4-12 shows the micro-benchmark results for our final sample configuration, that of a MicroBlaze soft processor implemented in a Xilinx Virtex II FPGA.

Memory Test	Transfer Rate
Stream (one-way)	10706KB
Stream (two-way)	4282KB
RAM-CPU-Stream	3536KB
Shared Mem (4B)	3536KB
Shared Mem+Signal (4B)	3528KB
Shared Mem (16B)	4784KB
Shared Mem+Signal (16B)	3881KB
Shared Mem (1024B)	5484KB
Shared Mem+Signal (1024B)	3892KB

Figure 4-12. Memory test results for the MicroBlaze processor in a Xilinx Virtex-II FPGA.

For this test, the FPGA was configured with a single OPB, embedded memory for program storage, and an external SDRAM. The first thing that stands out from these results is the large transfer rate obtained in the Stream (one-way) test. This result comes from the fact that Impulse C implements MicroBlaze-to-hardware streams using the Fast Simplex Link (FSL) provided for MicroBlaze. FSLs are dedicated FIFO channels connected directly to the MicroBlaze processor, thus avoiding the system bus altogether and providing single-cycle instructions to read and write data to and from hardware.

Although there is only one system bus, the MicroBlaze has dedicated instruction and data lines that can be connected to embedded memory for faster performance. Our sample configuration uses these dedicated connections and disconnects the embedded memory from the OPB to avoid interference from instruction fetching.

At first glance, the memory performance results from this test look good, but there are some additional considerations for this platform related to the use of Impulse C signals. Signals are implemented by the Impulse compiler as memory mapped I/O on the OPB. In applications making use of shared memory for process-to-process communication, a software process typically waits on a signal from the hardware process to know when the hardware process has completed and has finished writing to memory. Because this sample configuration uses one shared bus, the signal polling interferes with memory usage. For this reason we've shown two results for each shared memory test. While at smaller block sizes the performance was not adversely affected, we find that for larger block sizes, the signal polling significantly reduces performance.

The results also show that, for random access with small block sizes, using shared memory does not provide any advantage, and a streams-based approach would be preferable because it is simpler to program. However, as in the earlier tests using Altera Nios, the performance is significantly better using shared memory with large block sizes, provided that signal polling can be avoided.

4.15 SUMMARY

In this chapter we have introduced the Impulse C libraries and provided a brief set of examples demonstrating the essence of streams-oriented programming. We have seen how producer and consumer processes can be used to generate test data, and we have explored alternative methods of communicating between processes, including the use of signals and shared memories. We have also explored some of the important trade-offs to be made when considering streams-based and memory-based communications in specific FPGA platforms.

In the chapters that follow we'll delve into a few actual applications, describe the development and testing process, and demonstrate how to increase the performance of Impulse C applications through specific C-language coding techniques.

Find out more at www.ImpulseC.com

The remaining chapters present a series of examples demonstrating important concepts and techniques for programming FPGA-based applications in C. Additional examples may also be found on the companion website at www.ImpulseC.com/practical.

CHAPTER 5

Describing a
FIR Filter

In this chapter we present a simple digital filtering application in order to more fully introduce concepts of streams-based programming and to demonstrate the use of desktop simulation for design verification. In subsequent chapters we'll describe the process of actual hardware generation and discuss how to take advantage of statement-level and system-level parallelism for increased process throughput.

5.1 DESIGN OVERVIEW

A finite impulse response (FIR) filter is a type of signal processing element whose output is determined by a weighted sum of its past and present input values. There are many types of digital FIR filters, but all are defined primarily by the size of the window (number of discrete historical samples) that are used to perform a given filter calculation. This window size is often referred as the number of *filter taps*.

The filter we will describe here, using C language and the Impulse C libraries, is a 51-tap FIR filter. This filter will demonstrate a single hardware process and will provide us with opportunities to introduce important aspects of C programming for hardware—concepts that we will discuss in much greater detail in subsequent chapters.

5.2   THE FIR FILTER HARDWARE PROCESS

The hardware process representing our 51-tap FIR filter is shown in its entirety in Figure 5-1. As with the HelloFPGA example presented in the preceding chapter, this process accepts data on an input stream, filter_in, and performs a repetitive calculation on that stream to produce a corresponding filtered output stream, filter_out.

Stepping through the process, notice the following:

- The only interfaces into and out of the process are via the two streams filter_in and filter_out. An FIR filter such as this requires coefficients in order to perform its operation, so these values (which in the case of this filter are 32-bit integer values) are passed via the same input stream as the data to be filtered, presumably as a part of the system initialization. (A later design refinement might be to obtain the coefficients via a shared memory resource or to modify the process such that new coefficients could be dynamically loaded in response to an input signal.)

- Two local arrays, coef and firbuffer, are defined at the start of the process along with variables nSample, nFiltered, accum, and tap. You can see from the definition of TAPS (which is 51) and the data width (32 bits) that these arrays each equate to 204 bytes (1632 bits) of storage. In traditional C programming you know that such arrays declared within a subroutine are allocated on the stack, and that in all likelihood there is plenty of stack space available for such modestly sized arrays. When targeting hardware, however, you may need to consider (and perhaps specify) where such memory will be allocated and how that allocation relates to the trade-offs of the target platform. For now, however, we'll just take an abstract view of the application, leaving such considerations for a later discussion.

- After the input and output streams are opened (using modes O_RDONLY and O_WRONLY, respectively), the 51 coefficients are read from the input stream filter_in using co_stream_read. The assumption here is that the upstream process (the one feeding inputs to the FIR filter) is well-behaved and will send exactly 51 coefficients on the stream before it starts sending the data to be filtered. If the upstream process has been properly written (the programmer hasn't made an "off by one" or similar error), this method of counting the number of coefficients is quite safe. This is because the self-synchronizing nature of streams (which are FIFO buffers) guarantees that no data will be lost, regardless of whether the upstream and downstream processes are running at vastly different rates (such as when the producer process is running in software and the consuming process is running in hardware).

```
#define TAPS 51

void fir(co_stream filter_in, co_stream filter_out) {
    int32 coef[TAPS], firbuffer[TAPS];
    int32 nSample, nFiltered, accum, tap;

    co_stream_open(filter_in, O_RDONLY, INT_TYPE(32));
    co_stream_open(filter_out, O_WRONLY, INT_TYPE(32));

    // First fill the coef array with the coefficients...
    for (tap = 0; tap < TAPS; tap++) {
        co_stream_read(filter_in, &nSample, sizeof(int32));
        coef[tap] = nSample;
    }
    // Now fill the firbuffer array with the first n values...
    for (tap = 1; tap < TAPS; tap++) {
        co_stream_read(filter_in, &nSample, sizeof(int32));
        firbuffer[tap-1] = nSample;
    }

    // Now we have an almost full buffer and can start
    // processing the streaming waveform samples.
    //
    // Each time we process a sample we will "shift" the buffer
    // one position and read in a new sample.
    //
    while ( co_stream_read(filter_in, &nSample, sizeof(int32))
                              == co_err_none ) {
        firbuffer[TAPS-1] = nSample;
        accum = 0;
        for (tap = 0; tap < TAPS; tap++) {
            accum += firbuffer[tap] * coef[tap];
        }
        nFiltered = accum >> 2;
        co_stream_write(filter_out, &nFiltered, sizeof(int32));
        for (tap = 1; tap < TAPS; tap++) {
            firbuffer[tap-1] = firbuffer[tap];
        }
    }
    co_stream_close(filter_in);
    co_stream_close(filter_out);
}
```

Figure 5-1. A 51-tap FIR filter described using Impulse C.

- After the first loop has completed and the coefficients have been read, a second initialization loop (a loop that is outside of the main processing loop of this process) sets up the filter by filling the primary buffer (firbuffer) with enough values to begin the calculations, which take place in the subsequent main processing loop.

- The main processing loop follows the same pattern we observed in the HelloFPGA example of the preceding chapter, in which the return value of co_stream_read serves as the loop condition. This is the most efficient way to code an Impulse C inner code loop and should be used wherever possible.

- The actual calculations to perform the FIR filtering operation are contained in two inner code loops. The first of these two loops performs the filter calculation for the current value (which is an accumulation of products of the input buffer firbuffer and the corresponding coefficient). The second inner code loop (after a result value is written to the output stream) shifts the array by one in preparation for the next iteration of the main processing loop.

- When the filter process detects an end-of-stream condition (indicating that there are no more filter inputs to process), control drops out of the loop and the two streams (filter_input and filter_output) are closed. In an actual hardware implementation it can be assumed that this will never occur; the process will continue processing values until the system is powered down. (If the stream was closed for some reason, the hardware for the FIR filter would become inactive and useless until the system was restarted.) But for desktop simulation, such cleanup after a set of tests have been run against the process can be useful. In fact, if the streams are never closed, the process might never terminate itself, other processes might not close as a result, and the simulation might not complete as expected.

5.3 THE SOFTWARE TEST BENCH

To verify the correct function of this FIR filter prior to attempting any optimizations, we have written a simple software test bench that reads coefficient and source data from files, using the standard C fopen and related functions, and presents these values to the FIR filter process during desktop simulation. The producer and consumer processes representing the software test bench are shown in Figures 5-2 and 5-3, respectively. The configuration subroutine that ties these two test bench processes to the FIR filter process is presented in Figure 5-4.

How fast will it run?

What you see in Figure 5-1 is all that is required to define the operation of a generic 51-tap FIR filter.

What you might be asking yourself at this time (and any time you create a new Impulse C process of your own) is how the process will perform when generated as FPGA hardware. What kind of latencies and/or throughput can be expected? How large will the resulting logic be? These questions can be answered relatively quickly by doing some initial compilations with the C-to-hardware compiler and optimizer, which reports information on loop delays (measured as stages), pipeline latency and rate, and other information useful for determining cycle counts and a rough size of the generated logic. More precise information regarding performance can be obtained by performing hardware simulations (using an HDL simulator) on the generated outputs and by synthesizing the generated logic to obtain an FPGA netlist.

This design has some clear trade-offs that can be made either by doing some simple recoding (for example, by manually unrolling the inner code loops) or by using compiler pragmas provided with Impulse C to perform such unrolling automatically and, if desired, to generate a pipeline.

As written, this FIR filter will compile to produce a rather slow implementation of a FIR filter, but one that is reasonably compact in terms of the generated hardware. The filter will be slow (when measured in terms of clock cycles to complete one filter operation) because, by default, the Impulse C compiler will not attempt to unroll the inner code loops or pipeline the statements of the main processing loop.

Armed with your unique knowledge of the sytem requirements, you may choose to have these optimizations performed, and you might insert instructions to that effect in the code, in the form of compiler pragmas. You must, however, always remain cognizant of the likely results and learn how to apply such optimizations intelligently to balance size and speed. This balancing of goals is covered to some extent in Chapter 6 and is covered in much greater depth in Chapters 7 and 10.

```
void Producer(co_stream waveform_raw) {
   FILE *coefFile, *inFile;
   const char *CoefFileName = "coef.dat";
   const char *FileName = "waveform.dat";
   int32 coefValue, rawValue, coefcount = 0;
   cosim_logwindow log = cosim_logwindow_create("Producer");

   coefFile = fopen(CoefFileName, "r");
   if ( coefFile == NULL ) {
      fprintf(stderr, "Error opening file %s\n", CoefFileName);
      exit(-1);
   }
   // Read and write the coefficient data...
   co_stream_open(waveform_raw, O_WRONLY, INT_TYPE(32));
   cosim_logwindow_write(log, "Sending coefficients...\n");
   while (coefcount < TAPS) {
      if (fscanf(coefFile,"%d",&coefValue) < 1) {
         break;
      }
      coefcount++;
      co_stream_write(waveform_raw, &coefValue, sizeof(int32));
      cosim_logwindow_fwrite(log, "Coef: %x\n", coefValue);
   }
   fclose(coefFile);

   inFile = fopen(FileName, "r");
   if ( inFile == NULL ) {
      fprintf(stderr, "Error opening file %s\n", FileName);
      exit(-1);
   }
   // Now read and write the waveform data...
   co_stream_open(waveform_raw, O_WRONLY, INT_TYPE(32));
   cosim_logwindow_write(log, "Sending waveform...\n");
   while (1) {
      if (fscanf(inFile,"%d",&rawValue) < 1)
         break;
      co_stream_write(waveform_raw, &Value, sizeof(int32));
      cosim_logwindow_fwrite(log, "Value: %x\n", rawValue);
   }
   cosim_logwindow_write(log, "Finished writing waveform.\n");
   co_stream_close(waveform_raw);
   fclose(inFile);
}
```

Figure 5-2. Producer process for the FIR filter software test bench.

```
void Consumer(co_stream waveform_filtered) {
    int32 waveformValue;
    unsigned int waveformCount = 0;
    const char * FileName = "results.dat";
    FILE * outFile;

    cosim_logwindow log = cosim_logwindow_create("Consumer");

    outFile = fopen(FileName, "w");
    if ( outFile == NULL ) {
        fprintf(stderr, "Error opening file %s for writing\n",
                    FileName);
        exit(-1);
    }
    co_stream_open(waveform_filtered, O_RDONLY, INT_TYPE(32));

    cosim_logwindow_write(log, "Consumer reading data...\n");
    while ( co_stream_read(waveform_filtered, &Value,
                    sizeof(int32)) == co_err_none ) {
        fprintf(outFile, "%d\n", waveformValue);
        cosim_logwindow_fwrite(log, "Value: %x\n", waveformValue);
        waveformCount++;
    }
    cosim_logwindow_fwrite(log, "Consumer read %d datapoints.\n",
                    waveformCount);

    co_stream_close(waveform_filtered);
    fclose(outFile);
}

int main(int argc, char *argv[]) {
    co_architecture my_arch;
    void *param = NULL;

    printf("FIR filter software test bench.\n");
    my_arch = co_initialize(param);
    co_execute(my_arch);
    printf("\n\nTest complete.\n");
    return(0);
}
```

Figure 5-3. Consumer process and main function for the FIR filter software test bench.

Use the full power of C and C++ for test benches

There is no reason to limit yourself to ANSI C when writing Impulse C software test benches. If you wish, you can create a large, complex test bench using C++ and any number of useful class libraries, including the MinGW library, Microsoft MFC or ATL, or even SystemC. From an Impulse C processing standpoint, all that is required is that you maintain your co_initialize function, your configuration subroutine, and your hardware and software processes that are to be implemented in the actual target in distinct source files that can be analyzed by the Impulse C compiler. Also, be sure to restrict yourself to ANSI C syntax (or use ifdef statements as appropriate) to distinguish those portions that must be analyzed by the Impulse C hardware compiler.

The producer and consumer processes for this example are very much like the equivalent producer and consumer processes presented in Chapter 4, with the notable difference that they read test data from files (coef.dat and waveform.dat) and write the results of the test to file results.dat.

Because Impulse C is standard ANSI C (extended with C-compatible libraries), you can make use of the full power of C programming in your test bench. For example, you can do a direct, value-by-value comparison of the output values with a known-good set of results read by the test bench from another input file. For brevity, however, we have shown a minimal test bench that simply reads input values from one set of files and writes the filtered results to the output file indicated.

Let's look at the test bench processes (which are summarized in Figure 5-5) in more detail.

The Producer Process

The Producer process generates a stream of outputs representing (in sequence) the FIR filter coefficients and the data to be filtered (the input waveform). Starting at the top of this process, we find the following:

- C declarations, including declarations of the two input files containing the sample coefficients and the waveform values. These values may have been generated by some other test-related software, or they may be actual sampled data from some kind of instrument. Whatever their source, the values in these files are formatted quite simply, as one value (in decimal integer format) per file line.

```
extern void Producer(co_stream waveform_raw);
extern void Consumer(co_stream waveform_filtered);

void config_fir(void *arg) {
    co_stream waveform_raw;
    co_stream waveform_filtered;
    co_process producer_process;
    co_process fir_process;
    co_process consumer_process;

    IF_SIM(cosim_logwindow_init();)

    waveform_raw = co_stream_create("waveform_raw",
                    INT_TYPE(32), BUFSIZE);
    waveform_filtered = co_stream_create("waveform_filtered",
                    INT_TYPE(32), BUFSIZE);

    producer_process = co_process_create("producer_process",
                    (co_function)test_producer,
                    1, waveform_raw);

    fir_process = co_process_create("filter_process",
                    (co_function)fir,
                    2, waveform_raw, waveform_filtered);

    consumer_process = co_process_create("consumer_process",
                    (co_function)test_consumer,
                    1, waveform_filtered);

    // Assign FIR process to hardware element
    co_process_config(fir_process, co_loc, "PE0");
}

co_architecture co_initialize(int param) {
    return(co_architecture_create("fir_arch","Generic_VHDL",
                    config_fir,(void *)param));
}
```

Figure 5-4. Configuration subroutine for the FIR filter software test bench.

- A call to the **cosim_logwindow_create** function. This function call results in an Application Monitor window being created at runtime (during software simulation) and associated with the indicated name and local pointer (of type **cosim_logwindow**).

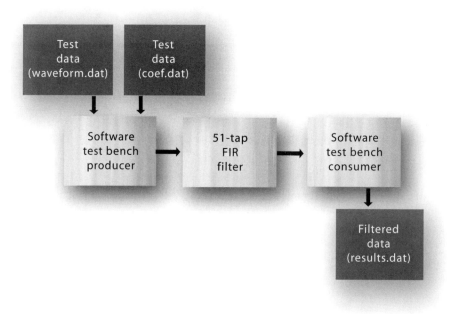

Figure 5-5. FIR filter test bench block diagram.

- A call to the standard C **fopen** function, followed by a loop that iteratively reads the coefficient values from the file **coef.dat**. These values are written to the filter's input stream (which in the **Producer** process is called **waveform_raw**) using **co_stream_write**. Notice the use of **cosim_logwindow_fwrite** to pass formatted text messages to the previously opened log window. The formatting supported in this function is identical to the formatting available in the standard C **fprintf** function.

- After all 51 coefficients have been read from the file and passed into the stream, the input file is closed (using **fclose**) but the stream is left open for the next step.

- The second **Producer** loop (which appears after the second call to **fopen** to open the **waveform.dat** file) follows the same format and outputs the test waveform data to the filter via the **waveform_raw** data stream. This loop continues as long as data remains in the input file.

The Consumer Process

The **Consumer** process accepts outputs from the FIR filter on a stream named **waveform_filtered** and writes these values both to an output file (again using one value per line) and to a log window. Examining this process in more detail, we see the following:

- A declaration of the output file (results.dat) and a call to the cosim_logwindow_create function.

- A co_stream_open Impulse C function call followed by a main processing loop. This loop iteratively reads values from the waveform_filtered stream and writes these values both to the results.dat file and to the log window using cosim_logwindow_fwrite.

- When the FIR filter stream has been closed, the Consumer function cleans up by closing the output file and writing a final count of the number of values read to the log window.

- Following the Consumer process is a main function for this application. When combined with the configuration function and the co_initialize function, this represents a complete C program ready for compilation for the purpose of software (or "desktop") simulation.

5.4 DESKTOP SIMULATION

The term *desktop simulation* refers to a method of debugging in which your complete Impulse C application, representing both hardware and software processes operating in parallel, is compiled under a standard C development environment (an IDE) for the purpose of functional verification. This is an exceptionally powerful technique because it allows you to bring in the full power of standard C language, and the many third-party libraries available for C and C++, to create comprehensive test fixtures—or test benches—and thereby validate the correct function of your application before you attempt to compile it to actual hardware.

When compiled in such a way, the various parallel processes of your application (in the case of the FIR filter application there are three such processes) are implemented using multiple threads. This allows you to emulate the parallel behavior of your application while still allowing you to make use of standard, widely available debugging tools such as Microsoft Visual Studio or open-source tools based on gdb.

Figure 5-6, for example, shows a design session in which the FIR filter example is being debugged using single-stepping and variable probing features of the Visual Studio debugger.

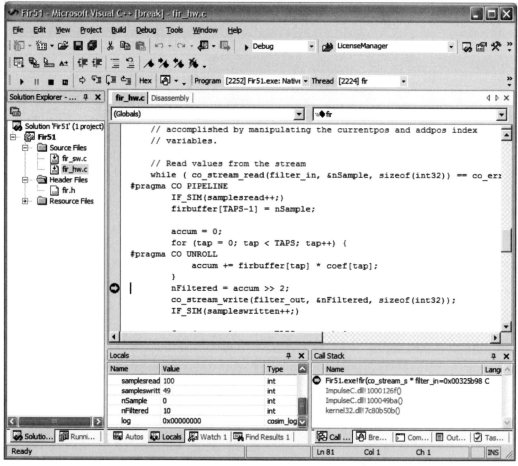

Figure 5-6. Using Visual Studio's debugger to validate the FIR filter example.

5.5 APPLICATION MONITORING

One aspect of the FIR filter software test bench that differs from the
HelloFPGA example of the preceding chapter is the presence of application
monitoring function calls, which are identified by the prefix cosim_. Application monitoring is a mechanism by which individual processes in an application may be selectively instrumented for the purpose of debugging.

The cosim_ instrumentation functions are compatible with all supported
desktop compiler environments. When included in your running Impulse C
application, these functions communicate with the Impulse C application

monitor to provide you with additional debugging capabilities that may be used in conjunction with standard desktop debugging environments.

Why use monitoring functions? Standard C debuggers (such as the debugger provided with Visual Studio) are useful for examining control flow in an application and for examining values of variables within specific processes. For highly parallel applications, however (applications consisting of many, perhaps hundreds, of interrelated processes), it is useful as well to add debug-related instrumentation directly within the application source code. Software instrumentation can be used to analyze data dependencies, to dynamically view the contents of streams, and to better understand how an application and its component algorithms might be optimized by the application programmer.

Monitoring with Log Windows

Log windows (which appear as child windows of the application monitor) are used to organize, format, and display messages and other information from within an executing Impulse C application. Any number of log windows may be created. To create and use a log window, do the following:

1. Call the **cosim_logwindow_init** function from within the configuration function of your application. (It is only necessary to call this function once, no matter how many log windows you will be creating.)

2. From within a process run function, declare a variable of type **cosim_logwindow**.

3. Create and name the log window using **cosim_logwindow_create**. Assign the return value of **cosim_logwindow_create** to the variable declared in step 2.

4. Write to the window any time after creating it by using **cosim_logwindow_write** or **cosim_logwindow_fwrite**.

5. Prior to executing the application in a desktop simulation environment (either as a stand-alone application or under the control of a debugger), launch the application monitor program.

During the course of the simulation, the log windows you have defined will display useful information on the state of each input and output stream, as well as messages generated as a result of calls to **cosim_logwindow_fwrite**, as shown in Figure 5-7.

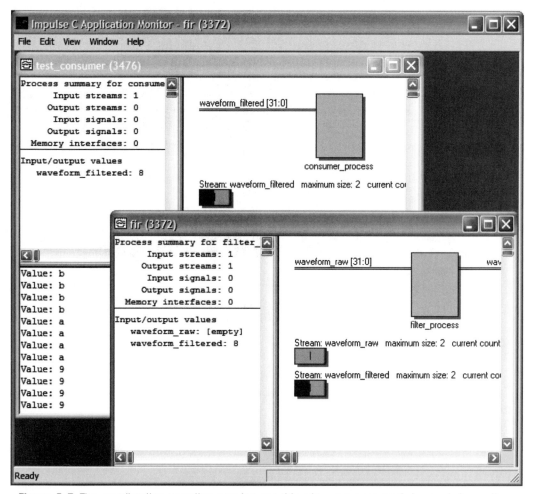

Figure 5-7. The application monitor can be used to view process and stream information.

5.6 SUMMARY

In this chapter, you have seen how Impulse C can be used to define a generic FIR filter. This application is representative of many types of filter applications, in which an input data stream must be processed to produce some corresponding output.

This chapter also showed you how desktop simulation and application monitoring can be used to debug an application at the level of C code. In the next chapter we'll spend more time with the FIR filter example and learn more about hardware generation tools. We'll investigate the compiler tool flow, learn about the kind of hardware that is produced, and learn as well how the generated hardware can be verified using a hardware simulator. We will also examine some alternative implementations of the FIR filter, with an eye toward achieving algorithm performance goals.

CHAPTER 6

Generating
FPGA Hardware

The preceding two chapters presented a method of programming, using the Impulse C libraries, that emphasizes a streams-oriented approach to application partitioning and that supports parallelism at the system level. This approach supports the concept of coarse-grained parallelism and allows us to abstract away the details of the actual target platform.

In actual practice, however, we need to be aware of the limitations of the target platform (the FPGA and any related IP components, such as embedded soft processors), and we need to understand how to write C code that can be efficiently compiled to either a processor or an FPGA, as appropriate. In this chapter we'll begin that process of understanding by describing the C-to-hardware compilation flow and by describing some of the C programming constraints you will encounter in current-generation tools.

When writing Impulse C processes that will be compiled to hardware using the Impulse C compiler, there are a number of such constraints you must observe, and there are methods of description that will result in better (or worse) results being generated. This chapter describes these constraints and programming methods in a general way; additional details will be presented in later chapters when we explore more specific examples.

As we investigate the processing flow, we will explore, to a limited extent, the hardware that has been generated in the form of hardware description language files.

6.1   THE HARDWARE GENERATION FLOW

When your application is processed by the Impulse C compiler, a series of steps are performed, some of which are dependent on the platform target you have selected. Figure 6-1 illustrates the design flow, beginning with C source files representing the hardware and software elements of the application.

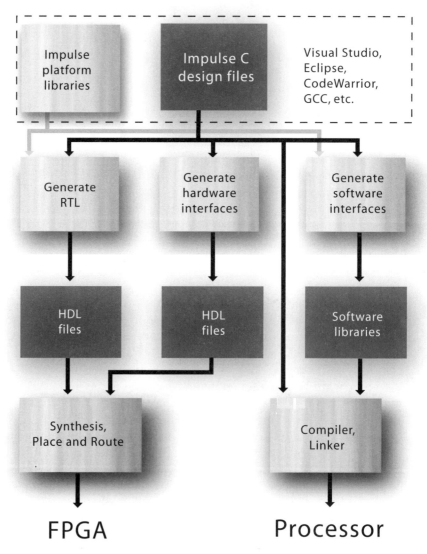

Figure 6-1. The C-to-hardware generation flow.

In this flow, design entry and initial desktop simulation and debugging are performed using common C development tools such as Visual Studio or gcc and gdb. The Impulse libraries provide the needed functions (as described in the preceding chapters) for software emulation of parallel processing.

Once an application has been developed and simulated in the context of standard C programming tools, it can then be targeted to a specific platform and compiled into hardware and software binary files. For software functions (those that will be compiled to an embedded processor, for example) the process is relatively straightforward and uses standard cross-compiler tools in conjunction with platform-specific runtime libraries provided with the Impulse compiler.

The hardware generation process is somewhat more complex and is summarized in the next section.

Optimization and Hardware Generation

Those portions of an Impulse C application that are intended for hardware (as specified using the co_process_config function) are analyzed by the RTL generator and optimized using a series of compiler passes. The optimization and hardware generation passes are summarized here and in Figure 6-2:

- *C preprocessing.* As in a traditional C compiler flow, the first step is a preprocessing pass that incorporates such things as include file references, macro expansion, and the like. If you have taken advantage of the preprocessor's #ifdef features, this is where C (or C++) statements that are related only to desktop simulation might be removed.

- *C analysis.* This compiler step is where, for example, the hardware and software processes of your application are identified. Specifically, this is where your application's co_initialize and configuration functions are examined to determine which processes you have configured as hardware processes, as defined by your use of the co_process_config function. Streams, signals, shared memories, and related elements are also identified here in preparation for hardware generation and for the generation of any needed hardware/software interfaces.

- *Initial optimization.* This phase performs various common optimizations on your hardware processes. (Software processes are not optimized or compiled at this point because they are processed later by a standard compiler environment for the target microprocessor.) Optimizations performed at this point might include constant folding, dead code elimination, and other such techniques. Certain compiler preoptimizations in support of later parallel optimizer passes are also performed at this point.

- *Loop unrolling.* This phase finds any uses of the UNROLL pragma and performs a corresponding expansion of loops into equivalent (typically larger) parallel statements. If the UNROLL pragma has been used on a loop that cannot be unrolled (for example, one with a nonstatic termination count), an error is generated.

- *Instruction stage optimization.* In this pass, a number of important optimizations are performed in order to extract parallelism at the level of individual C statements and at the level of blocks of statements. For example, two or more statements that appear in sequence but that have no interdependencies might be collapsed into a single clock cycle. Additionally, this optimizer pass performs loop pipelining if such pipelining has been requested via the PIPELINE pragma.

- *Hardware generation.* In this pass, the optimized and parallelized code is translated into equivalent hardware descriptions, resulting in a set of synthesizable (and simulatable) HDL files. These files contain descriptions of each hardware process, as well as references to the required stream, signal, and memory components. These latter components are referenced from a hardware library provided for a specific FPGA target.

For software processes, the compiler simply copies the relevant C source files to a location (typically a subdirectory of your project) and generates one or more related C and/or assembly language files that define the software-to-hardware interfaces, which are typically memory-mapped I/O routines.

This flow can change slightly for certain types of platforms, but it is representative of the work that is done by the compiler in preparing a C application (one consisting of both hardware and software elements) for mapping onto a programmable platform. It is important to understand, however, that more needs to happen before you can run your code on the FPGA target. These additional steps include the following:

- *Logic synthesis and FPGA technology mapping.* This is performed by the FPGA synthesis software provided by your FPGA vendor-supplied tools, or using third-party synthesis tools as appropriate. During this process, the hardware descriptions generated in the steps just described are further optimized (for example, to take advantage of FPGA-specific features such as multipliers and other "hard macros") to create an FPGA netlist, which is typically in EDIF format. (It should be noted here that some C-to-hardware design flows bypass this step and compile C code directly to EDIF netlists. This can provide the user and/or the optimizer with more opportunities for low-level hardware control.)

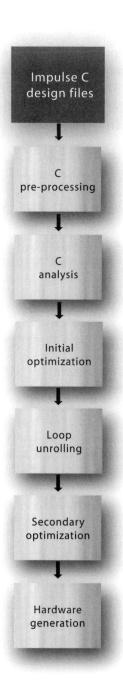

Figure 6-2. The C-to-hardware optimization and generation steps.

- *FPGA place-and-route*. In this step, the FPGA netlist created in the previous step is analyzed, and its many component references (which represent such things as registers and gate-level logic, as well as references to higher-level FPGA macros), are assigned to locations in the FPGA. The result of this step is an FPGA bitmap that can be downloaded to the device via a JTAG or other interface.

Due to the size of modern FPGAs, the complexity of the interconnect structures, and the need to decompose the logic into elements suitable for the particular structure of the FPGA, the preceding two steps can take many minutes, or even hours, to complete. These steps take even longer if it is necessary to synthesize, place, and route other elements of the design, such as an embedded soft processor and its peripherals, in order to map the complete design to hardware. For this reason it can be helpful to make use of hardware simulators (when practical) and verify the results of hardware generation at the level of the HDL files generated by the C compiler passes. Doing so can provide a much faster design and debug experience. We'll provide an example of hardware simulation a bit later in this chapter.

6.2   UNDERSTANDING THE GENERATED STRUCTURE

As we described in the preceding section, the outputs generated as a result of hardware generation will differ somewhat depending on the platform target you select, but at its core every generated hardware project includes both a top-level HDL module and one or more lower-level HDL modules representing the hardware process(es) that you have defined in C language and declared as hardware using **co_process_config**. When generated as VHDL, for example, the FIR filter presented in Chapter 5 is represented by two VHDL entity and architecture pairs, which are summarized in the listings shown in Figures 6-3 (the top-level file) and 6-4 (the lower-level file).

In the VHDL code shown (which has been modified to remove most of the details of the generated FIR filter logic), you can see that the FIR filter process has been converted into a VHDL top-level entity that in turn references a lower-level entity containing the actual function of the FIR filter as expressed in register transfer level (RTL) code.

Why these two levels in the generated VHDL files? First, if the application consists of multiple, parallel hardware processes, the top-level entity and its corresponding architecture are where these various processes are combined and interconnected.

The second reason for partitioning the generated hardware in this way is that the process itself (the FIR filter) is not directly connected to the outside world. Instead, the connections are made via buffered data streams, and these stream components must be referenced and instantiated in the top-level module. These stream interfaces are a critical aspect of the design and represent the hardware realization of the abstract streams used in the FIR filter process, and in fact for the majority of Impulse C applications.

In the generated VHDL top-level file, notice that the lower-level entity/architecture is referenced using a component declaration and is then instantiated using a component instantiation statement, along with two stream components from the Impulse hardware library. This represents the basic hierarchy of our application as it will exist in hardware: a top-level entity/architecture that references our generated FIR filter process and whatever stream components are needed to provide the I/O channels as specified in the original C code.

If you look carefully at this generated VHDL code (and you have some VHDL experience), you might notice a few other things. First, notice that the FIR process and the reference stream components have reset lines and clocks. Of course, no resets or clocks were defined in the original untimed C source code, but the hardware generation process has resulted in clocks and resets being generated. The compiler includes options that allow you to control this generation of clocks and resets to some extent. For example, if you will be clocking different hardware and software processes at different rates (as might be the case when one process must interface to an external device at a limited rate when compared to other processes), you can specify dual-clock streams. In this case, the compiler generates stream components with multiple clocks. There is one clock for each end of the stream, corresponding to the differing clock rates used by the producer and consumer processes. Understanding these various compiler options can help you better optimize your applications for specific types of platforms.

Also notice the use of generics to specify stream parameters—specifically, the width and depth of the streams. Generics are also used for generated hardware processes if parameters (of type co_parameter) are used in your Impulse C descriptions. This example does not make use of compile-time Impulse C parameters, so you do not see any corresponding generics applied to the filter_process component.

Last, notice that each of the two streams specified in the original C code has resulted in three control signals generated by the compiler in addition to the data input or output. These control signals are described in the next section. Note that signal interfaces (which are not used in this example) follow a similar pattern.

```
-- This file was automatically generated.
--
library ieee;
use ieee.std_logic_1164.all;

library impulse;
use impulse.components.all;

entity fir_arch is
  port (
    reset, sclk, clk : in std_ulogic;
    p_producer_process_waveform_raw_en : in std_ulogic;
    p_producer_process_waveform_raw_eos : in std_ulogic;
    p_producer_process_waveform_raw_data : in
                                    std_ulogic_vector (31 downto 0);
    p_producer_process_waveform_raw_rdy : out std_ulogic;
    p_consumer_process_waveform_filtered_en : in std_ulogic;
    p_consumer_process_waveform_filtered_data : out
                                    std_ulogic_vector (31 downto 0);
    p_consumer_process_waveform_filtered_eos : out std_ulogic;
    p_consumer_process_waveform_filtered_rdy : out std_ulogic);
end;

architecture structure of fir_arch is
  component fir is
    port (
      reset, sclk, clk : std_ulogic;
      p_filter_in_rdy : in std_ulogic;
      p_filter_in_en : inout std_ulogic;
      p_filter_in_eos : in std_ulogic;
      p_filter_in_data : in std_ulogic_vector (31 downto 0);
      p_filter_out_rdy : in std_ulogic;
      p_filter_out_en : inout std_ulogic;
      p_filter_out_eos : out std_ulogic;
      p_filter_out_data : out std_ulogic_vector (31 downto 0)
    );
  end component;

  signal p_filter_process_filter_in_rdy : std_ulogic;
  signal p_filter_process_filter_in_en : std_ulogic;
  signal p_filter_process_filter_in_eos : std_ulogic;
  signal p_filter_process_filter_in_data : std_ulogic_vector (31 downto 0);
  signal p_filter_process_filter_out_rdy : std_ulogic;
  signal p_filter_process_filter_out_en : std_ulogic;
  signal p_filter_process_filter_out_eos : std_ulogic;
```

Figure 6-3. Top-level generated entity for the FIR filter application. (*continues*)

```
        signal p_filter_process_filter_out_data : std_ulogic_vector (31 downto 0);
        signal local_reset : std_ulogic;
    begin
      local_reset <= reset;

      filter_process: fir
        port map (
          local_reset, sclk, clk,
          p_filter_process_filter_in_rdy,
          p_filter_process_filter_in_en,
          p_filter_process_filter_in_eos,
          p_filter_process_filter_in_data,
          p_filter_process_filter_out_rdy,
          p_filter_process_filter_out_en,
          p_filter_process_filter_out_eos,
          p_filter_process_filter_out_data
          );

      inst0: stream
        generic map (datawidth => 32, addrwidth => 1)
        port map (
          reset => local_reset, clk => clk,
          input_en => p_producer_process_waveform_raw_en,
          input_rdy => p_producer_process_waveform_raw_rdy,
          input_eos => p_producer_process_waveform_raw_eos,
          input_data => p_producer_process_waveform_raw_data,
          output_en => p_filter_process_filter_in_en,
          output_rdy => p_filter_process_filter_in_rdy,
          output_eos => p_filter_process_filter_in_eos,
          output_data => p_filter_process_filter_in_data);

      inst1: stream
        generic map (datawidth => 32, addrwidth => 1)
        port map (
          reset => local_reset, clk => clk,
          input_en => p_filter_process_filter_out_en,
          input_rdy => p_filter_process_filter_out_rdy,
          input_eos => p_filter_process_filter_out_eos,
          input_data => p_filter_process_filter_out_data,
          output_en => p_consumer_process_waveform_filtered_en,
          output_rdy => p_consumer_process_waveform_filtered_rdy,
          output_eos => p_consumer_process_waveform_filtered_eos,
          output_data => p_consumer_process_waveform_filtered_data);

    end;
```

Figure 6-3. *continued*

```
-- This file was automatically generated.
--

library ieee;
use ieee.std_logic_1164.all;

library impulse;
use impulse.components.all;

entity fir is
  port (
    reset, sclk, clk : std_ulogic;
    p_filter_in_rdy : in std_ulogic;
    p_filter_in_en : inout std_ulogic;
    p_filter_in_eos : in std_ulogic;
    p_filter_in_data : in std_ulogic_vector (31 downto 0);
    p_filter_out_rdy : in std_ulogic;
    p_filter_out_en : inout std_ulogic;
    p_filter_out_eos : out std_ulogic;
    p_filter_out_data : out std_ulogic_vector (31 downto 0)
  );
end;

architecture rtl of fir is
  type pipeStateType is (idle, run, flush);
  signal s_b1_state : pipeStateType;
  -- The remainder of the local signal delcarations have been removed for brevity
  -- . . .

  process (clk,reset,stateEn)
  begin
    if (reset='1') then
      thisState <= init;
    elsif (clk'event and clk='1') then
    -- The remainder of the hardware description has been removed for brevity.
    -- . . .
```

Figure 6-4. Lower-level generated entity and architecture for the FIR filter (abbreviated).

6.3 STREAM AND SIGNAL INTERFACES

When processes written using the Impulse C libraries are compiled to hardware, HDL files are generated representing (as you have seen) the various processes, streams, and other elements described in the C source files. The generated hardware may represent a complete hardware system (as might be the case when streams connect to an external microprocessor) or may be interfaced to other hardware elements, such as high-speed serial interfaces, through the use of well-defined stream, signal, and memory interfaces.

Depending on the application requirements, stream and signal interfaces may be used to directly connect different hardware processes (for example, to create a system-level hardware pipeline) or may be used to connect hardware processes running in the FPGA with other processes running as software in an embedded FPGA processor. For hardware-to-hardware connections the stream and signal protocols are relatively straightforward and are described in this section. Hardware-to-software connections are somewhat more complex and are specific to the platform being targeted.

Note: If you are using a predefined Impulse platform support package, the required hardware-to-software or hardware-to-hardware stream and signal interfaces will be generated automatically for you. If, however, you are interfacing Impulse C hardware processes to other components of your system (such as external VHDL hardware elements), you will need to study and understand the basic communication mechanisms used to implement streams and signals at a hardware level. Appendix B presents one such example, in which a stream is interfaced to an external hardware component.

Stream and Signal Protocols

The Impulse C compiler uses well-defined protocols when generating hardware interfaces for streams and signals. Figure 6-5 illustrates the relationship between the top-level system and a lower-level component with its associated stream and signal components. As shown in the figure, each stream is generated with one data line and three control lines on its input side and four corresponding lines on its output side. One of the control lines is controlled by the stream (the rdy signal), and one is controlled externally (the en signal) to allow a handshake method of control. The eos control line flows through the stream and indicates to the downstream process that the upstream process has closed the stream (typically by using the co_stream_close function). The data line, which has a width defined by the application programmer, accepts data on the input side of the stream and receives data on the output side.

During operation, a stream generates indications on its rdy lines indicating that the stream is ready for reading (output side) or writing (input side). When rdy is active, the system or the connected process (depending on the direction of the stream) may read or write data from or to the stream by moving data from or to the data lines and then asserting the en signal.

Top-level HDL module (entity)

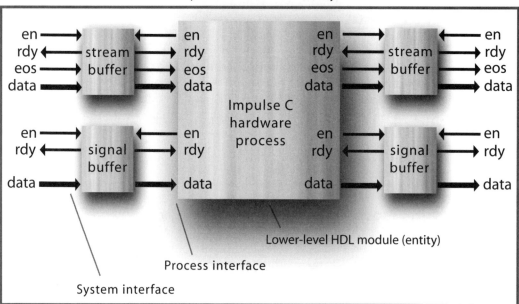

Figure 6-5. Streams and signals are implemented as components at the top level of the generated HDL.

Signals operate in a similar fashion, with the elimination of the **eos** control line and with different rules regarding how messages are posted. (Write operations on a signal component never block, even if an unserviced message has already been posted to the same signal.)

In the case of a hardware-to-hardware interface (as illustrated in Figure 6-6), the producer and consumer processes share a common stream or signal interface as shown.

Streams Used in Write Mode

When used in write mode (as indicated by the use of **O_WRONLY** in the **co_stream_open** function call), stream interfaces are generated with the following ports and corresponding signals:

```
<stream_name>_rdy : OUT    -- Ready to accept data.
<stream_name>_en : IN      -- Enable write.
<stream_name>_eos: IN      -- Write is EOS.
<stream_name>_data: IN     -- Write data.
```

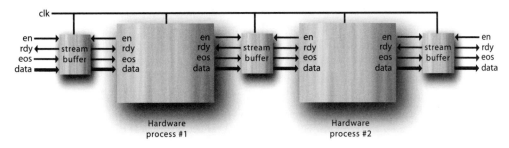

Figure 6-6. Multiple hardware processes connected via a shared stream buffer.

Streams Used in Read Mode

When used in read mode (as indicated by the use of O_RDONLY in the co_stream_open function call), stream interfaces are generated with these port names and corresponding signal names:

```
<stream_name>_rdy : OUT    -- Data is available.
<stream_name>_en : IN      -- Enable read.
<stream_name>_eos : OUT    -- Read is EOS
<stream_name>_data : OUT   -- Read data.
```

Signals Used in Post (Write) Mode

When used in write mode (as indicated by the use of the co_signal_post function call), signal interfaces are generated with the following ports and corresponding signals:

```
<signal_name>_en : IN      -- Enable write.
<signal_name>_data: IN     -- Write data.
```

Notice that there is no _rdy signal here, which reflects the fact that signal writes (using co_signal_post) are nonblocking operations.

Signals Used in Wait (Read) Mode

When used in wait mode (as indicated by the use of the co_signal_wait function call), signal interfaces are generated with the following ports and corresponding signals:

```
<signal_name>_rdy : OUT    -- Signal is posted.
<signal_name>_en : IN      -- Enable read.
<signal_name>_data : OUT   -- Read data.
```

6.4   USING HDL SIMULATION TO UNDERSTAND STREAM PROTOCOLS

The easiest way to understand the generated stream interfaces and their communication protocols is to examine them in the context of a simulation test
bench. Figures 6-7 presents such a test bench for the FIR filter example. When
used in a VHDL simulator, this test bench applies the coefficient and waveform data from the same files used in the software simulation of Chapter 5.

The important parts of this VHDL test bench are the two loops that
apply the data from the input files to the stream. One of these loops (the one
that applies the actual waveform data) appears as follows:

```
k := 0;
while true loop
    p_producer_process_waveform_raw_en <= '0';
    if (p_producer_process_waveform_raw_rdy = '1') then
        p_producer_process_waveform_raw_data <= testval_table(k);
        p_producer_process_waveform_raw_en <= '1';
        k := k+1;
    end if;
    wait until rising_edge(clk);
    if k=testval_count then
        exit;
    end if;
end loop;
```

This loop defines the following basic protocol, which applies to all stream inputs:

1. Set the _en control line low (value '0'). This tells the stream that the data
   currently appearing on the _data input is no longer valid.

2. Wait for the _rdy line (which is an output status signal from the stream)
   to be asserted. This can be done by using a VHDL wait statement sensi-
   tive to the _rdy input, or by checking the value of _rdy at a clock edge, as
   just shown above.

3. When _rdy is asserted, assign the desired data value to the _data input
   of the stream.

4. Set the _en line high (value '1'). This instructs the stream to accept the
   data. The stream will accept that data at the next rising clock edge.

5. After the clock edge, set the _en low (step 1).

To accept data from an output stream the steps are similar, with the _en signal being used to indicate to the stream that a value has been successfully
read and the stream may be updated with the next value. In the FIR filter test
bench, the _en control signal is simply held high, allowing values to be observed directly during the course of the simulation.

```
-- FIR filter test bench
--
library ieee;
use ieee.std_logic_1164.all;
use ieee.std_logic_arith.all;
use ieee.std_logic_textio.all;
use ieee.std_logic_arith.all;
use std.textio.all;
use work.config.all;

entity test_top is
end test_top;

architecture behavior of test_top is
  signal reset, clk : std_ulogic;
  signal p_producer_process_waveform_raw_en : std_ulogic;
  signal p_producer_process_waveform_raw_eos : std_ulogic;
  signal p_producer_process_waveform_raw_data :
                                      std_ulogic_vector (31 downto 0);
  signal p_producer_process_waveform_raw_rdy : std_ulogic;
  signal p_consumer_process_waveform_filtered_en : std_ulogic;
  signal p_consumer_process_waveform_filtered_data :
                                      std_ulogic_vector (31 downto 0);
  signal p_consumer_process_waveform_filtered_eos : std_ulogic;
  signal p_consumer_process_waveform_filtered_rdy : std_ulogic;
  constant PERIOD: time := 10 ns;
component fir_arch is
  port (
    reset, sclk, clk : in std_ulogic;
    p_producer_process_waveform_raw_en : in std_ulogic;
    p_producer_process_waveform_raw_eos : in std_ulogic;
    p_producer_process_waveform_raw_data :
                              in std_ulogic_vector (31 downto 0);
    p_producer_process_waveform_raw_rdy : out std_ulogic;
    p_consumer_process_waveform_filtered_en : in std_ulogic;
    p_consumer_process_waveform_filtered_data :
                              out std_ulogic_vector (31 downto 0);
    p_consumer_process_waveform_filtered_eos : out std_ulogic;
    p_consumer_process_waveform_filtered_rdy : out std_ulogic);
end component fir_arch;
begin
  reset_stimulus: process
  begin
    reset <= '1'; wait for PERIOD;
```

Figure 6-7. VHDL test bench for the FIR filter. (*continues*)

```
    reset <= '0'; wait;
end process;


clk_stimulus: process
begin
  clk <= '0'; wait for PERIOD/2;
  clk <= '1'; wait for PERIOD/2;
end process;

stimulus: process
 variable k : integer;
 begin
  p_producer_process_waveform_raw_eos <= '0';
  p_producer_process_waveform_raw_en <= '0';
  p_producer_process_waveform_raw_data <= X"00000000";
  wait for PERIOD;

  -- First output the coefficients
  k := 0;
  while true loop
    p_producer_process_waveform_raw_en <= '0';
    if (p_producer_process_waveform_raw_rdy = '1') then
      p_producer_process_waveform_raw_data <= coef_table(k);
      p_producer_process_waveform_raw_en <= '1';
      k := k+1;
    end if;
    wait until rising_edge(clk);
    if k=coef_count then
      exit;
    end if;
  end loop;

  -- Now output the waveform values
  k := 0;
  while true loop
    p_producer_process_waveform_raw_en <= '0';
    if (p_producer_process_waveform_raw_rdy = '1') then
      p_producer_process_waveform_raw_data <= testval_table(k);
      p_producer_process_waveform_raw_en <= '1';
      k := k+1;
    end if;
    wait until rising_edge(clk);
```

Figure 6-7. *continued*

```
            if k=testval_count then
                exit;
            end if;
        end loop;
    end process stimulus;


    results: process
    begin
        p_consumer_process_waveform_filtered_en <= '1';
        -- Now compare to the results of the software encryption
        while (p_consumer_process_waveform_filtered_eos = '0') loop
            -- results will appear on p_consumer_process_waveform_raw_data
            wait until rising_edge(clk);
        end loop;
        wait;
    end process;

    -- Instantiate the design under test
    DUT: entity work.fir_arch
        port map (
            reset, sclk, clk,
            p_producer_process_waveform_raw_en,
            p_producer_process_waveform_raw_eos,
            p_producer_process_waveform_raw_data,
            p_producer_process_waveform_raw_rdy,
            p_consumer_process_waveform_filtered_en,
            p_consumer_process_waveform_filtered_data,
            p_consumer_process_waveform_filtered_eos,
            p_consumer_process_waveform_filtered_rdy
        );
    end behavior;
```

Figure 6-7. *continued*

6.5 DEBUGGING THE GENERATED HARDWARE

Hardware that has been generated from C code should (in a perfect world) exhibit behavior that is exactly the same as is observed during a software simulation, such as when running under the control of a C debugger. In practice, however, there are many situations in which subtle coding errors made in the C code (such as relying on variables being initialized or making incorrect assumptions about process synchronization) can result in an application that operates perfectly during software simulation but fails in the actual hardware. To help guard against this, making use of hardware debugging techniques and hardware simulators can be an important part of your design efforts.

Note: Although debugging automatically generated HDL may seem daunting (particularly if you are a software engineer), it is actually not as bad as you might think. You will find that the generated outputs will be quite dense with intermediate, low-level signals (perhaps hundreds or even thousands of them, most of which will be optimized away in the hardware synthesis process). Fortunately, you will also find that the variables used in your C file are still, for the most part, intact and have their names preserved so they can be monitored during debugging.

To help in analyzing control flow and cycle-by-cycle synchronization issues, it's useful to know that the hardware generator implements each process in your application as a separate state machine, with symbolic state names that can be referenced back to specific blocks and stages of C code in the original application. Parallel operations are found within the state machine and/or within concurrent statements found elsewhere in the generated HDL module.

The following excerpts from the generated FIR filter hardware description help illustrate this point. First, notice that in the declarations section for the lower-level HDL file we have the following declaration:

```
type stateType is
        (init,b0s0,b0s1,b0s2,b0s3,b0s4,b0s5,b0s6,b0s7,b0s8,b0s9,b0s10,
        b0s11,b0s12,b0s13,b0s14,b0s15,b0s16,b0s17,b0s18,b0s19,b0s20,
        b0s21,b0s22,b0s23,b0s24,b0s25,b0s26,b0s27,b0s28,b0s29,b0s30,
        b0s31,b0s32,b0s33,b0s34,b0s35,b0s36,b0s37,b0s38,b0s39,b0s40,
        b0s41,b0s42,b0s43,b0s44,b0s45,b0s46,b0s47,b0s48,b0s49,b0s50,
        b0s51,b0s52,b0s53,b0s54,b0s55,b0s56,b0s57,b0s58,b0s59,b0s60,
        b0s61,b0s62,b0s63,b0s64,b0s65,b0s66,b0s67,b0s68,b0s69,b0s70,
        b0s71,b0s72,b0s73,b0s74,b0s75,b0s76,b0s77,b0s78,b0s79,b0s80,
        b0s81,b0s82,b0s83,b0s84,b0s85,b0s86,b0s87,b0s88,b0s89,b0s90,
        b0s91,b0s92,b0s93,b0s94,b0s95,b0s96,b0s97,b0s98,b0s99,b0s100,
        b0s101,b0s102,b1s0,b1s1,b2s0,finished);
    signal thisState, nextState : stateType;
```

The generated type **stateType** symbolically represents all the blocks and stages of the generated process. In the case of the FIR filter there are quite a few of these states in the machine (107 of them, to be exact) that represent two major blocks of functionality in the expanded code. One of these states (the first one, b0s0) is shown here, along with the clock logic that drives the machine:

```
if (clk'event and clk='1') then
    case thisState is
        when b0s0 =>
            if (stateEn = '1') then
                r_tap <= ni4126_tap;
            end if;
```

Comments found elsewhere in the generated HDL help identify which specific block and cycle a given operation is associated with. For example, the following concurrent multiply and accumulate operations are associated with stage one of block one, as indicated by the comment line preceding them:

```
-- b1s1
ni4130_nSample <= r_filter_in;
ni4131_firbuffer_50 <= ni4130_nSample;
ni4132_accum <= X"00000000";
ni4133_tap <= X"00000000";
ni4134_accum <= add(ni4132_accum, mul(r_firbuffer_0, r_coef_0));
ni4135_tap <= X"00000001";
ni4136_accum <= add(ni4134_accum, mul(r_firbuffer_1, r_coef_1));
```

Of the 107 states in the machine, those blocks identified by the b0 state name prefix represent the initialization section of the FIR filter, which consisted of two unrolled loops in the original C code. There are many stages in this block, but because this is only initialization code, the overhead of all those cycles is of little importance.

The key routine of the FIR filter, the inner code loop that actually processes the data stream, is represented by the states prefixed by b1 and b2, of which there are only two (b1s0 and b2s0) when pipelining has not been enabled and only one (b1s0) when pipelining has been enabled through the use of the PIPELINE pragma. You can use these symbolic states as an aid to hardware debugging with an HDL simulator, as shown in Figure 6-8.

Figures 6-9 and 6-10 show another hardware debugging session (again using the FIR application as an example) in which the expanded source code of the original example (in which the specific blocks and stages of the code have been identified, both graphically and in an expanded source listing) can, without too much difficulty, be related to specific lines of the generated HDL. Notice in the example shown that the variable firbuffer_50, which corresponds to one element of the scalarized firbuffer array, is easily identified in the HDL code during source-level hardware simulation and debugging. Comments embedded in the HDL code also help identify the specific blocks and stages of the original C code that correspond to the HDL statements being executed.

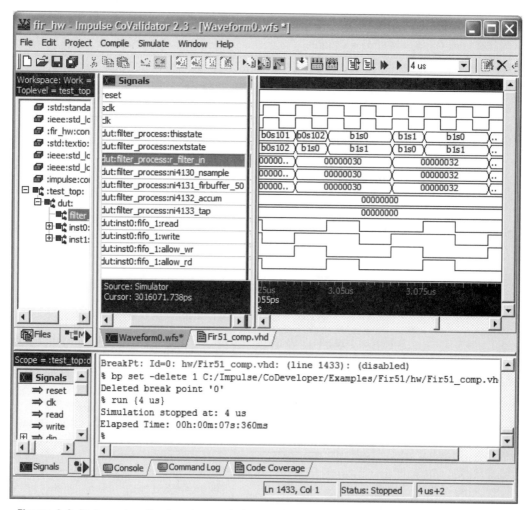

Figure 6-8. Debugging the hardware state machine using a VHDL simulator.

The goal of performing hardware simulations at this level (after compilation from C) is to identify and verify correct cycle-by-cycle behavior. An example of this kind of debugging is stepping through the design one clock cycle at a time (or through some defined number of cycles) to zero in on a specific problem area, as defined by both space (the area of code) and time (the clock cycle in which a problem manifests itself). The Impulse design flow has three fundamental ways in which to perform cycle-accurate hardware simulations of this type:

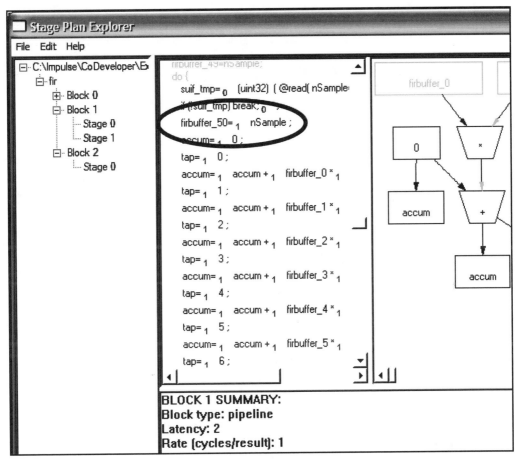

Figure 6-9. Viewing the expanded C code in the Impulse C Stage Master optimizer.

- The generated HDL files may be combined with an HDL test bench (as was done earlier in this chapter) for use in an HDL simulator, as you've just seen. This method generally requires at least a basic understanding of hardware design tools and requires access to an HDL simulator. The advantage of this method is that you have access to the many debugging and visualization features provided in hardware simulators.

- The generated HDL files may be translated (through the use of a utility provided by Impulse) to a cycle-accurate C-language model. This model may then be linked into the original application (replacing the original "untimed" C code) so that cycle-accurate behaviors may be observed in the context of a C debugger. An advantage of this method is that your original C-language test bench can be used to drive the simulation.

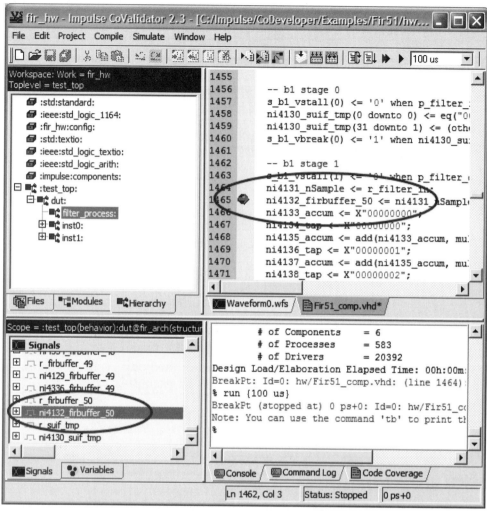

Figure 6-10. Debugging the same sequence of code in a hardware VHDL simulator.

- The synthesized design may be simulated at the level of the FPGA netlist, including incorporating post-route timing details through the use of timing-annotated HDL. This is the least desirable method if the goal is to verify function because the simulation can be extremely time-consuming and the generated netlist may be very difficult to relate back to the original C code. This is the only effective way, however, of debugging problems related to clock skew or to actual gate delays that are impacting your maximum operating frequency. (Such timing-related debugging is beyond the scope of this book, but many excellent books, application notes, and tools are available to help.)

Which of these methods you choose will depend on the nature of the problem you are attempting to debug, on your expertise as a hardware designer, and on your access to HDL simulation tools.

6.6   HARDWARE GENERATION NOTES

The following subsections describe a few important topics related to the generation of hardware. These topics will be expanded upon in subsequent chapters.

Instruction-Level Pragmas

Instruction- and block-level optimization features may be controlled at the level of your C source code through the use of certain predefined pragmas.

Pragma CO PIPELINE

Pipelining of instructions is not automatic and requires an explicit declaration in your C source code as follows:

```
#pragma CO PIPELINE
```

This declaration must be included within the body of the loop and prior to any statements that are to be pipelined. For example:

```
for (i=0; i<10; i++) {
#pragma CO PIPELINE
   sum += i << 1;
}
```

Note: The PIPELINE pragma must appear at the top of the loop to be pipelined, before any other statements within the loop, and the loop may not contain any nested loops.

Pragma CO UNROLL

Loop unrolling may be enabled with the use of the UNROLL pragma, which appears as follows:

```
for (tap = 0; tap < TAPS; tap++) {
#pragma CO UNROLL
   accum += firbuffer[tap] * coef[tap];
}
```

Unrolling a loop will result in that code within the loop being duplicated in hardware as many times as needed to implement the operation being described. It is therefore important to consider the size of the resulting hardware and unroll loops that have a relatively small number of iterations. The iterations of the loop must also not include interdependent calculations and/or assignments that would prevent the loop from being implemented as a parallel (unrolled) structure in hardware.

Note that the UNROLL pragma must appear at the top of the loop, before any other statements in the loop, and the loop must be a for loop with a loop variable of type int and constant bounds.

Pragma CO SET StageDelay

The general-purpose pragma CO SET is used to pass optimization information to the optimizers. One SET pragma is currently defined:

    #pragma CO SET stageDelay 32

The numeric argument refers to the maximum number of combinational gate delays permissible for an instruction stage. This pragma is described in more detail later in this chapter.

Understanding Latency and Rate

The latency and rate numbers reported by Stage Master apply to pipelines. In this context, latency refers to the number of cycles required for an input to reach the output of a pipeline, or, in other words, the length of the pipeline.

The rate is the number of cycles required for each input to the pipeline. (This is sometimes called the *input rate* or the *introduction rate*.) A rate of 1 means that the pipeline accepts inputs every cycle. A rate of 2 means that the pipeline accepts an input every other cycle.

Controlling Stage Delays

It was stated earlier in this chapter that all statements within a stage are implemented in a single clock cycle. One implication of this is that the number of individual statements (operations and assignments) being performed may have a direct (and potentially large) impact on the maximum clock rate of your application when synthesized to actual hardware. This impact can be mitigated by using the StageDelay parameter. This parameter specifies the maximum delay for the stages of the generated hardware. StageDelay parameters are specified using the generic CO SET pragma:

    #pragma CO SET StageDelay 32

The stage delay specified in the CO SET StageDelay pragma refers to the maximum number of combinational delays (levels of logic, or gate delays) that are allowed within a given pipelined stage. Note that optimizations performed by FPGA synthesis tools may further reduce (or in some cases expand) the number of combinational delays in the final implementation.

A combinational gate delay is roughly equivalent to the gate delay in the target hardware. Depending on the capabilities of the synthesis and routing tools being used, a logic operator such as an AND, OR, or SHIFT will require one delay unit, while an arithmetic operation or relational operation may require n or more delays, where n is the bit width.

Another way to explicitly control stage delays is to make use of the co_par_break function described in Appendix C. This function forces a new stage, allowing you to create sequential, multicycle operations in C code that would otherwise be generated as parallel logic.

6.7 MAKING EFFICIENT USE OF THE OPTIMIZERS

When writing Impulse C processes, it is important to have a basic understanding of how C code is parallelized by the compiler and optimizer during hardware generation so you can achieve the best possible results, both in terms of execution speed and the size of the resulting logic.

In later chapters we will explore specific techniques for writing good, optimizable C code for hardware generation. In this section we'll describe in general terms how the Impulse C optimizer performs its job and describe some of the current constraints of writing C for hardware generation.

The Stage Master Optimizer

The Impulse C optimizer (Stage Master) works at the level of individual blocks of expanded C code that are generated by the compiler, such as are found within a loop body. For each block of C code, the optimizer attempts to create the minimum number of instruction stages by scheduling instructions that do not have conflicting data dependencies.

If pipelining is enabled (via the PIPELINE pragma, which is added within the body of a loop and applies only to that loop), such identified stages occur in parallel. If no pipelining is possible, the stages are generated sequentially. Note that all statements within a stage are implemented in combinational logic operating within a single clock cycle.

To get maximum benefit from the optimizer, you should keep in mind the types of statements that will result in a new instruction stage being created. These include:

- A control statement such as an if test or a loop.

- Any access (read or write) to a memory or array that is already being addressed in the current stage.

In the case of a loop, pipelining attempts to execute multiple iterations of the loop in parallel without duplicating any code.

For example, this loop:

```
for (i=0; i<10; i++)
    sum += i << 1;
```

results in one stage (with one adder and one shifter) being executed ten times. When not pipelined, the computation time is at least ten multiplied by the sum of the shifter and adder delays, or 10 X (delay(shifter) + delay(adder)). If we enable pipelining, however:

```
for (i=0; i<10; i++) {
#pragma CO PIPELINE
    sum += i << 1;
}
```

the result for the same loop is that two stages representing the shifter and adder are executed concurrently, with the computation time being approximately 10 X max((delay(shifter), delay(adder)).

Note that pipelining is not automatic and requires an explicit declaration in your C source code, as described earlier:

```
#pragma CO PIPELINE
```

This declaration must be included within the body of the loop and prior to any statements that are to be pipelined.

Note: If used, the **PIPELINE** pragma should be inserted before the first statement in a loop. If you attempt to pipeline only part of a loop (for example by inserting the **PIPELINE** pragma in the middle of the loop) then the result will be an increase in the size of the generated hardware but will probably not result in increased performance when compared to the nonpipelined equivalent hardware.

Instruction Scheduling and Assignments

The instruction scheduler that organizes statements within a block or stage will always produce correct results (it will not break the logic of your C source code), if necessary by introducing stalls in the pipeline. To minimize such stalling, you should consider where you are making assignments and reduce the number of dependent assignments within a section of code. In some instances you may find that adding one or more intermediate variables to read and storing smaller elements of data (such as an array) at the start of a process may result in less stage delay. You should also make assignments to any recursive (self-referencing) variable as early as possible in the body of a loop and make other references to such variables as late as possible in the loop body so that introduced delays in one stage of the loop do not overly impact later stages.

Impacts of Memory Access

An important consideration when writing your inner code loops for maximum parallelism is to consider data dependencies. In particular, the optimizer cannot parallelize stages that access the same bank of memory (whether expressed as an array or using memory block read and block write functions). For this reason you may want to move subregions of a large arrays into local storage (local variables or smaller arrays) before performing multiple, otherwise parallel computations on the local data. Doing so will allow the optimizer to parallelize stages more efficiently, with a small trade-off of extra assignments that may be required. This aspect of C code optimization is explored in greater detail in Chapter 10.

6.8 LANGUAGE CONSTRAINTS FOR HARDWARE PROCESSES

Impulse C is compatible with standard ANSI C and allows you to take full advantage of C language (and third-party C libraries) for describing the software elements of your application. For those parts of your application that will target hardware, however, many constraints are placed on your ability to write and compile general-purpose C code. Some of these constraints are are obvious in the context of hardware generation (it doesn't really make sense to support recursive function calls, for example), while other constraints are more specific to the Impulse C hardware compiler. Some of the following constraints, in fact, are specific to the tools that were available as of this writing and may no longer be constraints when you read this. Check with Impulse Accelerated Technologies at www.ImpulseC.com for the latest updates.

No Support for Hardware Function Calls

Impulse C processes are intended to be created at compile time and exist statically (as persistent objects) at runtime. As such, it is not feasible to call an Impulse C process as a remote procedure (with argument lists and return values) as is normally done in traditional C programming. Similarly, it is not possible for an Impulse C process that will be compiled to hardware to call other functions or processes in the usual manner. The one exception to this is inline functions, which are resolved at compile time and do not require any special hardware (such as a stack) in order to be implemented.

Integer Math Operations

For integer datatypes, the standard C arithmetic operations +, -, *, ++ are supported as single-cycle operations. Division using the / operation results in a
multicycle implementation requiring one cycle for each bit in the result. The
algorithm is designed to be low in area and delay, with the goal of reducing
the overall system fMax.

Shift Operations

Shift operations are limited to constant values for the shift operand. For example, the following loop:

```
iFlags = 0;
for(i = 0; i < 8; i++)
{
    if(afCubeVal[i] <= FTARGETVALUE)
        iFlags |= 1 << i;   // Not supported in hardware compilation
}
```

should be rewritten as:

```
iFlags = 0;
bit = 1;
for(i = 0; i < 8; i++)
{
    if(afCubeVal[i] <= FTARGETVALUE)
        iFlags |= bit;
    bit = bit << 1;
}
```

Support for Datatypes Is Limited

As of this writing there is no support in the Impulse C compiler for complex
datatypes such as structs and unions, although support for these datatypes is
expected in a later release of the compiler. Floating-point types are also not
supported at this writing.

Limited Support for Pointers

The use of pointers to reference array data in your Impulse C hardware
processes is supported, but with some major limitations. Fundamentally, all
uses of pointers must be resolvable at compile time to static references to specific memory locations. Therefore, pointer support in Impulse C hardware
processes is limited to the following situations:

1. <pointer> = & <array element>

   For example:

   ```
   p = &(a[4]);   // Assign address of fifth element
   ```

2. <pointer> ++

   For example:

   ```
   p++;   // Increment pointer
   ```

3. <pointer> = <pointer> + <integer expression>

   For example:

   ```
   p = p + 2;   // Pointer addition
   ```

Also, pointers may not point to more than one array. The following, however, is allowed:

```
p = &(a[2]);
   ...
p = &(a[3]);
```

The second assignment in the following is not okay:

```
p = &(a[2]); // This is OK
   ...
p = &(b[3]); // This is not OK. Can't re-use the pointer for b
```

Using Pointers with Multidimensional Arrays

Rather than using nested loops to iterate through a two-dimensional array, it is more efficient to write the following:

```
p = &(a[0][0]);
for (i=0; i<4*5; i++) {
   ... // Access the array elements using p
  p++;
}
```

6.9   SUMMARY

In this chapter we have explored how FPGA hardware, in the form of hardware description language files, is created by the Impulse C compiler. We have also seen how these hardware processes may be connected to external hardware components, using a VHDL test bench for demonstration purposes. And lastly, we have explored some of the language-level optimizer controls available to you as an Impulse C program and have shown (briefly) how cycle-accurate debugging can be accomplished using a hardware simulator.

The chapters that follow spend more time exploring important topics related to hardware generation and optimization, using a series of examples as the basis for discussion. First, however, it is important for us to examine in more detail the statement-level optimizations available to you and to describe how certain programming techniques can help increase application performance.

Learn the latest about hardware generation

The structure of hardware descriptions created by the Impulse C compiler depends on many factors, including the coding style used (in particular, how stream reads and writes are specified), the target platform, and the version of the hardware generator and/or optimizer. Changes are being made to the software at a rapid pace. For a more complete understanding of the generated hardware structure, you may want to visit the Impulse C online discussion forums at www.ImpulseC.com/forums.

CHAPTER 7

Increasing Statement-Level Parallelism

The C language was designed to describe sequential operations on a sequential CPU. In fact, the semantics of C describe the exact order and manner in which each operation should occur. In this chapter we'll investigate the apparent contradiction between sequential programming in C and the desired result of creating highly parallel, accelerated hardware.

7.1 A MODEL OF FPGA COMPUTATION

Before we discuss specific techniques for extracting parallelism at the level of instructions—blocks of C statements—within a process, let's consider how computation is actually performed on an FPGA. This will help you understand the hardware execution model of your C code running in hardware. If you can understand the basic architecture of the modern CPU, it's not difficult to understand the hardware execution model. As in a traditional CPU, there must be some control logic, one or more arithmetic units, and a governing clock.

Actual computation on an FPGA is done by some number of predesigned arithmetic units such as adders, subtractors, and multipliers. As in a CPU, multiple instances of these units can operate in parallel, but in the FPGA the level of parallelism (the potential number of such computational

133

units) is limited only by the amount of available hardware and its interconnect structure.

In both the CPU and the FPGA, a clock controls the speed of operation of the arithmetic units. Each unit has one specific task to perform in each cycle. As a result, the maximum clock speed is limited by the speed of the basic arithmetic units. A CPU typically has a single clock controlling a relatively small number of such arithmetic units, while in the FPGA multiple clocks could be controlling different arithmetic units operating at different rates.

Finally, as in a CPU, an FPGA computing model must include a controller that maintains the order of execution, indicating what data should be supplied to each of the arithmetic units and where the results should go. But unlike in a CPU, the controller can (and should) be specific to the algorithm being performed.

Given that the structure of the generated hardware is much like that of a CPU, where does the speedup in computation come from? As it turns out, a number of limitations inherent in modern CPUs can be avoided by a generated model of computation.

First, a traditional CPU does not perform control and computation at the same time; that is, the CPU must spend noncomputational time decoding instructions, performing branches, and so on. In the case of FPGA-based computation these operations can be performed by a dedicated controller that operates in parallel with the arithmetic units. (It should be pointed out, however, that modern high-performance CPUs do introduce a great deal of parallelism—for example, through the use of instruction pipelining and prefetch operations.)

Second, by compiling a software process directly to hardware, it is in theory possible to generate any number of arithmetic units as needed, whereas the general-purpose CPU has only a fixed set of units from which to choose. (This fact alone has the potential to dramatically increase the processing efficiency of FPGA-based computations over a traditional processor, in which only a small fraction of the die area is actually engaged in computations at any given time.)

Third, the generated hardware may include custom single-cycle operations, which may be formed from multiple distinct statement lines in the original algorithm description, and highly optimized instruction pipelines.

Finally, a CPU generally has one connected memory resource, whereas the generated hardware for an FPGA implementation may take advantage of multiple blocks of RAM and other external resources to support multiple memory operations that are performed in parallel with each other and with other I/O operations.

7.2   C Language Semantics and Parallelism

Understanding the semantics of a given programming language allows the programmer to predict what each line of code will do during the design process and (equally important) to trace the observed failure of that code during debugging. (In fact, there would be little benefit to using a C syntax if we did not also obey the standard C semantics.) The result, however, is that the compiler is limited to performing operations in parallel only if it can guarantee the results will be the same as the sequential execution.

To generate efficient hardware from C language, then, the programmer should understand how to express detectable parallelism in the original algorithm and in the C statements used to describe that algorithm. As with any kind of optimization, the best approach is to design the algorithm in the simplest and most straightforward manner possible and then iteratively refine the C code for the desired performance.

This chapter helps you understand how to predict and control parallelizing within a single process by describing how computation is performed in hardware and how parallelism is extracted from C language. In the following three chapters we will present a specific example of how these techniques may be used to accelerate an actual design, and in later chapters we'll present other opportunities for expressing parallelism at the system level.

7.3   Exploiting Instruction-Level Parallelism

The Impulse C tools include C and RTL optimizers that provide general C-language optimizations as well as a parallel optimizer (called Stage Master) that is specific to FPGA targets. When writing Impulse C processes, it is important to have a basic understanding of how C code is parallelized by these optimizers so you can achieve the best possible results in terms of execution speed and size of the resulting logic.

The optimizer works at the level of individual blocks of C code, such as are found within a process or loop body. For each block, the optimizer attempts to create the minimum number of instruction stages, using instruction pipelining where possible and by scheduling instructions that do not have conflicting data dependencies.

Instruction Scheduling

Scheduling includes three major activities: parallelism extraction, control inference, and register insertion. Parallelism extraction is critical for obtaining optimal hardware performance from an inherently sequential algorithm description. While coarse-grained parallelism is often better expressed at the system level and even through the actual system partitioning (for example, using two processors implies a level of parallelism), fine-grained parallelism is often difficult for a designer to code or even recognize for a complex algorithm. C-to-RTL compilers automatically perform comprehensive data dependency analysis to find and exploit parallelism in a design. This parallelism is exploited during scheduling of the design. The use of other, user-specified optimizations such as pipelining and loop unrolling may allow the creation of additional parallelism.

Pipeline Generation

If pipelining is enabled (via the **PIPELINE** pragma, which is added within the body of a loop and applies only to that loop), such identified stages occur in parallel. If no pipelining is possible, the stages are generated sequentially. Note that all statements within a stage are implemented in combinational logic operating within a single clock cycle.

There are three main opportunities for instruction-level parallelizing in C code:

1. Suboperations of a complex expression may be performed in parallel. A simple example of this might be an expression such as x = (a + b) * (c + d). In a traditional processor the two add operations would require two distinct instruction cycles to determine the sums prior to the multiply being performed. In a parallel implementation, the two sums could take place in the same instruction cycle, in parallel. This is possible because the two operations a + b and c + d have no timing-sensitive interdependencies.

2. Multiple statements in a fragment of straight-line code may be performed in parallel. For sequences of instructions in which there are no interdependencies, such as in the following statements:

   ```
   x = a + b;
   y = a + c;
   ```

   These two statements (and more such statements) can be implemented as parallel hardware structures, potentially requiring only a single clock cycle for processing.

3. Multiple iterations of a loop may be performed in parallel, using a technique called pipelining. Pipelining allows the next iteration of a loop to begin executing before completing the current iteration in parallel with the execution of the current iteration.

Examples of all three of these types are presented in this and subsequent chapters. First, however, let's review how software-to-hardware compilation is performed and how the compiler looks for the sort of parallelism just described.

Optimizer Operation

The basic operation of the Impulse optimizer (Stage Master) is to divide a C procedure into basic blocks and then assign each operation of each basic block to one or more stages. Figure 7-1 illustrates how one stage of the FIR filter example (presented in Chapters 5 and 6) represents multiple parallel operations.

In general, each stage is performed in a single clock cycle in hardware. The exception is that if a particular stage is waiting for data from an external source (such as a stream or memory), that stage may stall for one or more clock cycles until the data arrives. Stages are performed one at a time and the generated hardware controller determines what stage is being performed and which stage should be performed next.

The optimizations performed include four basic passes. The first pass (optional and controlled by the **UNROLL** pragma) is to unroll inner loops and replicate their code blocks to create parallel structures. Next, a scheduling pass determines the minimum number of steps—or stages—possible to implement each basic block of statements. During this pass, each operation in a basic block is assigned a stage. All operations assigned to a given stage execute in parallel. A third optimization pass attempts to balance the delays of the individual stages. For each loop being pipelined (controlled by the **PIPELINE** pragma), a final pass analyzes resource constraints and interiteration dependencies to determine when and how much of the next iteration can begin in parallel with current iteration.

Expression-Level Optimizations

The optimizer automatically parallelizes computations both within and across multiple expressions and statements. For example, in the following code:

```
X = a + b + c;
X = X << 2;
Y = a - c;
```

the optimizer assigns the two additions of the first statement, the shift operation of the second statement, and the subtract operation of the third statement to a single stage.

Note that if many operations were to be chained together in this way within a single stage, the maximum clock speed could be seriously reduced. This is because each stage is to be performed in a single cycle, regardless of the amount of combinational logic required to implement the described

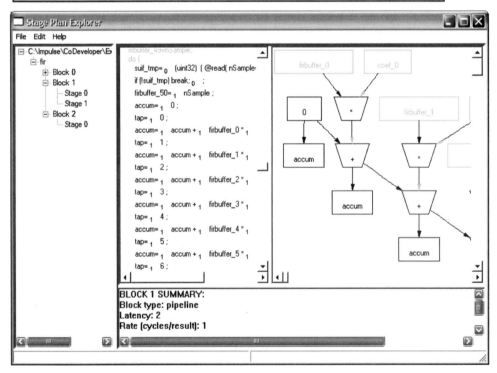

Figure 7-1. During optimization, C statements are are divided into blocks and stages.

operation. By using the StageDelay option in Impulse C, the programmer can control this by specifying the maximum delay (measured as an estimated number of gate delays) that the optimizer will allow in a single stage.

Optimization Within Basic Blocks

A basic block is a contiguous block of straight-line code with no intervening control statements such as if, while, and so on. For example, if the previous example were this instead:

```
if (odd) {
    x = a + b + c;
    x = x << 2;
}
y = a - c;
```

the optimizer would generate one stage to perform (a + b + c) << 2 and a second stage to compute a - c.

7.4   LIMITING INSTRUCTION STAGES

To get maximum benefit from the optimizer, you should keep in mind those types of statements that will result in a new instruction stage being created. These statements include

- A control statement such as an if test or a loop
- Any access (read or write) to a memory or array that is already being addressed in the current stage

---

Tip: To the greatest extent practical, you should reduce the use of unnecessary control statements and memory accesses to reduce the number of instruction stages.

Reduce Memory Accesses for Higher Performance

An important consideration when writing your inner code loops for maximum parallelism is to consider data dependencies. In particular, the optimizer will not be capable of parallelizing stages that access the same "bank" of memory (whether expressed as an array or using memory block read and block write functions). For this reason you may want to move subregions of a large array into local storage (local variables or smaller arrays) before performing multiple, otherwise parallel computations on the local data. Doing so will allow the optimizer to parallelize stages more efficiently, with a small trade-off of extra assignments that may be required.

Array Splitting

The way that memory (including local arrays) is accessed within a process can have a dramatic impact on the ability of the optimizer to limit instruction stages and to parallelize C statements. Consider the following example:

```
x = A[0] + A[1] + A[2]
x = x << 2;
```

This example involves an array **A** that is stored in a local RAM block. Only one element of the array can be read from the memory in a single cycle, so the computation must be spread out over four stages:

1. Read A[0]

2. Read A[1]

3. Read A[2], perform A[0]+A[1]

4. Rerform +A[2], perform <<2

One way to avoid this problem with memory is to use multiple arrays in multidimensional algorithms. For example, the following algorithm has the same problem as the preceding example:

```
int a[4][10];

for (i=0; i<10; i++) {
    a[3][i] = a[0][i] + a[1][i] + a[2][i];
}
```

However, suppose this algorithm is written using a separate array for each row of **a**, as follows:

```
int a0[10],a1[10],a2[10],a3[10];
for (i=0; i<10; i++) {
    a3[I] = a0[i] + a1[i] + a2[i];
}
```

In this example, each row is stored in a separate block of RAM, allowing each row to be read/written simultaneously. As a result, the body of this loop executes in a single stage instead of the four stages that would be required if the array were not split.

Tip: As this example demonstrates, array splitting is a useful technique to allow multiple simultaneous memory accesses and thereby increase parallelism. In Chapter 10 we'll explore this and other techniques in more detail.

7.5   UNROLLING LOOPS

As described in the previous section, the optimizer automatically parallelizes across expressions and statements within a basic block of C code. The optimizer does not, however, automatically parallelize operations across iterations of a loop. Loop unrolling is one technique that can be used to parallelize a loop by turning the loop into one basic block of straight-line code.

If the number of iterations of a loop is known at compile time, it is possible to unroll the loop to create dedicated logic for each iteration of the loop. Unrolling simply duplicates the body of the loop as many times as there are iterations in the loop.

For example, consider the following loop:

```
for (i=0; i<10; i++) {
    sum += A[i];
}
```

Without unrolling, this loop will generate logic to perform each iteration in two cycles. The first cycle will read from memory A, and the second cycle will calculate the addition. One adder is generated that will be used ten times during execution of the loop. Now consider the same loop with unrolling:

```
int i;    // Loop index must be type int
. . .
for (i=0; i<10; i++) {
#pragma CO UNROLL
    sum += A[i];
}
```

Unrolling simply duplicates the body of the loop for all values of i. In this example the result is equivalent to the following:

```
sum += A[0];
sum += A[1];
sum += A[2];
sum += A[3];
sum += A[4];
sum += A[5];
sum += A[6];
sum += A[7];
sum += A[8];
sum += A[9];
```

In this case, ten adders are generated, and each one is used only once during the execution of the loop. Most of the time, as in this case, loop unrolling alone has no specific benefit. Only one value of A can be read in a given cycle, so this example loop still requires ten cycles to execute, and ten adders have been generated, which requires a lot of logic. However, if the "scalarize array variables" optimizer option is used together with loop unrolling, the elements of the array are replaced with scalar variables. The results would then be equivalent to the following:

```
sum += A_0;
sum += A_1;
sum += A_2;
sum += A_3;
sum += A_4;
sum += A_5;
sum += A_6;
sum += A_7;
sum += A_8;
sum += A_9;
```

Instead of generating a memory for the array A, registers are generated for each of the ten elements of the array. All ten registers can be read simultaneously and thus this entire loop can be executed in a single cycle.

Note: The use of unrolling requires some care because a large amount of logic can be easily generated and the cycle delay can be greatly increased, with a corresponding decrease in maximum clock frequency. In the preceding example, ten adders will be generated, and because the result of each adder is used by the next adder, this cycle's delay is ten times the delay of a single adder's delay. You can keep this delay under control by placing a **StageDelay** pragma immediately preceding the loop.

7.6 PIPELINING EXPLAINED

When pipelining is enabled for inner code loops of your application (through the use of the **PIPELINE** pragma), the Impulse optimizer attempts to parallelize statements appearing within that loop with the goal of reducing the number of instruction cycles required to process the entire pipeline.

Pipelining is conceptually similar to a manufacturing assembly line, where the person at the first station of the assembly line can send a product to the next station for more assembly while he or she starts working on a second product. In this example, each station performs a portion of the overall assembly work. In a hardware pipeline, the body of a loop containing a sequence of operations is divided into stages that are analogous to the stations of the assembly line. Each stage performs a portion of the loop body. After the first stage has completed its portion of loop iteration i, the second stage begins its portion of iteration i, while the first stage starts processing iteration i+1 in parallel.

Pipelining reduces the number of cycles required to execute a loop by allowing the operations of one iteration to execute in parallel with operations of one or more subsequent iterations. Figure 7-2 illustrates a four-stage pipeline that is operating on three data packets (**A**, **B**, and **C**), each of which requires four clock cycles for the complete computation. By introducing a pipeline, the effective throughput rate of the process can be increased as shown. In this example it is assumed that each data packet requires two sequential operations (two pipeline stages) before proceeding to the next computation, resulting in an effective throughput rate of two cycles per data packet, which is an effective doubling of performance. Other pipelines may have greater or lesser performance gains, depending on the number of dependencies between different pipeline stages and other factors.

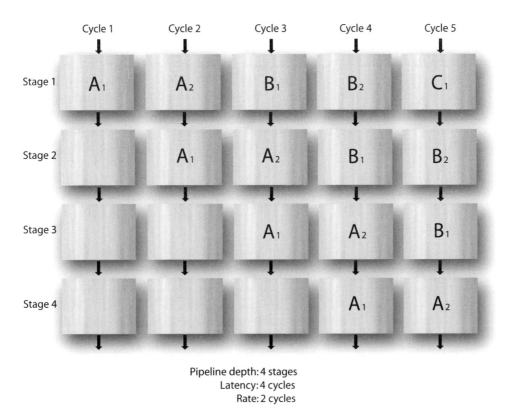

Pipeline depth: 4 stages
Latency: 4 cycles
Rate: 2 cycles

Figure 7-2. A four-stage pipeline with a throughput rate of two cycles.

For a more concrete example, consider the following loop:

```
while (1) {
   if (co_stream_read(istream,&data,sizeof(data)) != co_err_none)
      break;
   sum += data;
   co_stream_write(ostream,&sum,sizeof(sum));
}
```

Without pipelining, the body of this loop requires two cycles: the first cycle reads data from the input stream, and the second performs the addition and writes the result to the output stream. This example requires 200 cycles to process 100 inputs. Now, suppose that pipelining were used:

```
while (1) {
#pragma CO PIPELINE
   if (co_stream_read(istream,&data,sizeof(data)) != co_err_none)
      break;
   sum += data;
   co_stream_write(ostream,&sum,sizeof(sum));
}
```

This example results in a pipeline with two stages. The first stage reads data from the input stream, and the second stage performs the addition and writes the result to the output stream, similar to before. In the pipelined version, however, after the first data value is read, stage one immediately starts reading the next input in parallel with the computation and output of the sum using the first input. This example now requires only 101 cycles to process 100 inputs, or about one cycle per iteration.

The number of cycles required to complete one iteration of the loop is equal to the number of stages in the pipeline and is usually called the latency of the pipeline. In this example the latency is two.

Pipeline Rate

In most cases it is not possible to perform all stages of a pipeline in parallel. This can occur, for example, if two stages read from the same memory, such as from a local array. As a result, the pipeline will not be able to complete an iteration of the loop every cycle. The pipeline rate is the number of cycles between the time an iteration begins execution and the time the next iteration is allowed to begin. (This is sometimes called the *input rate* or *introduction rate*.) For example, if the rate were two, the pipeline would require about two times the number of iterations to complete the loop.

The fastest rate at which a loop can execute depends on inter-iteration dependencies and the use of sequential resources such as memories and streams. A common cause of reduced pipeline rates is multiple loads/stores of the same array variable.

The rate of a pipeline that contains multiple loads/stores of a single multidimensional array can sometimes be improved by using multiple arrays. For example, an image stored as RGB values might be implemented in C as

```
int8 img[16][3];
```

This array has three columns for the red, blue, and green components of each pixel. If img[i][0], img[i][1], and img[i][2] all appeared in the loop, the pipeline rate would be increased to permit three reads from the img array.

To improve performance, the image array might be repartitioned as follows:

```
int8 red[16],green[16],blue[16];
```

The loop body now references red[i], green[i], and blue[i]. Because these are separate arrays, they can be read simultaneously, and the pipeline rate will be improved.

7.7 SUMMARY

In this chapter we have described in general terms how C code is parallelized by the compiler, and we have provided some useful techniques for writing C code that is "optimizer-friendly." Unlike writing software code, writing efficient C for hardware requires some additional thought and a basic awareness of how parallelism is extracted by the compiler and optimizer. In the next three chapters we will apply this knowledge directly and show how the performance of a real-world C algorithm can be dramatically improved with relatively little effort. In subsequent chapters we'll apply these techniques to larger, more interesting applications.

CHAPTER 8

Porting a Legacy Application to Impulse C

In this and the following two chapters we will pull together some of the ideas presented in the preceding chapters and create a hardware/software implementation of a data encryption engine. While doing this we will demonstrate how legacy C code that was not originally written with a hardware implementation in mind can be iteratively ported, compiled, simulated, and refined to create a reasonably efficient hardware implementation.

The initial goal of this effort will not be to create the fastest possible or most compact implementation of this particular algorithm. Instead, our goal is to give you the skills and knowledge you'll need when analyzing and converting other types of algorithms that may already be implemented in standard C on standard processors.

An important aspect of any application porting (whether you are porting to an FPGA or to another, more traditional processor architecture, such as one supporting hyperthreading) is to analyze data movement in the application, in large part to ensure that your efforts at creating an efficient, high-performance implementation in hardware are not negated by simple bandwidth limitations.

To demonstrate how such an evaluation can be performed, consider the problem of data encryption, in which a stream of incoming data must be processed very quickly against a specified set of values (the key) to generate a resulting encrypted or decrypted data stream. Such a problem may involve

substantial computation but is also bandwidth-intensive: the final implementation must not compromise data throughput in order to increase overall performance.

In this chapter we will focus on the initial design process, including the
use of standard C debugging tools and application monitoring. We will also
generate some HDL (using a generic VHDL hardware platform) and perform
a simulation of the resulting hardware.

In Chapter 9 we'll target a specific FPGA and discuss as well the creation
of an embedded software test bench. This test bench will allow us to obtain
some quantifiable results by actually running the application in hardware,
under the control of an embedded processor.

In Chapter 10 we'll continue with this example, demonstrating through
a series of steps how an application not originally optimized for a hardware
implementation can be made to run faster and to require less hardware. As
we go through these stepwise optimizations, we hope to show some of the
techniques that can be used to accelerate many types of applications, both at
the level of the process (the coding of inner code loops and so on) and at the
level of the algorithm and its I/O requirements.

8.1   THE TRIPLE-DES ALGORITHM

In order to evaluate software versus hardware implementations of data encryption, we begin with publicly available source code for the triple-DES encryption algorithm. This source code was originally written by Phil Karn (of
Qualcomm) and is based on an algorithm described in *Applied Cryptography*,
written by Bruce Schneier and published in 1995 by John Wiley & Sons. This
original source code was written in standard C language and was not optimized for any specific processor target, nor was it written to take advantage
of algorithm-level parallelism.

We'll use Impulse C library calls to make the conversion from the original C code to a version suitable for hardware compilation in the selected
FPGA target, and to perform the required C-to-hardware compilation. Our
goal in this evaluation is to quickly evaluate the relative performance and
trade-offs of hardware versus software implementations for one specific algorithm (the triple-DES encryption function, represented by approximately 180
lines of C source code). Therefore, we have decided at the outset to make only
the minimum changes necessary to allow efficient hardware compilation and
to refrain from making non-obvious changes to the algorithm as a whole.
Those changes will come later, in Chapter 10, after we have generated a
working prototype.

The DES (Data Encryption Standard) algorithm was designed to encrypt/decrypt 64-bit blocks of data representing eight characters per block. Large amounts of data are processed by simply applying the same algorithm over and over to these 64-bit blocks, each of which represents the eight characters in an incoming ASCII text stream. In addition to the input data to be encrypted, the algorithm uses key schedule and SP Box data to perform the actual encryption/decryption. The key schedule data is generated from the encryption key and is constant for each stream of data. The SP Box data is a fundamental part of the algorithm and is constant for all streams.

The changes made to the encryption function in support of hardware compilation are as follows:

- The streaming model is a natural fit for this algorithm, so we use Impulse C library functions (including **co_stream_open**, **co_stream_read**, **co_stream_write**, and **co_stream_close**) to read the input data as a stream of data and output the data as a stream. The original algorithm assumed the data was in a global array, but the streaming implementation better reflects a real-world application where data would be processed as it is received from elsewhere. Streaming is also the preferred programming model for hardware/software interfaces when using Impulse C, because it more closely matches how data is most efficiently moved around in hardware.

- We create an additional configuration data stream as an input that accepts the encryption key (the key schedule) as well as the "SP box" static data specified by the encryption algorithm. (In the legacy C version these values were also accessed via global arrays.)

- We create top-level producer and consumer processes (also written in C, and again described using the Impulse C libraries) that serve as a test bench for the algorithm. This lets us stream random text characters into both the original, legacy C algorithm (which is compiled along with the test producer and consumer processes into native executable code on the embedded processor) and the hardware version, which is compiled directly to hardware and operates on the FPGA alongside the embedded processor.

- For debugging purposes, we will also create a more comprehensive test application (developed using Microsoft Visual Studio) that exercises the encryption algorithm in a desktop simulation environment. This test application combines the two encryption functions (the legacy C version and the Impulse C version) with corresponding decryption algorithms to verify the functional correctness of the application using various text inputs. This test will be set up and run, and the results verified, before going to the next step and compiling to the target FPGA platform.

8.2 CONVERTING THE ALGORITHM TO A STREAMING MODEL

The inner processing loop of the original (legacy) encryption function is shown in the code excerpt in Figure 8-1. Notice that this function repetitively processes blocks of characters, performing various operations on these blocks (using the SP box and key schedule data) to generate a resulting encrypted block.

As we indicated in the introduction to this chapter, our goal is to create a version of this algorithm that is compatible with hardware compilation, while at the same time making only the minimum necessary changes to the original function. In practice, many optimizations can be introduced right up front, but we will delay those optimizations for a later chapter. Instead we will focus primarily on the movement of data, which is often the most important aspect to consider when moving processing from a traditional processor to an FPGA.

As we've described in previous chapters, the most efficient way to move data from one hardware or software process to another is often via data streams. In fact, even if some other method of data transfer such as DMA will eventually be required (perhaps for performance reasons), the use of streams can greatly simplify the design and debugging of complex hardware/software systems.

To create a streaming implementation of this function, it's first necessary to wrap the core algorithm in a stream read/write loop. This outer loop will read characters from an input data stream, package these characters into the block array, perform the preceding calculations, and then write the results, eight characters at a time, to the output stream.

Before performing these operations on the data stream, however, the encryption algorithm must be initialized with the encryption keys (the key schedule) and the SP box data. In the original legacy C version of the algorithm these values were provided in global arrays. In an Impulse C implementation these values can either be streamed in or accessed via a shared memory. For maximum portability we will stream in the key schedule and SP box data via a separate stream called config_in. This will be done prior to the start of the main code loop.

Note: We could also use a shared memory and memory block read and write functions for this initialization data, but memory types are FPGA platform-specific and can introduce unwanted complexity during hardware simulation. For this reason it's often best to make use of streams during development and defer the use of shared memories for a later design optimization.

```
while ( blockCount < NumBlocks ) {
  for ( i = 0; i < DES_BLOCKSIZE; i++ ) {
    block[i] = inputBlocks[(blockCount * DES_BLOCKSIZE) + i];
  }
  // Process the block...
  // Read input block and place in left/right in big-endian order
  //
  left = ((unsigned long)block[0] << 24)
        |((unsigned long)block[1] << 16)
        | ((unsigned long)block[2] << 8)
        | (unsigned long)block[3];
  right = ((unsigned long)block[4] << 24)
        | ((unsigned long)block[5] << 16)
        | ((unsigned long)block[6] << 8)
        | (unsigned long)block[7];

  work = ((left >> 4) ^ right) & 0x0f0f0f0f;
  right ^= work;
  left ^= work << 4;
  work = ((left >> 16) ^ right) & 0xffff;
  right ^= work;
  left ^= work << 16;
  work = ((right >> 2) ^ left) & 0x33333333;
  left ^= work;
  right ^= (work << 2);
  work = ((right >> 8) ^ left) & 0xff00ff;
  left ^= work;
  right ^= (work << 8);
  right = (right << 1) | (right >> 31);
  work = (left ^ right) & 0xaaaaaaaa;
  left ^= work;
  right ^= work;
  left = (left << 1) | (left >> 31);

  // First key
  F(left,right,(*ks)[0]);
  F(right,left,(*ks)[1]);
  F(left,right,(*ks)[2]);
  F(right,left,(*ks)[3]);
  F(left,right,(*ks)[4]);
  F(right,left,(*ks)[5]);
  F(left,right,(*ks)[6]);
  F(right,left,(*ks)[7]);
  F(left,right,(*ks)[8]);
  F(right,left,(*ks)[9]);
```

Figure 8-1. Inner processing loop of the original encryption function. (*continues*)

```
                F(left,right,(*ks)[10]);
                F(right,left,(*ks)[11]);
                F(left,right,(*ks)[12]);
                F(right,left,(*ks)[13]);
                F(left,right,(*ks)[14]);
                F(right,left,(*ks)[15]);
                // Second key (must be created in opposite mode to first key)
                F(right,left,(*ks)[16]);
                F(left,right,(*ks)[17]);
                F(right,left,(*ks)[18]);
                F(left,right,(*ks)[19]);
                F(right,left,(*ks)[20]);
                F(left,right,(*ks)[21]);
                F(right,left,(*ks)[22]);
                F(left,right,(*ks)[23]);
                F(right,left,(*ks)[24]);
                F(left,right,(*ks)[25]);
                F(right,left,(*ks)[26]);
                F(left,right,(*ks)[27]);
                F(right,left,(*ks)[28]);
                F(left,right,(*ks)[29]);
                F(right,left,(*ks)[30]);
                F(left,right,(*ks)[31]);
                // Third key
                F(left,right,(*ks)[32]);
                F(right,left,(*ks)[33]);
                F(left,right,(*ks)[34]);
                F(right,left,(*ks)[35]);
                F(left,right,(*ks)[36]);
                F(right,left,(*ks)[37]);
                F(left,right,(*ks)[38]);
                F(right,left,(*ks)[39]);
                F(left,right,(*ks)[40]);
                F(right,left,(*ks)[41]);
                F(left,right,(*ks)[42]);
                F(right,left,(*ks)[43]);
                F(left,right,(*ks)[44]);
                F(right,left,(*ks)[45]);
                F(left,right,(*ks)[46]);
                F(right,left,(*ks)[47]);
                // Inverse permutation, also from Hoey
                // via Outerbridge and Schneier
                right = (right << 31) | (right >> 1);
                work = (left ^ right) & 0xaaaaaaaa;
                left ^= work;
```

Figure 8-1. *continued*

```
    right ^= work;
    left = (left >> 1) | (left  << 31);
    work = ((left >> 8) ^ right) & 0xff00ff;
    right ^= work;
    left ^= work << 8;
    work = ((left >> 2) ^ right) & 0x33333333;
    right ^= work;
    left ^= work << 2;
    work = ((right >> 16) ^ left) & 0xffff;
    left ^= work;
    right ^= work << 16;
    work = ((right >> 4) ^ left) & 0x0f0f0f0f;
    left ^= work;
    right ^= work << 4;
    // Put the block into the output stream with final swap
    block[0] = (co_int8) (right >> 24);
    block[1] = (co_int8) (right >> 16);
    block[2] = (co_int8) (right >> 8);
    block[3] = (co_int8) right;
    block[4] = (co_int8) (left >> 24);
    block[5] = (co_int8) (left >> 16);
    block[6] = (co_int8) (left >> 8);
    block[7] = (co_int8) left;

    for ( i = 0; i < DES_BLOCKSIZE; i++ ) {
        outputBlocks[(blockCount * DES_BLOCKSIZE) + i] = block[i];
    }
    ++blockCount;
}
```

Figure 8-1. *continued*

A summary of the required stream read/write loops appears in Figure 8-2. Note that, for brevity, only the key schedule read loop is shown.

In the complete application (which is listed in Appendix D), this encryption process is connected via its data streams to producer and consumer processes that generate and accept data for the purpose of testing or as part of the complete application that resides in the FPGA. This embedded test bench is described in Chapter 9.

```
// The key schedule and SP box data stream in via config_in,
// while the text data streams in via blocks_in.
//
void des_ic(co_stream config_in, co_stream blocks_in,
        co_stream blocks_out)
{
  co_int8 i, k;
  co_uint8 block[8];
  co_uint32 ks[KS_DEPTH][2];

  /* Get the key schedule and put it in a local array */
  co_stream_open(config_in, O_RDONLY, UINT_TYPE(32));
  for ( k = 0; k < 2; k++ ) {
    for ( i = 0; i < KS_DEPTH; i++ ) {
      if (co_stream_read(config_in, &ks[i][k],
                sizeof(uint32)) != co_err_eos ) {
      }
    }
  }
  co_stream_close(config_in);
  . . .

  // Now read in the data and process, one block at a time
  co_stream_open(blocks_in, O_RDONLY, UINT_TYPE(8));
  co_stream_open(blocks_out, O_WRONLY, UINT_TYPE(8));

  while (co_stream_read(blocks_in, &block[0],
                sizeof(uint8)) == co_err_none ) {
    for ( i = 1; i < BLOCKSIZE; i++ ) {
      if ( co_stream_read(blocks_in, &block[i],
                sizeof(uint8)) == co_err_none ) {
      }
    }

    // Now process the current block
    . . .
    // Done processing the current block.

  }
  co_stream_close(blocks_in);  // finished, close the streams
  co_stream_close(blocks_out);
}
```

Figure 8-2. Summary of the required stream read/write loops for the encryption process.

8.3 PERFORMING SOFTWARE SIMULATION

Before going through the process of choosing an FPGA-based platform target and compiling/synthesizing the encryption algorithm to that target, we will first use a standard C development environment to verify that the application, including both the legacy C code and the modified Impulse C version of the code, is correct in terms of the computations being performed. Because the Impulse C libraries are compatible with most popular C development environments, we can perform this test using any number of such compiler/debugger environments.

The test bench that we will create for this encryption process will allow us to directly compare the results of both encrypting and decrypting a stream of characters that originate from a text file, as shown in Figure 8-3.

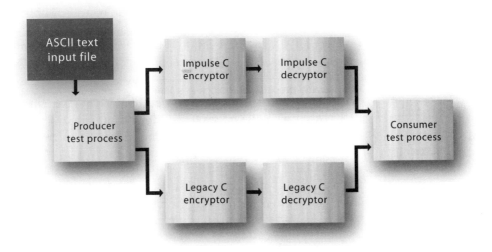

Figure 8-3. Block diagram of the complete encryption/decryption test bench.

As shown in the diagram, the software simulation consists of six distinct processes:

- A producer test process. This process reads characters of data from an input text file (a sequence of ASCII text characters) and streams the characters one byte at a time into the encryptor. The producer also moves those same characters into an array that will be used by the legacy C version of the encryptor to allow direct comparison of the results.

- An Impulse C process representing the encryption algorithm. This process (which was summarized in the preceding section) accepts configuration and text data from the consumer and writes the encrypted results one character at a time to its output stream.

- An Impulse C process representing a decryption algorithm. This process is actually the same process used for the encryption, but with different compile-time parameters.

- An Impulse C process representing the legacy C algorithm. This process does not accept data on a stream, but instead reads the input characters from the array populated by the producer process. The producer process communicates a "ready" status with this process using a signal to indicate that the array is filled and ready for processing.

- An Impulse C process representing the legacy C algorithm, with parameters necessary for decryption rather than encryption.

- A consumer test process. This process accepts the results of both the Impulse C (streaming) version of the encryption and decryption algorithms and the results of the legacy C version (which have been written to a results array) to verify that the new and old versions of the algorithm produce identical results.

This method of creating test benches in C, in which an Impulse C algorithm is tested in parallel with a legacy, known-good algorithm, is extremely useful, particularly when you will be making later C-language optimizations to the algorithm.

By using a standard IDE in conjunction with the Application Monitor provided with the Impulse tools, we can use standard C debugging techniques (including source-level debugging) while observing how data moves between the various processes in the system, which now includes two versions of both the encryption and decryption processes, plus the test consumer and producer processes. The result, using Visual Studio as the debugger, is shown in Figures 8-4 and 8-5.

8.4 COMPILING TO HARDWARE

After simulating its functionality using standard desktop tools, we are ready to implement the application on a mixed FPGA/processor target. We chose a Xilinx MicroBlaze-based FPGA target for this test, selecting the Virtex-II MicroBlaze Development Kit (available from Memec Design) as our reference system. The Memec kit includes a hardware reference board populated with a Virtex II FPGA and various peripheral interfaces, as well as all development tools needed to compile and synthesize hardware and software applications (consisting of HDL source files for hardware and C source files for software) to the FPGA target. When combined with the Impulse C compiler, this kit provided us with everything needed to compile and execute the test application from our C language source files.

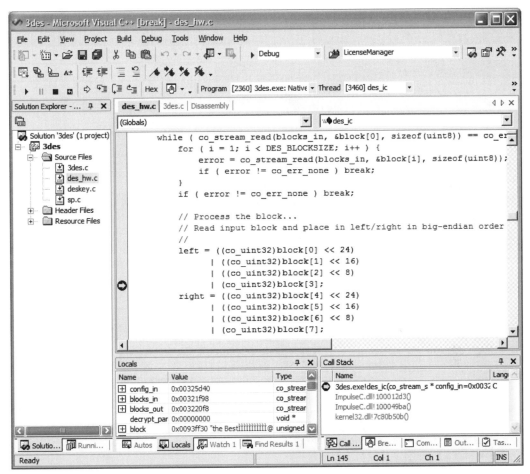

Figure 8-4. Debugging the Impulse C encryption processing using Visual Studio.

The following steps (which are presented in detail in the next chapter) were required to compile the encryption test application to the target:

1. The Xilinx MicroBlaze Platform Support Package was selected from within the Impulse tools, and C source files representing the application were processed, resulting in approximately 6,500 lines of generated RTL and related hardware/software interface source files.

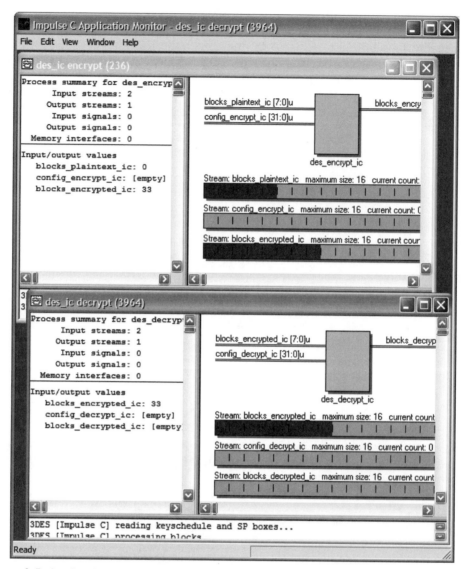

Figure 8-5. Application monitoring.

2. A new project was created using the Xilinx Platform Studio tools. A MicroBlaze platform was generated in Platform Studio, including the necessary peripherals such as a serial port, RAM controller, and so on.

3. The generated hardware and software files were exported from the Impulse tools to the newly created Platform Studio project.

4. Platform Studio was used to connect the generated hardware process (the Impulse C encryption process) to the MicroBlaze processor via the MicroBlaze FSL high-speed bus.

5. The complete system, including the MicroBlaze processor and the generated encryption-related hardware, was synthesized using the Xilinx tools, which are invoked from within Xilinx Platform Studio.

6. The software portions of the application (the legacy C version of the encryption algorithm and the test functions, including main) were imported into the Platform Studio project and compiled.

7. A bit file was generated using the Xilinx tools and downloaded to the platform (via a JTAG interface cable) for execution.

8.5 PRELIMINARY HARDWARE ANALYSIS

Once the algorithm has been ported to use the Impulse C library for streaming data and has been verified using desktop simulation, we can perform some preliminary analysis of the hardware processes before going all the way through hardware synthesis. By simply compiling the hardware processes in Impulse C, we can see some basic statistics in the output report from the optimization. This information includes stage counts for basic loops, and pipeline rates and latencies.

Using the optimizer outputs, we can look at all the basic blocks of the main loop in the algorithm and determine that each iteration requires 191 stages, which generally correspond to cycles. Using that information, we can determine an estimated throughput rate for a given clock frequency. For example, assuming a 50 MHz clock, we can compute 50e6/191, which equates to 261,780 blocks per second (at one block per process iteration). That in turn equates to 2,094,240 bytes per second, which at eight bytes per block is approximately 2MB per second performance. This is a rough estimate and assumes ideal conditions such as data always being available on the input stream.

While this analysis represents only a theoretical maximum throughput, it can be useful to determine the potential benefit of compiling to hardware and to suggest when further optimization is clearly necessary. Performing an initial analysis like this can help you avoid building actual hardware (whether that hardware is generated by the C compiler or hand-crafted using an HDL) for processes that turn out to be ill-suited for hardware implementation. This kind of analysis can also help when performing iterative language-level optimizations on an algorithm as it is being refined. Because the actual hardware synthesis process can be time-consuming, the design cycle can be significantly reduced and hardware/software partitioning decisions can be more accurately made.

Initial Results: 10.6X Performance Increase

The results of the test (expressed as the computation times for a specified number of data blocks) were generated using timers available on the MicroBlaze processor and invoked from within the C language test application. (This embedded test bench is described in more detail in Chapter 9.)

The results demonstrated that, for this algorithm running on the Virtex II, a hardware implementation would result in faster performance (a 10.6X speedup) than a software-only solution, even with the modest overhead of data communication between the processor and the FPGA-based encryption algorithm. This is due in part to the extremely low data communication overhead introduced by the Xilinx FSL bus. It is due as well to the compiler's ability to find and exploit low-level parallelism within the inner code loop of the algorithm. As you will learn in Chapter 10, however, the potential speedups for this algorithm are substantially greater.

8.6 SUMMARY

In this chapter we have taken a legacy C algorithm and made the changes necessary to create a streams-oriented hardware accelerator, one that could be connected to other hardware or software processes (using the streaming interface) and implemented directly in an FPGA. We made little or no attempt to optimize this algorithm. Instead, we have focused on creating and verifying the algorithm through the use of desktop simulation in a standard C development environment. We have also generated prototype hardware using the Impulse compiler and have obtained some initial performance numbers from the optimizer. These numbers have shown that, at a minimum, we should be able to accelerate this particular algorithm by an order of magnitude over its software equivalent, at least when that software equivalent is running on an embedded processor.

While 10.6X is a good start, and suggests that a hardware implementation for this algorithm may be appropriate, it is actually on the low end of what is possible when implementing software algorithms in programmable hardware. For this algorithm, further performance increases as well as reductions in gate count requirements can be obtained by optimizing the algorithm itself—for example, by reordering statements to better enable pipelining or by invoking the three stages of the triple-DES algorithm in parallel.

In the next chapter we will detail how to create an embedded, in-system test for this algorithm using the Memec FPGA prototyping board previously mentioned.

In Chapter 10 we will show, step-by-step, how an application such as this can be iteratively improved, resulting in enormous increases in performance for relatively little cost in added hardware.

CHAPTER 9

Creating an Embedded Test Bench

The latest generation of FPGAs featuring hard or soft embedded processors offer compelling platforms for hardware acceleration of computationally intensive software algorithms. Design teams taking advantage of these platforms are finding that a combination of traditional software applications running on the embedded processor and custom accelerators implemented within the FPGA fabric are an efficient way to create high-performance products and prototypes.

One important and often overlooked benefit of using a combination of embedded processors and hardware accelerators is the ability to create embedded software/hardware test fixtures, or *test benches*. By using such an approach in conjunction with a streaming programming model, it is possible to verify the correct operation of a hardware module (at least in terms of its functional behavior, if not its timing characteristics) with a high level of accuracy relative to the speed at which new tests can be developed.

Embedded test benches are therefore the subject of this chapter. To describe some potential ways in which these test benches may be written, we will continue with our discussion of the triple-DES algorithm presented in the preceding chapter.

9.1   A MIXED HARDWARE AND SOFTWARE APPROACH

For embedded system designers considering a mixed software/hardware implementation (whether that implementation is intended for testing of a hardware module or as part of a larger system), an important step is to evaluate the relative merits of various hardware and software targets for specific elements of the overall system. Often this evaluation is done based on the designer's past experience, by using "back of the envelope" calculations regarding raw instruction cycle counts for key calculations, or through an analysis of data throughput requirements. One specific step in such an evaluation may be to implement the same algorithm (which typically represents one component or software process within a larger hardware/software application) in both an embedded processor and in FPGA hardware to analyze the relative merits of a software versus hardware solution. In doing so, the designer must evaluate not only the resulting performance of the core algorithm but the relative overhead of setting up and managing the necessary software/hardware interfaces.

A critical factor in such an evaluation is the ability to compile algorithms to both a traditional microprocessor and to hardware, creating a test environment that allows both software and hardware versions to be tested and measured using identical inputs. As you've seen, the use of C-to-hardware compilation can assist in the process and allows a high degree of creativity on the part of the designer. This is because the software developer is free to quickly evaluate radically different ways of partitioning, describing, and implementing a mixed hardware/software application, without the need to write low-level hardware descriptions for those portions destined for hardware, or to make tedious hand calculations to determine relative performance numbers.

Considering Data Transfer Overhead

In algorithms involving streams of data (which represent a large percentage of the required processing in such domains as image processing and communications) there is a critical trade-off to be considered: the amount of computation required versus the data transfer overhead. More specifically, any evaluation of the merits of a hardware-based approach must consider, through direct measurement if possible, the cost of moving data between software components of the system (running on a traditional processor) and the dedicated hardware.

Every application evaluated in such a way will yield different cost/benefit results. The greatest benefits of hardware-based approaches are found in algorithms that are not unduly constrained by I/O, are computationally intensive, and include opportunities for parallelism to be exploited either at a low level (by scheduling statements within loops, for example) or at the level of pipelined or parallel processes. The ability to compile software algorithms directly to the FPGA makes such cost/benefit analysis a more efficient, less risky process.

9.2 THE EMBEDDED PROCESSOR AS A TEST GENERATOR

Even if you don't plan to use an embedded processor core in your final FPGA implementation, using an embedded FPGA processor can greatly enhance your ability to test and debug individual components of a complete FPGA-based application.

There are a number of ways in which C code running on an embedded processor can act as an in-system, software/hardware test bench, providing test inputs to the FPGA and (if necessary) validating the results and obtaining performance numbers. When used as part of a complete testing and debugging strategy, the embedded processor becomes an excellent vehicle for in-system FPGA verification and a good complement to hardware simulation.

By extending this approach to include not only C compilation to the embedded processor but C-to-hardware compilation as well, it is possible—with very little effort—to create high-performance, mixed software/hardware test benches that closely model real-world conditions.

One key to such an approach is to make use of standardized interfaces between test software (C-language test benches) running on the embedded processor and other test components (as well as the hardware under test) that are implemented in the FPGA fabric. These interfaces should be designed to take advantage of communication channels available in the target platform. For example, the MicroBlaze soft processor provided by Xilinx has access to high-speed serial interfaces (the FSL bus) that provide a high-performance data channel to the FPGA fabric. This interconnect channel provides much higher performance data transfers than the more general-purpose OPB bus, which is also available to MicroBlaze users. Other FPGA families and embedded processors have different mechanisms for processor-to-FPGA communication, and these mechanisms should be studied carefully before developing a mixed hardware/software application.

By using such interfaces along with a *unit test* philosophy, it is possible to quickly verify components of a larger application and to apply inputs that test the boundaries (or corner cases) of each component, thereby improving the quality of the application as a whole.

A Unit Test Philosophy

An important aspect of hardware/software testing is to create unit tests for critical modules in an application. Using HDL simulators, for example, it is relatively easy to create test benches that will exercise specific modules by providing stimulus (test vectors or their equivalents) and verifying results. For many classes of algorithms, however, such testing methods can result in very long simulation times or may not adequately emulate real-world conditions. Creating an in-system, prototype test environment is an excellent way to enhance simulation-based verification and insert more complex, real-world testing scenarios.

Unit testing is most effective when it focuses on unexpected or boundary conditions that might be difficult to generate when testing at a system level. In an imaging processing application that performs multiple convolutions in sequence, for example, you may want to just focus your efforts on one specific filter by testing pixel combinations that are outside the scope of what that filter would normally encounter in a typical image. For many such applications it would be impossible to test all permutations from the system perspective. The unit test allows you to build a suite of tests to validate specific areas of interest or test only the boundary/corner cases. By performing such tests using actual hardware (which may for testing purposes be running at slower-than-usual clock rates), it is possible to obtain real, quantifiable performance numbers for specific application components.

Key to this approach is the ability to quickly generate the test bench software that will run on the embedded processor and to generate any necessary prototype hardware that will operate within the FPGA to generate sample inputs. By using C-to-RTL compilation it is possible to generate such mixed software/hardware test routines with a minimum of effort.

Desktop Simulation Versus Embedded Test Benches

As you've seen in the preceding chapters, a major advantage of using a C-based approach to hardware design is the ability to create software/hardware models for the purpose of algorithm debugging in software, using standard development environments such as Visual Studio. In this approach, the complete application (including the unit under test, the producer and consumer test functions, and any other needed test bench elements) is described using C, compiled under a standard desktop compiler, and executed, perhaps under the control of a source-level debugger.

When we apply a similar philosophy to embedded testing benches, we are somewhat limited by the constraints of the platform target. It may be difficult, for example to create a test in which files are read from and written to a disk drive, or to create interactive tests, as is possible in a desktop simulation environment. As you will see in a later chapter, however, the addition of a lightweight embedded operating system to the embedded processor can greatly simply the creation of complex and powerful embedded tests.

Data Throughput and Processor Selection

When considering processors for in-system testing, there are a few important considerations. First, consider the fact that an embedded soft processor requires a certain amount of area in the target FPGA device. If you are only using the embedded processor as a test generator for a relatively small part of your complete application, this added resource usage may be of no concern. If, however, the unit under test already pushes the limits of utilization in the FPGA, you may need to move into a larger device during the testing process and/or choose an FPGA that includes an embedded hard processor such as the Xilinx Virtex-II Pro or Virtex-4. These FPGAs include built-in PowerPC processors that add very little cost to the overall FPGA platform.

Tip: Another factor to consider when choosing an embedded processor is the synthesis time. Depending on the synthesis tool chain being used, adding a soft processor core to your complete application may add substantially to the time required to synthesize and map your application component to the FPGA, which might be a factor if you are performing iterative compile, test, and debug operations. Again, the use of a hard processor core has an advantage over the soft processor core when design iteration times are a concern.

If data throughput from the processor to the FPGA is your primary concern, you will probably find that using a soft processor (such as the Xilinx MicroBlaze with its FSL bus) provides the highest performance for streaming data between software and hardware components. By using the Impulse C libraries for hardware/software communication, the differences between these different FPGA-based processors and bus protocols can be abstracted (to a great extent) using a common set of stream read and write functions.

Moving Test Generators to Hardware

If the performance of test generation software routines must be further increased, the obvious way to do this is by moving critical test functions (such as stimulus generators) into hardware. Normally this is a tedious process of re-implementing using VHDL or Verilog, but by using automated C-to-RTL compilation you can very quickly generate hardware representing test producer and/or consumer functions that interact with the unit under test using streaming interfaces to implement test inputs and outputs.

9.3 THE ROLE OF HARDWARE SIMULATORS

Although the use of in-system testing and debugging (using embedded
FPGA processors) is a compelling addition to the complete application testing
and debugging environment, there is no reason to abandon the use of hardware simulators. Hardware simulation has many important benefits, not the
least of which is the avoidance of the compile/synthesize/map times that are
required before a given hardware module can be tested in-system. Hardware
simulators also provide much more visibility into a design under test and
allow single-stepping and other methods to be used for zeroing in on errors.

   If your tests require very specific timing, you will probably find that
using an embedded processor to create test data will result in data rates that
are only a fraction of what is needed to obtain timing closure. In simulation
(using post-route simulation models) you can observe the system running at
any simulated clock speed, although it may require a very long time—perhaps hours or even days—to complete that simulation if the application being
tested is of any significant size.

Tip: In-system testing using embedded processors is a good complement to existing simulation-based testing methods. It allows
hardware elements to be tested (typically at lowered clock rates)
in a highly efficient way, using actual hardware interfaces and potentially more accurate real-world input stimulus. This helps aug-
ment simulation, because even at reduced clock rates the
hardware under test will operate substantially faster than is possible
in RTL simulation. And when C-to-hardware compilation tools are
used, the entire system (test bench included) can be modeled in C
and then iteratively ported to hand-coded HDL or optimized from
the level of C code to create increasingly high-performance unit
and system tests.

9.4 TESTING THE TRIPLE-DES ALGORITHM IN HARDWARE

In Chapter 8, we presented a triple-DES algorithm, showing how an iterative
process of software simulation, compilation, and hardware simulation could
be used to verify the proper function of the algorithm and to get some initial
performance numbers in terms of cycle delays. To evaluate that performance
we used a combination of information obtained from the compiler tools (the
latency and rate) and hardware simulation (using a VHDL simulator). But
while hardware simulation is an excellent way to both debug an application

and analyze performance, there is really no substitute for programming an application in hardware and watching it work.

As we mentioned briefly in the preceding chapter, one way to test individual Impulse C processes (or collections of processes as appropriate) is to set up a hardware/software test environment. We'll discuss this procedure in more detail here, and show the complete path from software down to the mixed processor and hardware implementation.

Platform Selection

A number of different FPGA and processor combinations would allow us to create a mixed hardware/software test for our prototype encryption algorithm. To simplify the creation of such a platform, we have selected a widely available FPGA prototyping board, the V2MB1000 board available from Memec Design (www.memec.com). This board includes a Xilinx Virtex II device as well as a variety of useful on-board peripheral interfaces, including a network interface, two serial ports, a USB port, and other such hardware. The V2MB1000 board is pictured in Figure 9-1.

To create the embedded software test bench, we will make use of the MicroBlaze soft processor core and its FSL (Fast Simplex Link) interconnects.

Figure 9-1. Insight/Memec V2MB1000 development board.

This platform combination will provide us with a highly efficient software-to-hardware communication channel, allowing us to stream character data from MicroBlaze to the FPGA-based algorithm at a relatively high rate.

Software and Hardware Algorithm Comparison

The goal of this test is to compile the same algorithm (the encryption process) on both the MicroBlaze processor as a standard C function and as hardware on the FPGA. This will allow us to compare the results, both in terms of accuracy of the algorithm and performance, for the hardware and software implementations. In addition, the MicroBlaze will be used to run test producer and consumer processes that will pass text data into the algorithm and accept the results, as shown in Figure 9-2.

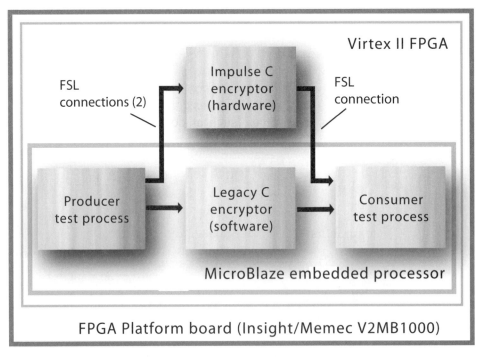

Figure 9-2. Embedded test bench block diagram.

The test that we will create for in-system validation will be much simpler than the test bench described in the previous chapter. Recall that in that test we created a software test bench that read a large number of characters from a text file, encrypted those characters, and then decrypted them to produce the identical characters for output. The test bench also performed the same encryption and decryption operations using legacy C function calls for comparison purposes. One aspect of that earlier software test bench was that it made use of multiple processes, including a producer, consumer, and processes that controlled the legacy encryption and decryption. In fact, the earlier test bench had six processes in total, which (when simulated under the control of Visual Studio) were implemented as six distinct threads. This allowed us to emulate the parallel behavior of those processes.

For this test we will need to create a simpler test bench, one that does not rely on threads to implement multiple processes. Although we could make use of a threading library (such as pthreads) or an embedded operating system, for this example we would like to keep this test bench as simple as practical. The resulting test bench process, which is compatible with both the MicroBlaze processor and Windows desktop compilation (by virtue of #ifdef statements) is shown in Figure 9-3. Starting from the top of this file, notice the following:

- The file begins by defining a TIMED_TEST macro, which indicates that the test bench will include the necessary interfaces to invoke a timer object in the target platform. This is also reflected in the use of the Xilinx--provided include file xtmrctr.h.

- The test bench also includes the co.h include file. This file is provided by Impulse and includes declarations for the Impulse C library functions. The functions themselves (such as co_execute and the various stream read and write functions and macros) are provided in a library that is specific to the chosen platform.

- The BLOCKSIZE and KS_DEPTH macros specify the size of an encryption block (eight characters) and the depth of the key schedule. These must not be changed.

- For compilation to the MicroBlaze target, the standard printf function is replaced by a MicroBlaze-specific function called xil-printf. This function maps to a UART peripheral that will be included when we build the FPGA platform.

- extern declarations are included that reference functions co_initialize and deskey. The co_initialize function is defined with the configuration function, as was described in Chapter 8. The deskey function is used to generate key schedule data from a set of 24 encryption keys. The resulting key schedule and SP box data are stored in the extern arrays Ks and Spbox, respectively.

- A static array of character values is declared as Block. These test values will be used as the test inputs (representing one block of character data) for the purpose of testing. A larger number of input characters could be used in this test bench with minor modifications.

```
#define TIMED_TEST
#include "xparameters.h"
#ifdef TIMED_TEST
#include "xtmrctr.h"
#endif
#include <stdio.h>
#include "co.h"
#include "des.h"
#define BLOCKSIZE 8      // Unsigned characters per block
#define KS_DEPTH 48      // Key pairs

#ifndef WIN32
#define printf xil_printf
#ifdef TIMED_TEST
XTmrCtr TimerCounter;
#endif
#endif

extern co_architecture co_initialize(void *);
extern void deskey(k,key,decrypt);  // Generates a key schedule for testing
extern unsigned long Spbox[SPBOX_X][SPBOX_Y];      // Combined SP boxes
extern DES3_KS Ks;        // Key schedule generated by deskey()

// Sample data, this could be read iteratively from a file.
static unsigned char Blocks[]={0x6f,0x98,0x26,0x35,0x02,0xc9,0x83,0xd7};
void des_test(co_stream config_out, co_stream blocks_out,
              co_stream input_stream) {
   int i, k;
   unsigned char block[8];
   uint8 blockElement;
   unsigned long data,err;
#ifdef TIMED_TEST
   Xuint32 counter;
#endif

   // Send the keyschedule and SPbox data
   HW_STREAM_OPEN(des_test,config_out, O_WRONLY, UINT_TYPE(32));
   for ( k = 0; k < 2; k++ ) {
      for ( i = 0; i < KS_DEPTH; i++ ) {
         data = Ks[i][k];
         HW_STREAM_WRITE(des_test,config_out,data);
      }
   }
   for ( i = 0; i < SPBOX_X; i++ ) {
```

Figure 9-3. Embedded test bench for the triple-DES hardware process. (*continues*)

```
            for ( k = 0; k < SPBOX_Y; k++ ) {
                data = Spbox[i][k];
                HW_STREAM_WRITE(des_test,config_out,data);
            }
        }
        HW_STREAM_CLOSE(des_test,config_out);

        // Send the test block of data to the encryption process
        HW_STREAM_OPEN(des_test,blocks_out, O_WRONLY, UINT_TYPE(8));
        HW_STREAM_OPEN(des_test,input_stream,O_RDONLY,UINT_TYPE(8));

#ifdef TIMED_TEST
        XTmrCtr_Reset(&TimerCounter,0);
#endif
        for ( k = 0; k < BLOCKSIZE; k++ ) {
            blockElement = Blocks[k];
            HW_STREAM_WRITE(des_test,blocks_out,blockElement);
        }

        for ( k = 0; k < BLOCKSIZE; k++ ) {
            HW_STREAM_READ(des_test,input_stream,blockElement,err);
            block[k] = blockElement;
        }

#ifdef TIMED_TEST
        counter = XTmrCtr_GetValue(&TimerCounter,0);
#endif

        HW_STREAM_CLOSE(des_test,blocks_out);
        HW_STREAM_CLOSE(des_test,input_stream);

#ifdef TIMED_TEST
        xil_printf("FPGA processing done (%d ticks).\n\r",counter);
#else
        xil_printf("FPGA processing done.\n\r");
#endif

        printf("FPGA block out:");
        for (i=0; i<BLOCKSIZE; i++) {
            printf(" %02x",block[i]);
        }
        printf("\n\r");
}
```

Figure 9-3. *continued*

```
int main(int argc, char *argv[])
{
   // The key is 24 bytes
   unsigned char * key = (unsigned char *) "Gflk jqo40978J0dmm$%@878";
   co_architecture my_arch;

#ifdef IMPULSE_C_TARGET
#ifdef TIMED_TEST
   XTmrCtr_Initialize(&TimerCounter, XPAR_OPB_TIMER_0_DEVICE_ID);
   XTmrCtr_SetResetValue(&TimerCounter,0,0);
   XTmrCtr_Start(&TimerCounter,0);
#endif
#endif

   printf("Impulse C 3DES DEMO\n\r");

   des3key(Ks, key, 0);  /* Create a keyschedule for encryption */

   printf("Running encryption test on FPGA ...\n\r");

   my_arch = co_initialize((void *)Iterations);
   co_execute(my_arch);

   return(0);
}
```

Figure 9-3. *continued*

9.5 SOFTWARE STREAM MACRO INTERFACES

Once an algorithm has been implemented in hardware and optimized (see Chapter 10), it may be capable of sustaining extremely fast data rates, possibly consuming and processing data every one or two clock cycles. Consequently, the software portion that supplies the data to the hardware should be able to output data as efficiently as practical in order to take advantage of the hardware implementation.

One way to reduce overhead in a software application is to reduce the use of function calls, which are inherently inefficient due to their need for stack-related overhead.

The standard method provided for opening, reading, writing, and closing streams is represented by the **co_stream_open**, **co_stream_read**, **co_stream_write** and **co_stream_close** functions. These functions may be used in hardware or software processes to set up and manage the communication of data across stream interfaces. These stream interfaces are subsequently implemented as buffered channels (FIFOs) on the hardware side of the interface and as low-level (memory-mapped) bus interfaces on the software side.

The overhead of making a simple procedure call to the Impulse C stream functions can require many cycles, potentially resulting in poor performance to the point of negating any potential benefits of hardware acceleration. The macro versions of these stream read and write functions therefore eliminate this function call overhead by using inline assembly instructions for a more efficient implementation.

For hardware processes you should always use the standard Impulse functions and observe the stream guidelines described in Chapter 4. For software processes, however, you may increase the performance of your stream I/O by replacing the co_stream_open, co_stream_read, co_stream_write, and co_stream_close function calls (which are *procedural interfaces*) with more direct calls that are provided in the form of the following Impulse C macros:

HW_STREAM_OPEN(proc,stream,mode,type)

HW_STREAM_READ(proc,stream,var,evar)

HW_STREAM_WRITE(proc,stream,var)

HW_STREAM_CLOSE(proc,stream)

In general, you should defer the use of these macros until late in the development process in order to preserve hardware/software compatibility for your Impulse C processes. It is not possible or desirable to use these macros for hardware processes.

9.6 BUILDING THE TEST SYSTEM

To run the hardware/software test using the embedded test bench just described, we need to set up a test platform using an FPGA prototyping board and configure that platform using the appropriate FPGA vendor tools. For the Xilinx MicroBlaze target, we have chosen to use the Virtex II prototyping board supplied by Insight Electronics (Memec Design). This board, which was shown in Figure 9-1, has the necessary I/O resources and other features useful for hardware/software prototyping and testing.

In this section we will go through the process, step-by-step, of creating a platform using the Xilinx Embedded Development Kit (EDK) tools. As we go through this process, keep in mind that the Xilinx and Impulse tools are in a near-constant state of change. By the time you read this description the process may be somewhat different, and quite possibly easier.

Specifying the Platform Support Package

The first step, before generating hardware, is to select a target platform in the Impulse tools. The compilation process for both hardware and software elements of the application is defined by a number of platform-specific factors, including the type of FPGA being targeted (in this case a Xilinx Virtex II device), the type of processor (the MicroBlaze), and the bus architecture to be used for data streams (FSL). Additional factors might include an embedded operating system or an unusual (board-specific) memory or I/O interface.

To specify a platform target appropriate for the V2MB1000 board we are using, we open the Generate Options dialog as shown in Figure 9-4 and choose the platform Xilinx MicroBlaze FSL. We also specify hw and sw for the hardware and software directories as shown, and specify EDK for the hardware and software export directories. Also, we ensure that the "Generate dual clocks" option is checked, which indicates that we expect to clock the FPGA component (the triple-DES encryption core) at a different clock rate than the MicroBlaze processor.

Generating HDL for the Hardware Process

To generate hardware in the form of HDL files, and to generate the associated software interfaces and library files, we select Generate HDL from the Impulse C Project menu. This invokes the compilation steps for hardware generation (see Figure 9-5). This series of operations results in the generation of

- hw/3des_comp.vhd, which represents the encryption process
- hw/3des_top.vhd, which represents the top-level HDL file
- hw/core, which is a directory containing files required by the Xilinx EDK tools
- hw/lib, which includes the Impulse library files required for FPGA synthesis and/or VHDL simulation
- sw/co_init.c, which is linked with the software portion of your application to start up and manage the stream interfaces and other software elements
- sw/driver, which is a directory containing runtime libraries implementing the low-level stream, signal, and memory interfaces and the various Impulse C function calls

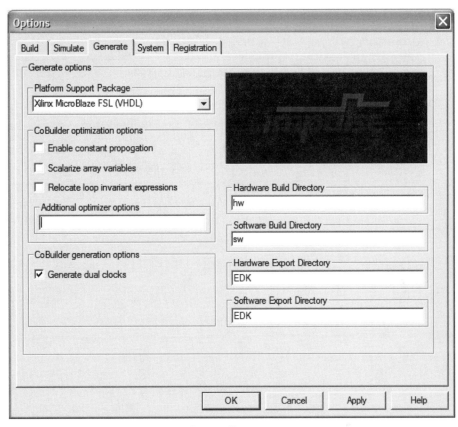

Figure 9-4. Setting the hardware generation options.

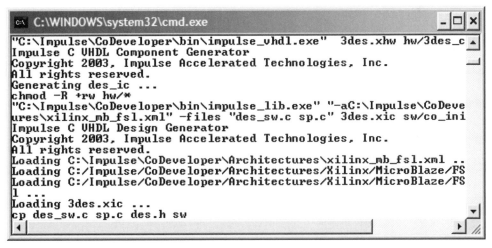

Figure 9-5. Generating hardware for the selected target.

Creating the Platform Using the Xilinx Tools

As described earlier, the result of compilation is a number of hardware and software-related output files that must all be used to create a complete hardware/software application on the target platform. Creating that platform requires a number of additional steps, most of which are performed in the Xilinx tools.

Note: Although this section focuses primarily on the use of the Xilinx tools and Xilinx devices, the process of creating mixed hardware/software applications for Altera-based platforms is conceptually the same. Altera provides a tool called SOPC Builder (described in Chapter 11) that provides capabilities similar to the EDK and Platform Studio tools described here. The Impulse compiler generates outputs compatible with either Xilinx or Altera platforms.

To create the platform, we need to create a new Xilinx EDK project then export the files generated by the Impulse C compiler to this new project. We will do this using the EDK System Builder (Platform Studio) tool environment.

We begin by creating a subdirectory within our project directory that will contain the new Xilinx Platform Studio (EDK) project files. For this subdirectory we have chosen the name EDK, which is also the directory name we specified earlier in the Generate Options dialog.

Creating a Platform Studio Project and Choosing a Board

Now we'll move into the Xilinx tool environment. We launch Xilinx Platform Studio and create a new project (called system.xmp, specifying that we want to use the Xilinx Base System Builder Wizard to define the platform.

The Base System Builder Wizard prompts us to choose our target board (Figure 9-6). We choose the Memec Design V2MB1000 (revision 3) board from the list. Note that board support packages such as this one are provided by the board vendors, typically on their support websites.

We click Next to continue with the System Builder Wizard, and in the next wizard page, we select MicroBlaze as the processor (see Figure 9-7). The other choice, PowerPC, is not available for this particular board, which does not include a PowerPC-equipped Virtex II Pro or Virtex-4 device.

We click Next to continue with the Base System Builder Wizard. The next steps (and corresponding series of Base System Builder screens) are summarized next. They involve configuring the MicroBlaze processor and creating the necessary I/O interfaces for the 3DES test application.

Base System Builder - Select Board

Select a target development board:

⊙ I would like to create a system for the following development board

Board Vendor   Memec Design

Board Name   Virtex-II V2MB1000 Development Board with P160 Comm. M

Board Revision 3

Vendor's Website   Contact Info

Download Third Party Board Definition Files

○ I would like to create a system for a custom board

Figure 9-6. Selecting the target development board.

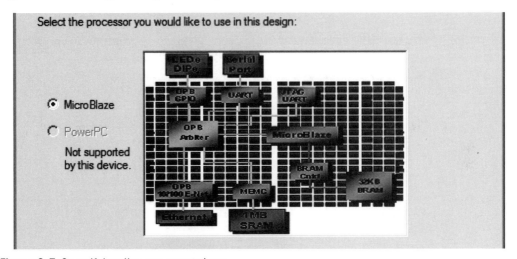

Figure 9-7. Specifying the processor type.

Configuring the MicroBlaze Processor

Now that we have created a basic MicroBlaze project in the System Builder Wizard, we need to specify additional information about the platform to support the requirements of the 3DES hardware/software application. Continuing with the steps required in the Base System Builder Wizard, we specify the following information in the Configure processor page, making sure to increase the local data and instruction memory, as shown in Figure 9-8.

System Wide Settings
 Reference Clock Frequency: 100 MHz
 Processor-Bus Clock Frequency: 100 MHz

Processor Configuration
 On-chip H/W debug module (default setting)
 Local Data and Instruction Memory: 64KB

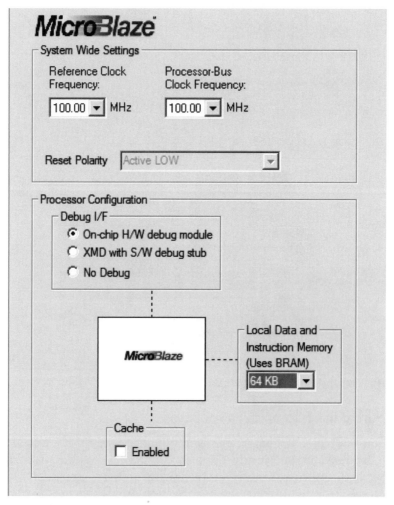

Figure 9-8. Configuring the MicroBlaze processor.

We click Next to continue with the Wizard. We are now presented with a series of wizard pages specifying the I/O peripherals to be included with the MicroBlaze processor. We select one RS232 device and the SRAM_256Kx32 peripheral by setting the following options:

I/O Device: RS232
Peripheral: OPB UARTLITE
Baudrate: 9600
Data Bits: 8
Parity: NONE
Use Interrupt: disabled

I/O Device: SRAM_256Kx32
Peripheral: OPB EMC

We click Next and disable all other I/O devices, with the exception of DDR_SRAM_16Mx16 in the Additional IO Interfaces dialogs that follow.

On the Software Configuration dialog (see Figure 9-9), we select the Generate Sample Application and Linker Script option (the default) and ensure that STDIN and STDOUT are mapped to the selected RS232 interface.

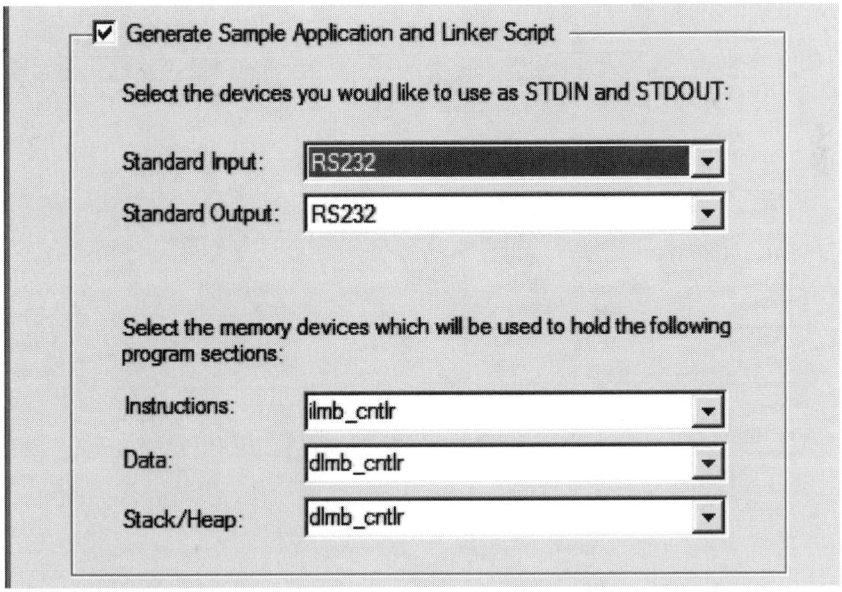

Figure 9-9. Specifying the STDIN and STDOUT devices and the processor memories.

At this point we have configured the platform and processor features. The Base System Builder Wizard displays a summary of the system we have created, as illustrated in Figure 9-10. We click Generate to generate the system and project files and click Finish to close the Wizard. We exit Xilinx Platform Studio completely before beginning the next step, which is exporting our generated Impulse C hardware to the newly created platform.

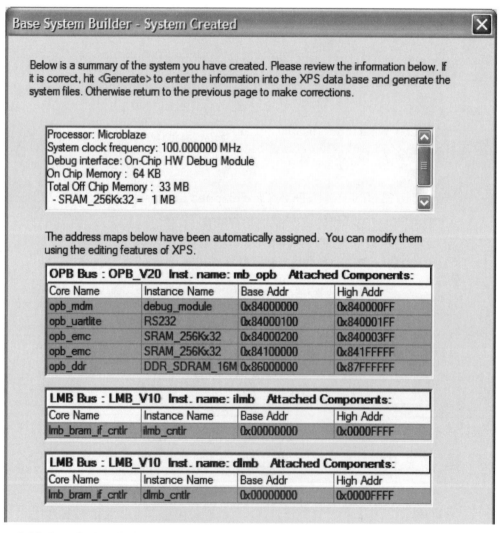

Figure 9-10. Base System Builder system summary.

Exporting Files from the Impulse Tools

With Xilinx Platform Studio closed, we return to the Impulse C tools. Recall that at the start of this section we specified the directory EDK as the export target for hardware and software (in Figure 9-4). This is also the directory where we created the new Platform Studio project. These export directories specify where the generated hardware and software processes are to be copied when the Export Software and Export Hardware features of the Impulse tools are invoked. Within these target directories (in this case EDK), the specific destination subdirectory for each file previously generated is determined from the Impulse Platform Support Package architecture library files. It is therefore important that the correct Platform Support Package (in this case Xilinx MicroBlaze FSL) be selected prior to starting the export process.

To export the files from the build directories (in this case hw and sw) to the export directories (in this case the EDK directory), we select the Export Generated Hardware and Export Generated Software options from the Impulse tools. This results in the generated hardware and software files being copied into the appropriate subdirectories of our new Platform Studio project. As a result, the next time we open that project we will have available the generated Impulse C hardware in the form of a standard FSL peripheral.

Importing the Generated Hardware

We now return to the Xilinx Platform Studio application to finish creating and configuring the target platform. This next part of this procedure is somewhat lengthy but needs to be done only once for any new project. In fact, it is useful to create one template platform (using the following methods) and save that platform for later reference and use.

After navigating to the EDK subdirectory and reopening the new project (system.xmp), we will add the generated and exported hardware module (which is called fsl_des) to the platform and create the necessary bus connections. We'll begin by adding the new IP core.

Adding the 3DES Hardware IP Core

To add the new IP core, we select Project -> Add/Edit Cores to bring up the dialog box shown in Figure 9-11. This dialog lists all the available IP cores, most of which are provided as standard with the Xilinx EDK distribution. We will be making use of two additional cores in this example (apart from the MicroBlaze and related cores that were already added by Base System Builder): The newly generated encryption core (fsl_des) and a timer core (opb_timer) that will be used to obtain actual timing information for our test.

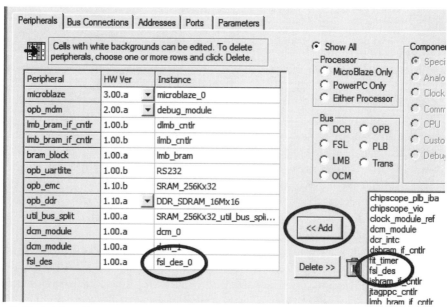

Figure 9-11. Adding the generated hardware block as a peripheral.

To add the **fsl_des** core, we select it from the list as shown and click the Add button. The **fsl_des** module now appears in the list of peripherals on the left side of the dialog box.

Setting the FSL Bus Parameters and Connections for the Core

Next, we need to set some parameters related to this hardware process, setting up the communication with the FSL bus. We click the Parameters tab in the Add/Edit dialog, then add the **C_FSL_LINKS** parameter by selecting IP instance **microblaze_0** and selecting parameter **C_FSL_LINKS**. We then click the Add button to add the parameter. After adding the **C_FSL_LINKS** parameter, we set its value to 3, indicating that the hardware process will require three FSL connections, one for each stream as shown in Figure 9-12.

Next, we click the Bus Connections tab to open the Bus Connections page. We select **fsl_v20_v1_00_b** and click Add three times to create three new bus connections. We specify the new bus connections as follows:

microblaze_0 mfsl0 connects to fsl_v20_0

microblaze_0 mfsl1 connects to fsl_v20_1

microblaze_0 sfsl2 connects to fsl_v20_2

fsl_des_0 sfsl0 connects to fsl_v20_0

fsl_des_0 sfsl1 connects to fsl_v20_1

fsl_des_0 mfsl2 connects to fsl_v20_2

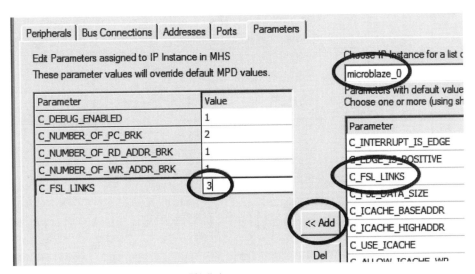

Figure 9-12. Specifying the number of FSL links.

The completed bus connections are shown in Figure 9-13.

We now select the Ports page by clicking the Ports tab. We add the clk and co_clk ports as internal ports under fsl_des_0. We also change the net name for port co_clk to sys_clk_2 as shown in Figure 9-14. This reflects the fact that we are creating a dual-clock system, in which the FPGA component (our hardware encryption algorithm) will run at a different clock rate than the MicroBlaze processor. Synchronization across clock domains is handled automatically by the dual-clock FIFO interfaces that were generated for us by the Impulse compiler. (Note: If co_clk is missing from the fsl_des_0 section at this point, we have probably set the hardware generation options incorrectly in the Impulse tools and have generated a single-clock interface.)

Now we need to add the external port for the secondary clock, which in this case of the Memec development board chosen is a 24MHz clock. We do this by selecting the co_clk entry in the Internal Ports Connections and clicking the Make External button, as shown in Figure 9-15. As specified, the MicroBlaze processor runs at 100MHz while the FPGA logic implementing the encryption algorithm operates at 24MHz.

Remaining on the Ports page, we now connect the clocks for the three created FSL bus connections (fsl_v20_0, fsl_v20_1, and fsl_v20_2). For each of these bus connections, we add the FSL_Clk and SYS_Rst ports, as shown in Figure 9-16.

For each FSL link, we set FSL_Clk to sys_clk_s. We also set SYS_Rst to net_gnd as shown, and we set the clk port for fsl_des_0 to sys_clk_s, as shown in Figure 9-17.

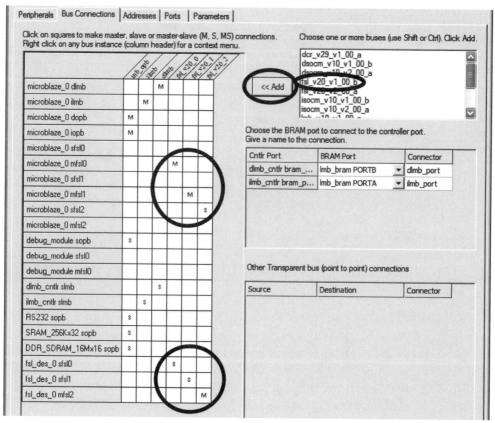

Figure 9-13. Connectng the new peripheral to the FSL links.

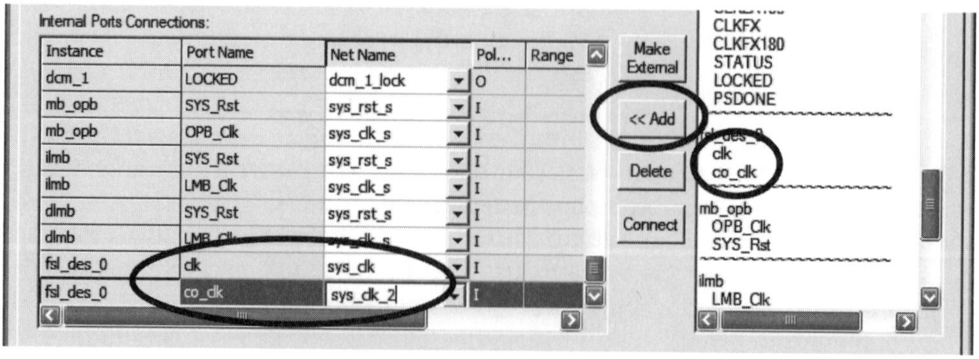

Figure 9-14. Adding the clock and reset lines for the new peripheral.

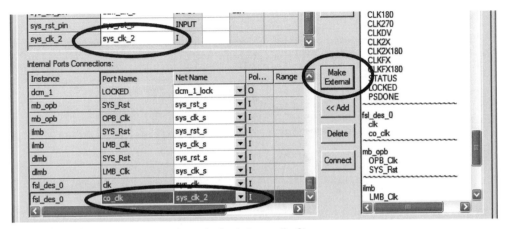

Figure 9-15. Adding a second external clock (**sys_clk_2**).

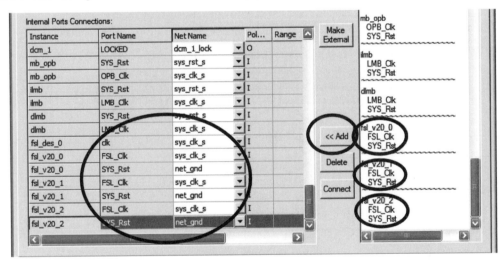

Figure 9-16. Adding clock and reset signals for the FSL links.

Adding and Configuring an OPB Timer Core

Now we need to add the OPB timer core. This timer will be used to time the software and hardware implementations of the 3DES algorithm so we can do a rough performance comparison. To add the timer core, we select the Peripherals tab of the Add/Edit dialog and choose opb_timer from the list of IP cores as shown in Figure 9-18. We then select the Bus Connections tab and connect the opb_timer_0 core to the mb_opd bus, as shown in Figure 9-19.

Last, we select the Ports tab and add the OPB_Clk port under opb_timer_0, setting OPB_Clk to sys_clk_s as was done for the fsl_des_0 clock ports.

Figure 9-17. Renaming the **fsl_des_0** clock net name.

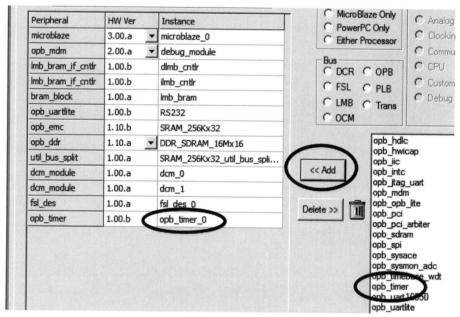

Figure 9-18. Adding an OPB timer peripheral.

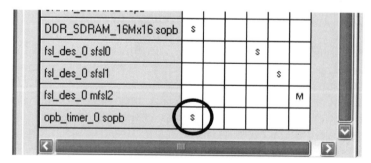

Figure 9-19. Connecting the OPB timer peripheral.

Specifying the Peripheral Addresses

Now we need to set the addresses for each of the peripherals specified for the platform, including the **opb_timer** just added. This is done by choosing the Addresses tab of the Add/Edit dialog and clicking the Generate Addresses button. The addresses are assigned automatically by Platform Studio (see Figure 9-20). Notice that the **fsl_des** peripheral is not listed here because it is not connected to the OPB.

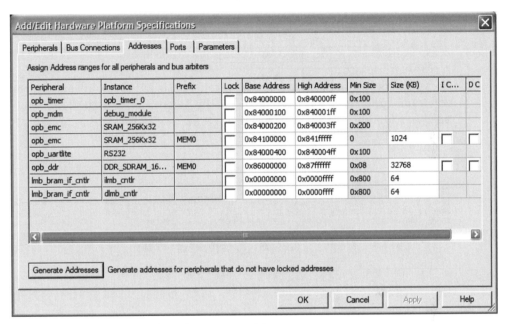

Figure 9-20. Setting peripheral and memory addresses.

Specifying the System Clock Pin

Now we need to add a second system clock pin to the system, representing the 24MHz clock that will drive our encryption peripheral. In the system view of the Platform Studio project, we double-click the **system.ucf** file (located under Project Files) then add the following line:

    Net sys_clk_2 LOC=A11;

The specified pin (**A11**) represents the 24MHz clock provided on the Memec board used in this example. (The clock pin on other target boards may be different.)

We have now exported all necessary hardware files from the Impulse tools to the Xilinx tools environment and have configured the new platform.

The next step is to import the software source files representing the test bench that will run on the MicroBlaze processor.

Importing the Application Software

Now we will import the relevant software source files to the Platform Studio project. To do this, we click the Applications tab of the project to view the sample project named TestApp (see Figure 9-21). This sample application was generated automatically by Base System Builder when we created the project.

We delete the automatically generated sample source file (TestApp.c) from the Project Sources and then right-click Sources under microblaze_0 and select Add Files. A file selection dialog appears. We navigate to the code directory and select the two files shown in Figure 9-22 (des_sw.c and co_init.c), clicking Open to add the source files to the project. These files comprise the software application that will run on the MicroBlaze CPU. (The des.h and sp.c files are included indirectly through the use of include statements in the des_sw.c file.)

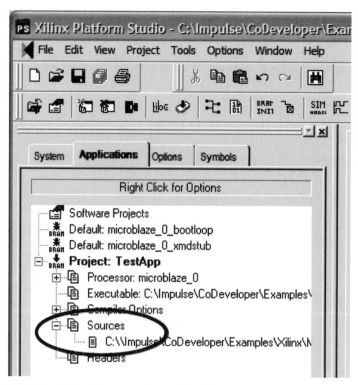

Figure 9-21. Modifying the sample application generated by Base System Builder.

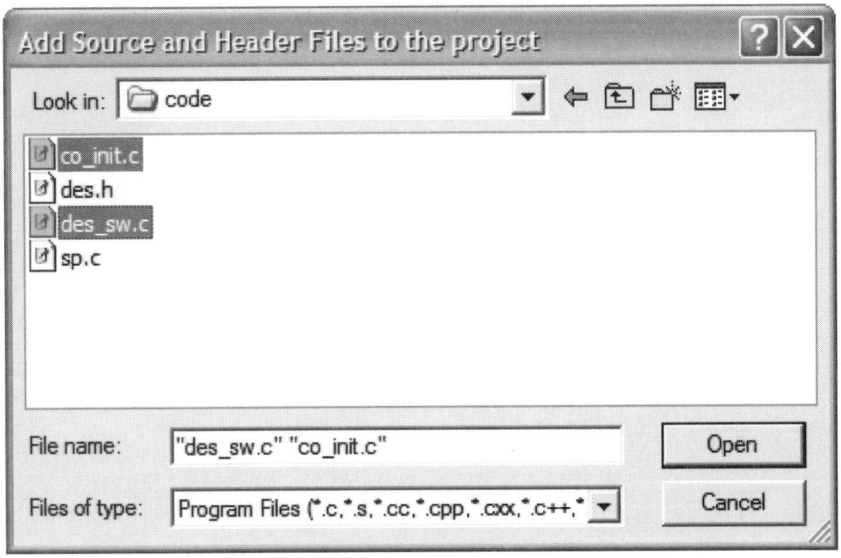

Figure 9-22. Importing the software test bench files.

The complete hardware/software application is now ready to be synthesized, placed, and routed to create a downloadable FPGA bitmap. The next section describes how to create the final bitmap, download the bitmap to the device, and run the test application.

Generating the FPGA Bitmap

We are now ready to generate the bitmap and download the complete application to the target platform. First, from within Platform Studio we select the Generate Bitstream item from the Tools menu (see Figure 9-23). This process requires about one hour to complete on our Windows development system.

The generated bitstream includes the FPGA image representing our encryption processes (fsl_des) as well as the MicroBlaze processor core and its various peripherals, including the UART core needed for the RS232 port, the SRAM controller, and the OPB timer peripheral. It does not yet, however, include the binary image for the software that will run on the MicroBlaze processor.

To generate the required software binary image, we must first select the Build All User Applications menu item, shown in Figure 9-24. This triggers a compile and link of des_sw.c, co_init.c, and related library files.

After the compilation is complete, we add the compiled binary image to the FPGA image by selecting the Update Bitstream menu item, as shown in Figure 9-25. This menu item adds the software binary image to the FPGA bitmap file. We are now ready to download and execute the application.

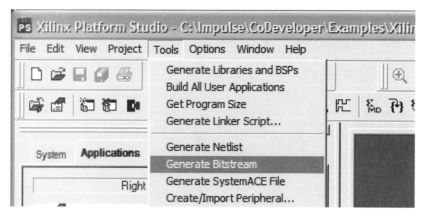

Figure 9-23. Generating the FPGA bitmap.

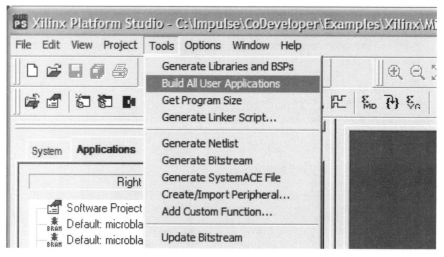

Figure 9-24. Building the MicroBlaze software application.

Downloading and Running the Application

Before downloading the application to the target board and FPGA, we open a Windows HyperTerminal session connecting our development machine to the serial port connected to the RS232 peripheral on the main FPGA development board. We use the same communication settings we chose when defining the RS232 peripheral in Base System Builder (9600 baud, 8 bits, no parity).

The RS232 cable is attached, as is the parallel port FPGA download cable. We turn on the power to the board, then select the Download function from the Tools menu in Platform Studio. A series of messages appears in the Platform Studio transcript, and after the FPGA has been successfuly programmed we see (on the Hyperterminal window) a series of messages generated by our software test bench through the use of printf statements. These messages indicate that the test was successful, and report the time required to perform the test encryption in both software and hardware, as shown in Figure 9-26.

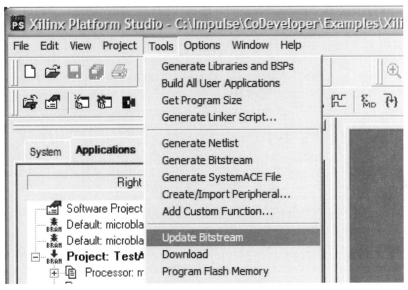

Figure 9-25. Updating the bitstream with the compiled application files.

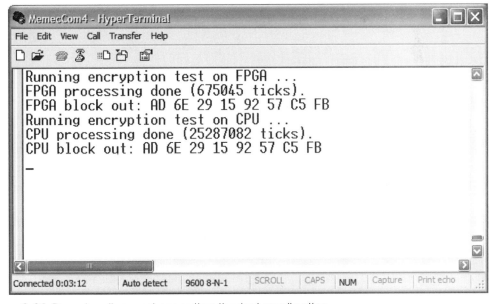

Figure 9-26. Downloading and executing the test application.

9.7 SUMMARY

In this chapter we have created a mixed hardware/software application in which the software portion communicates directly with the hardware process using streams-based communications. We have shown the steps needed to generate the FPGA hardware, combine this with a soft processor and download the combined application to an actual FPGA prototyping board.

This process has allowed us to determine the correctness of the algorithm in hardware and set up an in-system test environment to directly evaluate the results of subsequent design optimizations. In effect, what we have created by doing this is a hardware/software test bench that supports unit testing of this particular algorithm. This test bench lets us validate the algorithm quickly and apply a wide variety of test inputs simply by changing and recompiling the software application, which runs on the FPGA's embedded processor.

Armed with this software/hardware test bench, we can now proceed to optimize this application for performance, secure in the knowledge that the results will be fully testable at the process level, in actual hardware.

CHAPTER 10

Optimizing C for FPGA Performance

In Chapters 8 and 9 we presented a triple-DES encryption algorithm, showing how a process of software simulation, using standard C tools coupled with in-system testing, could be used to validate a legacy C algorithm that had been converted to an FPGA implementation. As we pointed out in those chapters, however, the performance obtained from this particular algorithm was not especially good relative to what could be achieved by a reasonably skilled VHDL or Verilog programmer.

C code that has not been written with hardware compilation in mind can result in less-than-optimal results when those results are measured in terms of process latencies, data throughput, and size of the generated logic. This is because the fundamental target of compilation (an FPGA and its constituent hardware resources) is quite different from a traditional processor, and the C language is, by design, optimized for processor-based architectures. By using some relatively simple C programming techniques, however, it is possible to dramatically accelerate the performance of many types of algorithms. This chapter presents some of these important techniques.

In Chapter 6 and 7 you learned that certain programming techniques can be applied to improve the performance of processes written in C and compiled to hardware. In this chapter we'll apply some of those techniques directly.

10.1 RETHINKING AN ALGORITHM FOR PERFORMANCE

Chapter 9 presented an example of data encryption using the triple-DES encryption algorithm, using a legacy C implementation of the algorithm as a
basis for the application. As indicated in that chapter, the purpose of the example was to demonstrate how to prototype a hardware implementation in
an FPGA, using C code that had previously been written for and compiled to
an embedded processor. When hardware simulation of the resulting RTL was
performed, it was determined that the rate at which the algorithm could generate encrypted blocks of data was approximately one block approximately
every 191 cycles. This rate is faster in terms of clock cycles than what could be
achieved in a software implementation (using an embedded processor for
comparison), but given the much higher clock rates possible in modern
processors, this is unlikely to be a satisfactory result for FPGA-based hardware acceleration.

To increase the performance of this application, we begin by taking a
closer look at the algorithm itself and how it was initially implemented as an
Impulse C process. To simplify the example for the purpose of discussion, we
will focus on only one of the three single-DES passes of this algorithm. The
following are excerpts from the original legacy C code that processes one 64-
bit block of eight characters:

```
left = ((unsigned long)block[0] << 24)
   | ((unsigned long)block[1] << 16)
   | ((unsigned long)block[2] << 8)
   | (unsigned long)block[3];
right = ((unsigned long)block[4] << 24)
   | ((unsigned long)block[5] << 16)
   | ((unsigned long)block[6] << 8)
   | (unsigned long)block[7];

work = ((left >> 4) ^ right) & 0x0f0f0f0f;
right ^= work;
left ^= work << 4;
work = ((left >> 16) ^ right) & 0xffff;
right ^= work;
left ^= work << 16;
work = ((right >> 2) ^ left) & 0x33333333;
left ^= work;
right ^= (work << 2);
work = ((right >> 8) ^ left) & 0xff00ff;
left ^= work;
right ^= (work << 8);
right = (right << 1) | (right >> 31);
work = (left ^ right) & 0xaaaaaaaa;
left ^= work;
```

```
right ^= work;
left = (left << 1) | (left >> 31);
```

The preceding block prepares the data by arranging the 64-bit block into two 32-bit values and performing some initial permutations on the data. Simple bit-level manipulations such as this will be easily converted by the Impulse C compiler into hardware logic requiring only a single cycle. Reading the data from the block array, however, will require at least eight cycles because only one value per cycle can be read from memory. Reducing the number of accesses required to memory is one key optimization that we will consider for this algorithm.

Next, consider the following excerpts, in which a relatively complex macro has been defined as part of the calculation:

```
#define F(l,r,key){\
        work = ((r >> 4) | (r << 28)) ^ key[0];\
        l ^= Spbox[6][work & 0x3f];\
        l ^= Spbox[4][(work >> 8) & 0x3f];\
        l ^= Spbox[2][(work >> 16) & 0x3f];\
        l ^= Spbox[0][(work >> 24) & 0x3f];\
        work = r ^ key[1];\
        l ^= Spbox[7][work & 0x3f];\
        l ^= Spbox[5][(work >> 8) & 0x3f];\
        l ^= Spbox[3][(work >> 16) & 0x3f];\
        l ^= Spbox[1][(work >> 24) & 0x3f];\
    }
```

This macro is subsequently used in the following section of the algorithm:

```
F(left,right,Ks[0]);
F(right,left,Ks[1]);
F(left,right,Ks[2]);
F(right,left,Ks[3]);
F(left,right,Ks[4]);
F(right,left,Ks[5]);
F(left,right,Ks[6]);
F(right,left,Ks[7]);
F(left,right,Ks[8]);
F(right,left,Ks[9]);
F(left,right,Ks[10]);
F(right,left,Ks[11]);
F(left,right,Ks[12]);
F(right,left,Ks[13]);
F(left,right,Ks[14]);
F(right,left,Ks[15]);
```

This code performs the main operation of the encryption/decryption. Some simple bit manipulations are performed, but there are also many references to both one and two-dimensional array elements. Each instantiation of the macro **F** performs eight loads from the **Spbox** array, and **F** itself is instantiated 16 times for a total of 128 loads. From that you can easily see that this block of code will require at least 128 cycles.

The final block of code performs some final permutations on the two 32-bit values and arranges the result in the output array:

```
right = (right << 31) | (right >> 1);
work = (left ^ right) & 0xaaaaaaaa;
left ^= work;
right ^= work;
left = (left >> 1) | (left << 31);
work = ((left >> 8) ^ right) & 0xff00ff;
right ^= work;
left ^= work << 8;
work = ((left >> 2) ^ right) & 0x33333333;
right ^= work;
left ^= work << 2;
work = ((right >> 16) ^ left) & 0xffff;
left ^= work;
right ^= work << 16;
work = ((right >> 4) ^ left) & 0x0f0f0f0f;
left ^= work;
right ^= work << 4;

block[0] = (unsigned char) (right >> 24);
block[1] = (unsigned char) (right >> 16);
block[2] = (unsigned char) (right >> 8);
block[3] = (unsigned char) right;
block[4] = (unsigned char) (left >> 24);
block[5] = (unsigned char) (left >> 16);
block[6] = (unsigned char) (left >> 8);
block[7] = (unsigned char) left;
```

Again, the simple bit manipulations being shown here will be easily compiled into a single-cycle operation. Storing the data into the block array will require at least eight cycles, however, because only one data value can be written to memory per cycle.

Using this original, unmodified code, the CoDeveloper tools generate a hardware implementation that uses close to 3,000 slices in a Xilinx Virtex-II FPGA and perform just 10.6 times faster than an equivalent algorithm running in software on an embedded MicroBlaze processor. In the following sections we will demonstrate a number of refinements that can be applied in this example to obtain better results, both in terms of clock cycles and in the size of the resulting logic.

10.2 REFINEMENT 1: REDUCING SIZE BY INTRODUCING A LOOP

As described earlier, a macro (F) was used to repeat the 16 steps of the core processing. In software, this might run a little faster than using a loop. In hardware, though, using the macro in such a way means that we are duplicating that code 16 times, creating a potentially much larger implementation. The obvious solution here is to reduce code repetition by introducing a loop.

Without making major modifications, we can introduce a loop as follows:

```
for (i=0; i<16; i++) {
    F(left,right,Ks[i]);
    i++;
    F(right,left,Ks[i]);
}
```

Regenerating hardware using the Impulse C tools, we obtain an implementation using about 1,500 slices in the Xilinx device. The looping instructions introduce some extra delay, but the performance is still 9.4 times faster than the unmodified software implementation. In short, for a small hit in performance the design size has been cut roughly in half. Refinement four will shed a somewhat different light on this performance difference.

Tip: Evaluate the use of straight-line code versus loops to help balance cycle delays with hardware size.

10.3 REFINEMENT 2: ARRAY SPLITTING

One of the most significant performance advantages of hardware implementations is the ability to access multiple memories in the same cycle. In the typical software implementation, each CPU is connected to one or more memories by a single bus. In that scenario, the software can only access data from, at most, one memory per bus transaction (one or more cycles).

When generating hardware, we have the opportunity to generate any topology necessary, including separate connections to multiple memories. The Impulse C tools generate a separate, individually connected memory for each array used in the process. For example, consider the following:

```
void run() {
    co_int32 i,A[4],B[4],C[4];
    for (i=0; i<4; i++) {
        A[i]=B[i]+C[i];
    }
```

This process generates three separate memories for the arrays A, B, and C, each with a dedicated connection to the computation hardware. The assignment statement that would require at least three bus transactions to execute in software can now be performed in a single cycle by performing reads from B and C and writing to A simultaneously.

Returning to our DES example, notice that two relatively large arrays are involved in the core computation, Spbox and Ks. These are each implemented in their own memory, but they still involve multiple accesses to the same array and thus require multiple cycles when used in this algorithm. Let's look at the memory access involved in the core computation as specified in macro F and in the main code loop:

```
#define F(l,r,key){\
    work = ((r >> 4) | (r << 28)) ^ key[0];\
    l ^= Spbox[6][work & 0x3f];\
    l ^= Spbox[4][(work >> 8) & 0x3f];\
    l ^= Spbox[2][(work >> 16) & 0x3f];\
    l ^= Spbox[0][(work >> 24) & 0x3f];\
    work = r ^ key[1];\
    l ^= Spbox[7][work & 0x3f];\
    l ^= Spbox[5][(work >> 8) & 0x3f];\
    l ^= Spbox[3][(work >> 16) & 0x3f];\
    l ^= Spbox[1][(work >> 24) & 0x3f];\
}

    ...

    for (i=0; i<16; i++) {
         F(left,right,Ks[i]);
         i++;
         F(right,left,Ks[i]);
    }
```

Each iteration of this loop requires four loads from Ks and eight loads from Spbox. The need for eight loads from Spbox implies that at least eight cycles will be required because (at most) one value can be read from Spbox per cycle. Notice, however, that Spbox is a multidimensional array and that the row indices are all constants. Because the indices are constant it is possible to actually split this array into individual arrays for each row.

Likewise, notice that Ks is a multidimensional array and the column indices are all constant. Ks can therefore be split into individual arrays for each column. The resulting code looks like this:

```
#define F(l,r,key0,key1){\
    work = ((r >> 4) | (r << 28)) ^ key0;\
    l ^= Spbox6[work & 0x3f];\
    l ^= Spbox4[(work >> 8) & 0x3f];\
    l ^= Spbox2[(work >> 16) & 0x3f];\
```

```
    I ^= Spbox0[(work >> 24) & 0x3f];\
    work = r ^ key1;\
    I ^= Spbox7[work & 0x3f];\
    I ^= Spbox5[(work >> 8) & 0x3f];\
    I ^= Spbox3[(work >> 16) & 0x3f];\
    I ^= Spbox1[(work >> 24) & 0x3f];\
}

. . .

    i=0;
    do {
        F(left,right,Ks0[i],Ks1[i]);
        i++;
        F(right,left,Ks0[i],Ks1[i]);
        i++;
    } while (i<16);
```

This is only a minor change to the code, but the result is that each of the new **Spbox** and **Ks** arrays are accessed only twice per iteration, reducing the total cycle count required to execute the process. After regenerating the hardware with CoDeveloper, we obtain an implementation that is slightly smaller in size and 19.5 times faster than the unmodified software implementation running on MicroBlaze. In this case, array splitting has doubled the performance over our first refinement. The result is also smaller because some address calculations have been eliminated by reducing the dimension of **Spbox** and **Ks**.

Tip: Array splitting is a useful technique for reducing both size and cycle counts.

10.4 REFINEMENT 3: IMPROVING STREAMING PERFORMANCE

Sometimes the most significant bottleneck in a system is the hardware/software communication interface. In the 3DES example, the data is assumed to be both produced and consumed by the connected CPU (a MicroBlaze processor connected via FSL). This results in a significant amount of data crossing over the software/hardware interface. When the code was initially ported to Impulse C hardware, no attention was given to the communication overhead, so the resulting stream implementation is very inefficient. Consider the following code, which is located at the start of the main processing loop:

```
while (co_stream_read(blocks_in,&block[0], sizeof(uint8))==co_err_none) {
    for ( i = 1; i < BLOCKSIZE; i++ ) {
        co_stream_read(blocks_in, &block[i], sizeof(uint8));
    }
}
```

Here, each 64-bit block is being transferred eight bits at a time (one character) over an 8-bit stream even though the hardware is connected to the CPU via a 32-bit bus. As with memories, streams require at least one cycle per read/write operation, so this code requires at least eight cycles. Furthermore, consider the code immediately following the stream reads:

```
left = ((unsigned long)block[0] << 24)
     | ((unsigned long)block[1] << 16)
     | ((unsigned long)block[2] << 8)
     | (unsigned long)block[3];
right = ((unsigned long)block[4] << 24)
      | ((unsigned long)block[5] << 16)
      | ((unsigned long)block[6] << 8)
      | (unsigned long)block[7];
```

After reading the data eight bits at a time, this code rearranges the data into two 32-bit values, which requires eight loads from the block array and therefore at least eight more cycles. The same situation is also present in the output communications at the end of the main processing loop.

Rewriting the streams interface to use 32-bit streams significantly improves performance. The input and output communication can be rewritten as follows:

```
while (co_stream_read(blocks_in, &left, sizeof(left))==co_err_none) {
    co_stream_read(blocks_in, &right, sizeof(unsigned long));
    . . .
    co_stream_write(blocks_out, &right, sizeof(unsigned long));
    co_stream_write(blocks_out, &left, sizeof(unsigned long));
}
```

Notice that the **block** array has been completely eliminated. Obviously this change to the streams specification requires a corresponding change to the producer and consumer processes (which in this example are represented by a single software test bench process running on the embedded MicroBlaze processor), but this is a simple change.

Regenerating hardware with this new communication scheme, we obtain system performance 48 times faster than the software implementation, which was also modified to access the block data in 32-bit chunks. Thus, the new result is nearly 2.5 times faster than the previous result (refinement 2).

Tip: Consider packing multiple data values into a single stream packet to increase process throughput.

10.5 REFINEMENT 4: LOOP UNROLLING

Refinement 1 began by reducing the size of the generated hardware by recoding some repeated code as a loop. Up to refinement 3, we considered performance improvements that kept the size fairly constant and in the end achieved an overall system speedup of 48X over the initial software-based prototype. In the remaining sections we will abandon our attempts to maintain the size and push instead for the maximum performance, as measured in cycle counts.

Refinement 1 introduced a loop and consequently some overhead to the core DES computation in favor of the reduced size. The performance loss was quite modest when considered in isolation; however, that was before optimizing the statements that make up the body of the new loop. Let's see what impact loop unrolling will have now, after the substantial optimizations of refinements 2 and 3. The following is how the code appears after unrolling the inner loop introduced in refinement 1:

```
F(left,right,Ks0[0],Ks1[0]);
F(right,left,Ks0[1],Ks1[1]);
F(left,right,Ks0[2],Ks1[2]);
F(right,left,Ks0[3],Ks1[3]);
F(left,right,Ks0[4],Ks1[4]);
F(right,left,Ks0[5],Ks1[5]);
F(left,right,Ks0[6],Ks1[6]);
F(right,left,Ks0[7],Ks1[7]);
F(left,right,Ks0[8],Ks1[8]);
F(right,left,Ks0[9],Ks1[9]);
F(left,right,Ks0[10],Ks1[10]);
F(right,left,Ks0[11],Ks1[11]);
F(left,right,Ks0[12],Ks1[12]);
F(right,left,Ks0[13],Ks1[13]);
F(left,right,Ks0[14],Ks1[14]);
F(right,left,Ks0[15],Ks1[15]);
```

Regenerating hardware using the Impulse C tools on this new revision produces a hardware implementation that requires a little over 2,000 slices and performs nearly 80 times faster than the software implementation.

Tip: As this example shows, the overhead introduced by a loop can be significant when the loop body is small.

Note that, rather than duplicating the body eight times and substituting constants for array index I as we have done here, Impulse C also has an **UNROLL** pragma that essentially does the same thing for loops with constant values for their iteration values. The **UNROLL** pragma performs unrolling for you as a preprocessing step before other optimizations are performed. For this example, however, it is more convenient to unroll the loop by hand as a prelude to refinement 5.

10.6 REFINEMENT 5: PIPELINING THE MAIN LOOP

Along with memory optimization, pipelining is also one of the most significant types of optimizations to consider when generating hardware. Pipelining can be applied to loops that require two or more cycles per iteration and that do not have dependencies between iterations. The DES loop processes independent blocks of data with no dependencies between iterations, so it should be possible to apply pipelining. To evaluate the potential use of pipelines, let's first look in more detail at how the process is currently being parallelized.

Using Stage Master Explorer

The Stage Master Explorer tool (available from Impulse) provides a way to see in detail how hardware is being generated from the C code. In refinement 5, we will try generating a pipeline to improve performance, but we first want to use the Stage Master Explorer to better understand how the current version is being mapped to hardware.

The screen image of Figure 10-1 shows the Stage Master Explorer display for the core DES computation. The center view in the application shows the input C code annotated with the cycle number of each operation. For example, line one in the display shows that the read of Ks[0] will be available in cycle two and that the new value of work will be assigned in cycle two. The second line shows the computation of left based on Spbox6[work & 63]. We saw that work is assigned in cycle two, so the index work & 63 is not available until cycle two, and thus the result of the read will be available in cycle three.

Because all arrays have been split, we see that the loads from Spbox and Ks can be done simultaneously. As a result, each instantiation of the F macro can be done in a single cycle. There are 16 instantiations of the F macro, and each one depends on the previous one because work is computed from the value of left computed in the previous instance of F. Due to this inherent dependency, there is no way to reduce the cycle count below 16 cycles per block.

For any loop that requires more than one cycle per iteration, it is worth considering the possibility of using a pipeline. The next section takes a look at introducing a pipeline in the main DES loop.

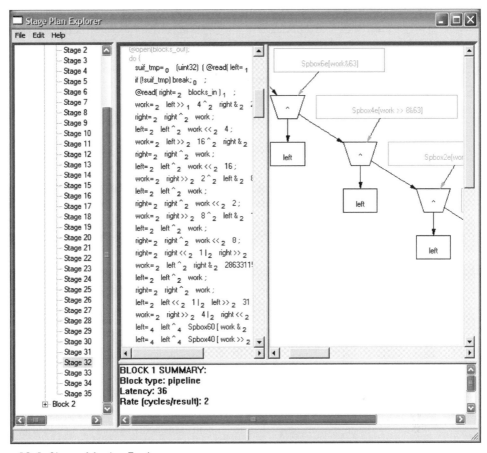

Figure 10-1. Stage Master Explorer.

The Goal of Pipelining

The goal of pipelining is to execute multiple iterations of a loop in parallel, much like the assembly line of a factory. The core DES computation contains 16 instantiations of the F macro, and these instantiations can be called F0, F1, …, F16. Iteration F16 is dependent on F15, which is in turn dependent on F14, and so on so that F1 through F16 must be executed sequentially. Imagine that F1 through F16 are stations in an assembly line processing a sequence of 64-bit blocks (the "raw material") and producing a sequence of 64-bit block "products" on the other end. At the start of the line, F1 receives block 1, does some processing, and passes the result to F2. While F2 is processing block 1, F1 is idle, so it can go ahead and start processing block 2. By the time block 1 makes it to station F16, station F1 is processing block 16 and all 16 stations are busy operating in parallel. The result of building such an assembly line is that we can generate one block of output every cycle, even though it still takes 16 cycles to process one block. The cost (continuing the assembly line metaphor) is that each station needs its own equipment for all of them to operate in parallel.

To generate a pipeline using Impulse C, we must insert the CO PIPELINE pragma inside the main loop:

```
while (co_stream_read(blocks_in, &left, sizeof(unsigned long)) == co_err_none) {
    #pragma CO PIPELINE
    co_stream_read(blocks_in, &right, sizeof(unsigned long));
    . . .
```

Unfortunately, simply adding the pipeline pragma to the refinement 4 code generates a pipeline that produces only one result every 16 cycles. Why did the pipeline fail? The reason is each instance of the F macro requires access to the Spbox and Ks arrays, which prevents the macros from being executed in parallel.

In order to generate an effective pipeline in this application, each instance of F must have its own copy of the Spbox arrays. The Ks array could also be duplicated, but since the array is small we use another technique, which is to convert the Ks array into a register bank. Registers can be read by many sources in the same cycle. To implement Ks as a register bank, all indices to Ks must be constants. The resulting code appears as follows:

```
#define F(l,r,key0,key1,sp0,sp1,sp2,sp3,sp4,sp5,sp6,sp7){\
    work = ((r >> 4) | (r << 28)) ^ key0;\
    l ^= sp6[work & 0x3f];\
    l ^= sp4[(work >> 8) & 0x3f];\
    l ^= sp2[(work >> 16) & 0x3f];\
    l ^= sp0[(work >> 24) & 0x3f];\
    work = r ^ key1;\
    l ^= sp7[work & 0x3f];\
    l ^= sp5[(work >> 8) & 0x3f];\
    l ^= sp3[(work >> 16) & 0x3f];\
    l ^= sp1[(work >> 24) & 0x3f];\
}


    . . .


F(left,right,Ks0[0],Ks1[0],
    Spbox00,Spbox10,Spbox20,Spbox30,Spbox40,Spbox50,Spbox60,Spbox70);
F(right,left,Ks0[1],Ks1[1],
    Spbox01,Spbox11,Spbox21,Spbox31,Spbox41,Spbox51,Spbox61,Spbox71);
F(left,right,Ks0[2],Ks1[2],
    Spbox02,Spbox12,Spbox22,Spbox32,Spbox42,Spbox52,Spbox62,Spbox72);
F(right,left,Ks0[3],Ks1[3],
    Spbox03,Spbox13,Spbox23,Spbox33,Spbox43,Spbox53,Spbox63,Spbox73);
```

```
F(left,right,Ks0[4],Ks1[4],
    Spbox04,Spbox14,Spbox24,Spbox34,Spbox44,Spbox54,Spbox64,Spbox74);
F(right,left,Ks0[5],Ks1[5],
    Spbox05,Spbox15,Spbox25,Spbox35,Spbox45,Spbox55,Spbox65,Spbox75);
F(left,right,Ks0[6],Ks1[6],
    Spbox06,Spbox16,Spbox26,Spbox36,Spbox46,Spbox56,Spbox66,Spbox76);
F(right,left,Ks0[7],Ks1[7],
    Spbox07,Spbox17,Spbox27,Spbox37,Spbox47,Spbox57,Spbox67,Spbox77);
F(left,right,Ks0[8],Ks1[8],
    Spbox08,Spbox18,Spbox28,Spbox38,Spbox48,Spbox58,Spbox68,Spbox78);
F(right,left,Ks0[9],Ks1[9],
    Spbox09,Spbox19,Spbox29,Spbox39,Spbox49,Spbox59,Spbox69,Spbox79);
F(left,right,Ks0[10],Ks1[10],
    Spbox0a,Spbox1a,Spbox2a,Spbox3a,Spbox4a,Spbox5a,Spbox6a,Spbox7a);
F(right,left,Ks0[11],Ks1[11],
    Spbox0b,Spbox1b,Spbox2b,Spbox3b,Spbox4b,Spbox5b,Spbox6b,Spbox7b);
F(left,right,Ks0[12],Ks1[12],
    Spbox0c,Spbox1c,Spbox2c,Spbox3c,Spbox4c,Spbox5c,Spbox6c,Spbox7c);
F(right,left,Ks0[13],Ks1[13],
    Spbox0d,Spbox1d,Spbox2d,Spbox3d,Spbox4d,Spbox5d,Spbox6d,Spbox7d);
F(left,right,Ks0[14],Ks1[14],
    Spbox0e,Spbox1e,Spbox2e,Spbox3e,Spbox4e,Spbox5e,Spbox6e,Spbox7e);
F(right,left,Ks0[15],Ks1[15],
    Spbox0f,Spbox1f,Spbox2f,Spbox3f,Spbox4f,Spbox5f,Spbox6f,Spbox7f);
```

As a result of this change, each instance of **F** has its own copy of the **Spbox** data, and **Ks0** and **Ks1** will be converted to registers by the compiler.

Running the compiler now results in a 20-stage pipeline that produces one block every two cycles. Two cycles are required because the 32-bit communication stream requires two cycles to transfer the blocks to and from the CPU. If a 64-bit interface were available, the pipeline would generate a single block every cycle.

The overall system performance is now 425 times faster than the software implementation. The size of the hardware has also increased to over 3,000 slices due to the duplicated resources required to implement a pipeline.

Tip: Pipelining can, when combined with other optimizations, result in significantly increased process throughput. Pipelining will, however, increase the size and complexity of the generated hardware. Pipelining should therefore be deferred until the application has been verified and other source-level optimizations have been used.

Need to assert more control over optimization?

Although this chapter focuses on the use of C language coding styles to improve the outcome of C code optimization and hardware generation, there are also methods (such as the **StageDelay** pragma described elsewhere) that can be used to more directly specify how logic is generated for a given set of C statements. In particular, the Impulse C compiler includes various flags for specifying hardware generation options and the Impulse library itself includes additional functions, including the **co_par_break** and **co_array_config** functions that allow you to specify actual clock cycle boundaries, specify memory types, and otherwise direct the operation of the optimizer and HDL generator. Refer to the latest Impulse C documentation for more information. Also be sure to visit the Impulse C discussion forums at www.ImpulseC.com/forums.

10.7 SUMMARY

This chapter has described a number of C coding techniques that may be used to increase the performance and reduce the size of C code. Although you can in many cases obtain satisfactory, prototype-quality results without considering the trade-offs of memory access, array specifications, instruction-level dependencies, loop coding styles, and instruction pipelines, you should consider these factors carefully when creating applications that require the highest practical performance and throughout. The techniques described in this chapter are a good starting point and are applicable to a wide variety of FPGA-based applications.

CHAPTER 11

Describing System-Level Parallelism

In the preceding chapters, we have focused primarily on the creation of individual processes and described how these processes may be optimized for performance using techniques such as array splitting, loop pipelining, and loop unrolling. We have also seen how additional processes may be used for the purpose of testing by creating both desktop and embedded software test benches that interact with Impulse C hardware processes during simulation or during actual operation within an FPGA.

In this chapter, we will return to the topic of communicating processes and show how the use of system-level parallelism can improve throughput for many types of applications. Applications designed in this way might reside entirely in hardware, as a collection of hardware processes communicating via streams that are implemented as FIFOs, or they might involve a combination of hardware processes that communicate with one or more software processes residing on an embedded or discrete microprocessor. Furthermore, these multiple, independent processes may reside on a single FPGA or be spread across multiple FPGAs to form a larger parallel computing grid.

When designing for system-level parallelism, it is important to create processes that can operate in parallel or in a pipelined sequence to take full advantage of the available hardware resources.

The example we will use for this purpose is an image-processing filter. This filter accepts a stream of pixels (which are assumed to be 24-bit RGB values) and performs an edge-detection operation, streaming the resulting modified pixel values to the output. Many similar image filters may be created using the methods described in this example. The general method of creating multiple parallel instances of hardware processes can be used for a wide variety of applications. We'll present three different ways of implementing this image filter, each of which demonstrates different aspects of system-level parallelism. In the latter half of this chapter, we will take one version of the image filter example all the way to hardware, using an Altera FPGA-based prototyping board.

11.1 Design Overview

Edge-detection algorithms are used to identify and enhance areas of an input image (whether a single image or a real-time video stream) that have particularly high contrast between adjacent pixels. There are many different edge-detection algorithms, each of which is optimized for particular requirements and for a particular hardware or software implementation.

Virtually all image-detection algorithms, and many other types of imaging filters, share a common attribute: they operate iteratively on specific "windows" of an image, where a window is defined as a collection of neighboring pixels spanning a number of rows and columns, with the target pixel in the center. In this example, we use a 3-by-3 window.

If we wish to create an image filter that operates on a stream of pixel data (see Figure 11-1), one pixel at a time, it is clear that the process must store at least enough pixel values to "look ahead" one row (plus one pixel) and "look behind" one row (plus one pixel). To do this, we will describe a circular streaming image buffer using some simple array indexing techniques in C:

```
while ( co_stream_read(pixels_in, &nPixel, sizeof(co_uint24)) == co_err_none ) {
        bytebuffer[addpos][REDINDEX] = nPixel & REDMASK;
        bytebuffer[addpos][GREENINDEX] = (nPixel & GREENMASK) >> 8;
        bytebuffer[addpos][BLUEINDEX] = (nPixel & BLUEMASK) >> 16;
        addpos++;
        if (addpos == BYTEBUFFERSIZE)
                addpos = 0;
        currentpos++;
        if (currentpos == BYTEBUFFERSIZE)
                currentpos = 0;
```

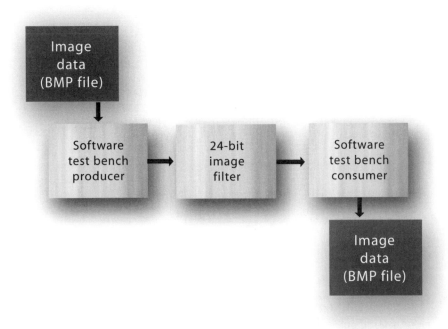

Figure 11-1. A single-process edge-detection filter and test bench.

The result of reading the 24-bit input data into the local storage area and unpacking the color values is three distinct buffers, one for each color, containing exactly enough pixel data at any given time to provide access to the required 3-by-3 window over a single scan line.

Once we have access to a window of pixel data, we can then perform the necessary calculations to determine the new value of the target pixel. We could make many possible calculations to produce the desired results. The edge-detection algorithm described in this example performs a relatively primitive set of calculations on the image window to enhance the edges: For each pixel n, four pairs of pixels, each pair surrounding n on one axis, are compared, and their absolute difference is calculated for each pixel color (red, green, and blue). For each color, a new value for n is assigned that represents the greatest observed difference between any enclosing pair of pixels.

This logic is expressed in an inner code loop (nested within a larger loop running over each target pixel), as shown in Figure 11-2. Let's examine this loop in detail:

```
for (clr = 0; clr < 3; clr++) { // Red, Green and Blue
  pixelN  = bytebuffer[B_OFFSETADD(currentpos,WIDTH)][clr];
  pixelS  = bytebuffer[B_OFFSETSUB(currentpos,WIDTH)][clr];
  pixelE  = bytebuffer[B_OFFSETADD(currentpos,1)][clr];
  pixelW  = bytebuffer[B_OFFSETSUB(currentpos,1)][clr];
  pixelNE = bytebuffer[B_OFFSETADD(currentpos,WIDTH+1)][clr];
  pixelNW = bytebuffer[B_OFFSETADD(currentpos,WIDTH-1)][clr];
  pixelSE = bytebuffer[B_OFFSETSUB(currentpos,WIDTH-1)][clr];
  pixelSW = bytebuffer[B_OFFSETSUB(currentpos,WIDTH+1)][clr];

  // Diagonal difference, lower right to upper left
  pixelMag = 0;
  pixeldiff = ABS(pixelSE - pixelNW);
  if (pixeldiff > pixelMag)
    pixelMag = pixeldiff;

  // Diagonal difference, upper right to lower left
  pixeldiff = ABS(pixelNE - pixelSW);
  if (pixeldiff > pixelMag)
    pixelMag = pixeldiff;

  // Vertical difference, bottom to top
  pixeldiff = ABS(pixelS - pixelN);
  if (pixeldiff > pixelMag)
    pixelMag = pixeldiff;

  // Horizontal difference, right to left
  pixeldiff = ABS(pixelE - pixelW);
  if (pixeldiff > pixelMag)
    pixelMag = pixeldiff;

  if (pixelMag < EDGE_THRESHOLD)
    pixelMag = 0;

  nByteMag[clr] = (co_uint8) pixelMag;
}
```

Figure 11-2. Edge-detection processing loop for one pixel and three colors.

- For each color (red, green, and blue) in the target pixel, the values for the eight adjacent pixels are copied from the scan line buffer (**bytebuffer**). Note that the **B_OFFSETADD** and **B_OFFSETSUB** macros are defined elsewhere (in an include file) and provide the circular buffer offsets needed to access the desired pixels.

- After the adjacent pixel values have been determined (and stored in local variables PixelN, PixelNE, PixelW, and so on), a set of four calculations is performed to determine the maximum difference in intensity between any two bounding pixels—for example, between PixelN and PixelS. The absolute value operation, in the form of the ABS macro defined elsewhere in an include file, is used in these calculations.

- After all four possible edge directions (the vertical, the horizontal, and the two diagonals) have been analyzed, the maximum intensity difference is compared to the edge threshold value (EDGE_THRESHOLD). If it is greater than the threshold value, it becomes the new value of the pixel. If the difference does not meet the threshold value, the color is given a value of zero, which corresponds to black. (This calculation can be easily modified to produce different levels of contrast in the output image or to produce a black-on-white image, as shown in Figure 11-4.)

After the new values for the three colors are determined, they are repackaged into a single 24-bit value and written to the output buffer using co_stream_write:

```
nPixel = nByteMag[REDINDEX] |
    (nByteMag[GREENINDEX] << 8) |
    (nByteMag[BLUEINDEX] << 16);
co_stream_write(pixels_out, &nPixel, sizeof(co_uint24));
```

The result is a single image filter process that accepts 24-bit pixels, caches just enough pixels locally (using a circular array) to assemble a window of eight pixels around each of the pixels in the scan line of interest, splits these eight required pixels into their component colors, and performs the edge-detection function. As you will see, there are more efficient ways to implement a filter such as this, but as a starting point (perhaps as a working prototype) this filter will produce the desired outputs and can be compiled to hardware.

11.2 PERFORMING DESKTOP SIMULATION

If we compile this application for desktop simulation (along with an appropriate software test bench), we can observe the results of the edge detect operation on a sample 800-by-800 pixel image, as shown in Figures 11-3 and 11-4.

Note that the software test bench created for this application makes use of standard C library functions to read and write sample image data from a Windows bitmap-format (.BMP) image file. Each pixel in the sample image is transmitted by the software test bench to the edge-detection process using a stream. The resulting convoluted pixels are then output to a second file, which appears as shown in Figure 11-4.

Figure 11-3. Image prior to edge-detection filtering.

11.3 REFINEMENT 1: CREATING PARALLEL 8-BIT FILTERS

As written, the inner code loop of this edge-detection process requires a total of 18 instruction stages (six stages per color), which may or may not be fast enough to meet the requirements. We can attempt to speed up this process using the techniques described in Chapter 9, but these optimizations are unlikely to accelerate the loop dramatically due to the number of memory accesses being performed on the bytebuffer array.

In particular, consider the following statements at the beginning and end of the inner code loop:

```
while ( co_stream_read(pixels_in, &nPixel, sizeof(co_uint24)) == co_err_none ) {
    bytebuffer[addpos][REDINDEX] = nPixel & REDMASK;
    bytebuffer[addpos][GREENINDEX] = (nPixel & GREENMASK) >> 8;
    bytebuffer[addpos][BLUEINDEX] = (nPixel & BLUEMASK) >> 16;
```

Figure 11-4. Image after edge-detection filtering.

These statements unpack the 24-bit value obtained from the stream to create three distinct references to the red, green, and blue pixel values, consuming three clock cycles in doing so. After unpacking, the edge detection is performed for each of the three colors in the inner code loop, as summarized here again:

```
for (clr = 0; clr < 3; clr++) {  // Red, Green and Blue
    pixelN  = bytebuffer[B_OFFSETADD(currentpos,WIDTH)][clr];
    pixelS  = bytebuffer[B_OFFSETSUB(currentpos,WIDTH)][clr];
    . . .
}
```

One approach to accelerating this algorithm might be to perform these three edge-detection operations in parallel, by creating three instances of the same image filter. This should, in theory, reduce the amount of time needed to process one pixel by a factor of three, plus whatever overhead is required to pack and unpack the 24-bit pixel values. In this application, which reads its inputs from an RGB-encoded source file, the pixels are already available as distinct color values. Therefore, the test bench (the C function that streams pixel data to the edge-detection process for the purpose of testing) can be easily modified to generate three streams of pixel color values, rather than one stream of whole pixel values, as in the original version.

As an initial attempt to speed up this edge-detection filter, we will create parallelism at the algorithm level by partitioning the algorithm into three distinct edge detection filters, one for each color, as illustrated in Figure 11-5.

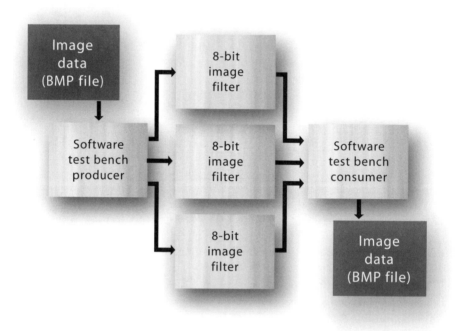

Figure 11-5. Creating three parallel instances of an 8-bit edge-detection process.

The resulting eight-bit, one-color filter (which is replicated three times) is shown in Figure 11-6. In this listing, notice that

- The input stream (pixels_in) is now used to transfer unsigned eight-bit values, as specified in the co_stream_open function.

- As in the original version, the bytebuffer array is used to store pixel color values. In the single-color version, bytebuffer is only a one-dimensional array.

```
void edge_detect(co_stream pixels_in, co_stream pixels_out) {
  int currentpos, addpos, idx = 0;
  short pixeldiff1, pixeldiff2, pixeldiff3, pixeldiff4;
  short pixelN, pixelS, pixelE, pixelW, pixelNE, pixelNW, pixelSE, pixelSW;
  co_uint8 pixelMag;
  co_uint8 bytebuffer[BYTEBUFFERSIZE];

  co_stream_open(pixels_in, O_RDONLY, UINT_TYPE(8));
  co_stream_open(pixels_out, O_WRONLY, UINT_TYPE(8));

  // Begin by filling the bytebuffer array...
  idx = 0;
  while (co_stream_read(pixels_in,&bytebuffer[idx],sizeof(co_uint8))
                                      == co_err_none ) {

    idx++;
    if (idx == BYTEBUFFERSIZE - 1)
      break;
  }

  // Now we have an almost full buffer and can start processing pixel
  // windows. But first, we need to write out the first line of pixels and
  // the first pixel of the second line (they don't get processed):

  for(idx=0; idx < WIDTH+1; idx++) {
    co_stream_write(pixels_out, &bytebuffer[idx], sizeof(co_uint8));
  }

  // Now, each time we process a window we will "shift" the buffers
  // one position and read in a new pixel. We will continue this until
  // the input stream is closed (eos). "Shifting" is accomplished by
  // manipulating the currentpos and addpos index variables.

  addpos = BYTEBUFFERSIZE - 1;
  currentpos = WIDTH;

  // Read pixel values from the stream...
  while ( co_stream_read(pixels_in, &bytebuffer[addpos],
          sizeof(co_uint8)) == co_err_none ) {
#pragma CO PIPELINE
    addpos++;
    if (addpos == BYTEBUFFERSIZE)
      addpos = 0;
    currentpos++;
    if (currentpos == BYTEBUFFERSIZE)
      currentpos = 0;
```

Figure 11-6. Eight-bit edge-detection process. (*continues*)

```
    // At this point we are guaranteed to have enough pixels in
    // our array to process a window, so let's do it...

    pixelN  = bytebuffer[B_OFFSETADD(currentpos,WIDTH)];
    pixelS  = bytebuffer[B_OFFSETSUB(currentpos,WIDTH)];
    pixelE  = bytebuffer[B_OFFSETADD(currentpos,1)];
    pixelW  = bytebuffer[B_OFFSETSUB(currentpos,1)];
    pixelNE = bytebuffer[B_OFFSETADD(currentpos,WIDTH+1)];
    pixelNW = bytebuffer[B_OFFSETADD(currentpos,WIDTH-1)];
    pixelSE = bytebuffer[B_OFFSETSUB(currentpos,WIDTH-1)];
    pixelSW = bytebuffer[B_OFFSETSUB(currentpos,WIDTH+1)];

    // Diagonal difference, lower right to upper left
    pixeldiff1 = ABS(pixelSE - pixelNW);

    // Diagonal difference, upper right to lower left
    pixeldiff2 = ABS(pixelNE - pixelSW);

    // Vertical difference, bottom to top
    pixeldiff3 = ABS(pixelS - pixelN);

    // Horizontal difference, right to left
    pixeldiff4 = ABS(pixelE - pixelW);

    pixelMag = (co_uint8) MAX4(pixeldiff1,pixeldiff2,pixeldiff3,pixeldiff4);
    if (pixelMag < EDGE_THRESHOLD) {
        pixelMag = 0;
    }
    co_stream_write(pixels_out, &pixelMag, sizeof(co_uint8));
}

// Write out the last line of the image (plus one extra pixel
// representing the last pixel on the end of the second to last
// line.)
for(idx = 0; idx < WIDTH+1; idx++) {
    co_stream_write(pixels_out, &bytebuffer[currentpos], sizeof(co_uint8));
    currentpos++;
    if (currentpos > BYTEBUFFERSIZE)
        currentpos = 0;
}

co_stream_close(pixels_in);
co_stream_close(pixels_out);
}
```

Figure 11-6. *continued*

- Because we no longer require the innermost code loop, which previously processed each color in turn, we can now use the PIPELINE pragma. This optimization will improve this loop's throughout rate by two clock cycles but is of limited benefit due to the manner in which bytebuffer is accessed. As before, memory access continues to be a limiting factor in increasing the speed of this algorithm.

To complete this version of the application, we will need to create and interconnect three instances of the single-color filter process. The configuration function that describes this structure is shown in Figure 11-7. In this configuration function (which also refers to the software test bench function) we can see declarations for six streams—three colors for the input side and three colors for the output side—as well as declarations and corresponding calls to co_process_create for the five processes: the three edge-detection processes, plus the producer and consumer test processes.

This version of the edge-detection algorithm will run substantially faster than the original version but will of course require more hardware when synthesized.

11.4 REFINEMENT 2: CREATING A SYSTEM-LEVEL PIPELINE

The partitioning of the image filter algorithm into three subprocesses (which actually represents a single process replicated three times) reduces the complexity of the filter and allows the algorithm's performance to be improved at the expense of increased hardware requirements. As we have mentioned, however, both versions of the filter are limited in performance because of how the bytebuffer array is accessed. At issue is the fact that both of the edge-detection algorithms just described must store locally (in the bytebuffer array) enough pixels to allow a look-ahead and look-behind of one scan line plus one pixel to sample the desired pixel window. If we rethink the algorithm and apply techniques of pipelined system-level parallelism, however, we can increase the filter's effective throughput dramatically.

The system-level pipeline that we will describe operates in much the same way as the instruction-level pipelines described in Chapter 7, but at a higher level. A system-level pipeline is designed by the application programmer, not automatically generated by the compiler, and consists of two or more processes connected in sequence. Each process in the sequence performs some transformation on the data and passes the results to the next process.

```
void config_edge_detect(void *arg) {
  co_stream blue_source_pixeldata, green_source_pixeldata,
            red_source_pixeldata, blue_result_pixeldata,
            green_result_pixeldata, red_result_pixeldata;
  co_signal header_ready;
  co_process producer_process;
  co_process blue_process, green_process, red_process;
  co_process consumer_process;

  blue_source_pixeldata = co_stream_create("blue_source_pixeldata",
                                      UINT_TYPE(8), BUFSIZE);
  green_source_pixeldata = co_stream_create("green_source_pixeldata",
                                      UINT_TYPE(8), BUFSIZE);
  red_source_pixeldata = co_stream_create("red_source_pixeldata",
                                      UINT_TYPE(8), BUFSIZE);
  blue_result_pixeldata = co_stream_create("blue_result_pixeldata",
                                      UINT_TYPE(8), BUFSIZE);
  green_result_pixeldata = co_stream_create("green_result_pixeldata",
                                      UINT_TYPE(8), BUFSIZE);
  red_result_pixeldata = co_stream_create("red_result_pixeldata",
                                      UINT_TYPE(8), BUFSIZE);
  header_ready = co_signal_create("header_ready");

  producer_process = co_process_create("producer_process",
                          (co_function)test_producer,
                          4, blue_source_pixeldata, green_source_pixeldata,
                          red_source_pixeldata, header_ready);
  blue_process = co_process_create("blue_process",
                          (co_function)edge_detect,
                          2, blue_source_pixeldata, blue_result_pixeldata);
  green_process = co_process_create("green_process",
                          (co_function)edge_detect,
                          2, green_source_pixeldata, green_result_pixeldata);
  red_process = co_process_create("red_process",
                          (co_function)edge_detect,
                          2, red_source_pixeldata, red_result_pixeldata);
  consumer_process = co_process_create("consumer_process",
                          (co_function)test_consumer,
                          4, blue_result_pixeldata, green_result_pixeldata,
                          red_result_pixeldata, header_ready);

  co_process_config(blue_process, co_loc, "PE0");
  co_process_config(green_process, co_loc, "PE0");
  co_process_config(red_process, co_loc, "PE0");
}
```

Figure 11-7. Parallel edge-detection configuration function.

As in the previous examples, this implementation of the edge-detection function requires a 3-by-3 window for processing each pixel in the source image. In this version of the algorithm, however, a separate upstream process generates the pixel window, in the form of three streams of pixel information on which the subsequent filter process can operate. This removes the need for **bytebuffer** and its corresponding circular buffer overhead.

For this example, we'll create a pipeline of four separate processes. Two of these processes will replace the edge-detection function, as previously described, while the other two will allow us to apply the image filter to an image buffer (an external memory), rather than rely on data streaming from an embedded processor or some other source. Each process performs one step of the filter, as described here and shown in Figure 11-8:

1. Image data is loaded from an image buffer using the Impulse C **co_memory_readblock** function and then are converted into a stream of 24-bit pixels using **co_stream_write**.

2. The stream of pixels is processed by a row generator to produce three output streams, each representing a row in three marching columns of pixels.

3. The marching columns of pixel data are used to compute the new value of the center pixel. This value is streamed out of the filter using **co_stream_write**.

4. The resulting pixel data is read (using **co_stream_read**) and stored to a new image buffer through the use of **co_memory_writeblock**.

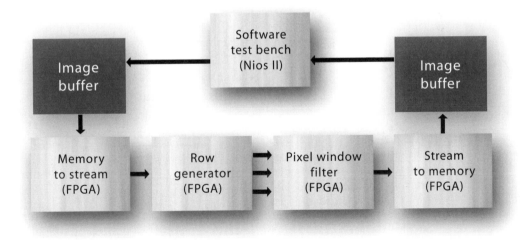

Figure 11-8. Four pipelined processes.

Because these four processes operate in parallel, in a system-level pipeline, the filter can generate processed pixels at a rate of one pixel every two clock cycles. For a standard-size image this is fast enough for real-time video image filtering at quite leisurely FPGA clock rates.

The following sections describe the four processes in more detail.

The DMA Input Process

One aspect of the image filter application that we ignored in discussing the first version of this edge detect algorithm was where the input and output images are actually stored. In a real-world example, pixel data most likely originates from some image buffer (external RAM) or is streamed via a direct hardware interface. The output image may similarly be stored in memory or sent via a hardware channel. In either case, there needs to be a process on either end of our image filter: one that converts the incoming data to an Impulse C stream, and another that converts the resulting filtered pixel stream for output.

Impulse C offers two mechanisms for moving high-volume data: streams and shared memory. Thus far, our examples have primarily used streams, but using shared memory can be a useful way to increase application performance for certain types of streaming applications. It can also be useful for applications that need to operate on nonstreaming data located in memory.

The factors influencing a decision to use streams versus shared memory are highly platform-specific. There are many issues to consider:

- How many cycles are required for each stream transaction? This is dependent on whether a processor is involved in the transfer and on the architecture of the specific CPU and bus combination. In particular, if the bus must be polled to receive data sent on a stream, there will be significant cycle-count overhead.

- How many cycles are required for a memory transfer? This number is dependent on the memory and bus architecture used, as well as the CPU/bus combination.

- Does the CPU have a cache? Is the data likely to already be in the cache?

- Is the memory on the same bus as the Impulse C hardware processes? If so, the hardware processes and memory could compete for access to the CPU, negatively impacting performance.

Some of these issues were discussed in Chapter 4, with specific benchmark examples for various platform configurations. For this example, we have chosen to use the shared memory approach, which proves to be more efficient than data streaming on the platform selected for this project, an Altera Stratix FPGA paired with a Nios II embedded soft processor.

In the Altera platform, the FPGA code that we will generate from this image filter and the shared memory are both accessed via Altera's Avalon bus. (The methods for accessing shared memory are similar on other platforms.) For the Altera Nios II processor, it is more efficient to move large blocks of data between the embedded processor and a hardware function such as our image filter using DMA transfer rather than making use of streams. On other platforms, such as the Xilinx MicroBlaze processor with its FSL bus, the use of software-to-hardware streams may provide faster performance than DMA transfers.

One disadvantage of using DMA with a shared memory is that the hardware process is blocked from doing any computation during the transfer. As a result, the use of shared memory often implies a separate process for handling shared memory operations, so that data transfer can be overlapped with computation. Thus, the first process in the pipeline serves only to read image data from the image memory (which is the heap0 memory defined by the selected platform) and send it out on a data stream. In the listing for this process in Figure 11-9, notice that

- The to_stream process takes three arguments: a signal, go; a shared memory, imgmem; and an output stream, output_stream.

- In the process run function itself, the co_memory_readblock and co_stream_write functions are used to read pixel data from the shared memory (one entire scan line at a time) and to write those pixels to the output stream, respectively. A signal (go) is used to synchronize with the CPU and ensure that the image memory is ready for processing.

- The algorithm accepts 16-bit pixel data even though our filter algorithm was designed to operate on 24-bit data. Consequently, this process also converts the 16-bit pixel value into a 24-bit value. Also, notice that a 32-bit memory is used even though our image data values are logically stored as 16-bit unsigned integers. The DMA transfers one element at a time to the array, so an array of 16-bit values would require twice as many bus transactions.

This process could be modified (or replaced by a hand-crafted hardware block) for use with many types of input sources.

The Column Generator Process

Image filters generally operate on a moving window of image data, as described previously. In this example, the filter operates on a 3-by-3 window. Notice that as the filter moves from one pixel to the next, the computation uses six of the same values used in the previous pixel computation plus three new values. A marching column generator generates a stream of three pixel rows that represent (at any given cycle) one three-pixel column corresponding to the three new values the filter requires for the next target pixel's filter window as it proceeds from left to right. The prep_run process buffers enough data to produce the three-pixel column data just as the original algorithm did, but in a much simpler way.

```
void to_stream(co_signal go, co_memory imgmem, co_stream output_stream) {
    int16 i, j;
    uint32 offset, data, d0;
    uint32 row[IMG_WIDTH / 2];
    co_signal_wait(go, &data);
    co_stream_open(output_stream, O_WRONLY, INT_TYPE(32));
    offset = 0;
    for ( i = 0; i < IMG_HEIGHT; i++ ) {
        co_memory_readblock(imgmem, offset, row, IMG_WIDTH * sizeof(int16));
        for ( j = 0; j < (IMG_WIDTH / 2); j++ ) {
#pragma CO PIPELINE
            d0 = row[j];
            data  = ((d0 >> 8) & 0xf8) << 16;
            data |= ((d0 >> 3) & 0xf8) << 8;
            data |= (d0 << 2) & 0xfc;
            co_stream_write(output_stream, &data, sizeof(int32));
            d0 = d0 >> 16;
            data  = ((d0 >> 8) & 0xf8) << 16;
            data |= ((d0 >> 3) & 0xf8) << 8;
            data |= (d0 << 2) & 0xfc;
            co_stream_write(output_stream, &data, sizeof(int32));
        }
        offset += IMG_WIDTH * sizeof(int16);
    }
    data = 0;
    co_stream_write(output_stream, &data, sizeof(int32));
    co_stream_write(output_stream, &data, sizeof(int32));
    co_stream_close(output_stream);
}
```

Figure 11-9. Shared memory to stream process for pipelined image filter.

As described previously, a limiting factor in the performance of the original algorithm was its method of accessing pixel values from a single buffer, which contained all the pixels needed to "look ahead" and "look behind" one entire row of the image. To increase the performance of the image filter, we therefore need to reduce the number of accesses to the same memory and thereby decrease the number of cycles required to perform the desired calculations. To do this, we use the array splitting technique described in Chapters 7 and 10 to eliminate the need to access the same array twice in the body of the loop. As a result, this process can achieve very high throughput, generating a pixel column once every two clock cycles. The source listing for this process is shown in Figure 11-10. Notice that this process accepts one stream of pixels, caches two scan lines into arrays B and C, and generates three output streams using a circular buffer technique similar to that in the original example. By taking care not to make unnecessary accesses to the same array, this process has a latency of just two clock cycles for each column output.

The Image Filter Process

The most important process in this pipeline is the filter itself, which is represented by the filter_run process listed in Figure 11-11. This process accepts the three pixel streams (the marching column) generated by the prep_run process and performs a convolution similar to the original algorithm.

In this version of the convolution, a number of statement-level optimizations have been made to reduce the number of cycles required to process the inputs:

- The adjacent pixels are captured in local variables p01, p02, p03, and so on rather than being accessed from an array, as was done in the original. This eliminates the need for simultaneous accesses to an array, which would prevent the optimizer from combining multiple calculations into single stages.

- The calculation of a difference between points in the horizontal, vertical and diagonal directions has been replaced by a simpler and more parallelizable version that eliminates the need for an abs (absolute value) function or macro. Notice how in this version the center pixel is repeatedly compared to its neighboring pixels to obtain an average difference.

The result of these changes is an edge-detection algorithm that can be accomplished with a throughput rate of just two clock cycles. This calculation can of course be modified as needed to perform other types of convolutions.

The Stream to Memory Process

This process, from_stream, provides the opposite function from the to_stream process described earlier. In this process, a single stream of input pixels is read, one scan line at a time, and written to the output memory using co_memory_writeblock. This process also swaps the order of the bytes being read from the stream and placed in memory, reflecting the requirements of the image format being used (which in this case is the TIFF format). The stream-to-memory process (from_stream) is shown in Figure 11-12.

```
void prep_run(co_stream input_stream, co_stream r0, co_stream r1, co_stream r2) {
  int32 i, j;
  int32 B[IMG_WIDTH], C[IMG_WIDTH];
  int32 A01, A02, p02, p12, p22;

  co_stream_open(input_stream, O_RDONLY, INT_TYPE(32));
  co_stream_open(r0, O_WRONLY, INT_TYPE(32));
  co_stream_open(r1, O_WRONLY, INT_TYPE(32));
  co_stream_open(r2, O_WRONLY, INT_TYPE(32));

  co_stream_read(input_stream, &A01, sizeof(int32));
  co_stream_read(input_stream, &A02, sizeof(int32));

  for ( j = 0; j < IMG_WIDTH; j++ )
        co_stream_read(input_stream, &B[j], sizeof(int32));
  for ( j = 0; j < IMG_WIDTH; j++ )
     co_stream_read(input_stream, &C[j], sizeof(int32));

  co_stream_write(r0, &A01, sizeof(int32));
  co_stream_write(r1, &B[IMG_WIDTH - 2], sizeof(int32));
  co_stream_write(r2, &C[IMG_WIDTH - 2], sizeof(int32));
  co_stream_write(r0, &A02, sizeof(int32));
  co_stream_write(r1, &B[IMG_WIDTH - 1], sizeof(int32));
  co_stream_write(r2, &C[IMG_WIDTH - 1], sizeof(int32));

  for ( i = 2; i < IMG_HEIGHT; i++ ) {
        j = 0;
        do {
              p02 = B[j];
              p12 = C[j];
              co_stream_read(input_stream, &p22, sizeof(int32));
              co_stream_write(r0, &p02, sizeof(int32));
              co_stream_write(r1, &p12, sizeof(int32));
              co_stream_write(r2, &p22, sizeof(int32));
              B[j] = p12;
              C[j] = p22;
              j++;
        } while ( j < IMG_WIDTH );
  }

  co_stream_close(input_stream);
  co_stream_close(r0);
  co_stream_close(r1);
  co_stream_close(r2);
}
```

Figure 11-10. Pixel window row generator for the pipelined image filter.

```
    void filter_run(co_stream r0, co_stream r1, co_stream r2,
            co_stream output_stream) {
    uint32 data,res, p00, p01, p02, p10, p11, p12, p20, p21, p22;
    uint16 d0;

    co_stream_open(r0, O_RDONLY, INT_TYPE(32));
    co_stream_open(r1, O_RDONLY, INT_TYPE(32));
    co_stream_open(r2, O_RDONLY, INT_TYPE(32));
    co_stream_open(output_stream, O_WRONLY, INT_TYPE(32));

    p00 = 0; p01 = 0; p02 = 0;
    p10 = 0; p11 = 0; p12 = 0;
    p20 = 0; p21 = 0; p22 = 0;
    while ( co_stream_read(r0, &data, sizeof(int32)) == co_err_none ) {
#pragma CO PIPELINE
#pragma CO set stageDelay 256
        p00 = p01; p01 = p02;
        p10 = p11; p11 = p12;
        p20 = p21; p21 = p22;

        p02 = data;
        co_stream_read(r1, &p12, sizeof(int32));
        co_stream_read(r2, &p22, sizeof(int32));

        d0 = RED(p11) << 3;
        d0 = d0 - RED(p00);
        d0 = d0 - RED(p01);
        d0 = d0 - RED(p02);
        d0 = d0 - RED(p10);
        d0 = d0 - RED(p12);
        d0 = d0 - RED(p20);
        d0 = d0 - RED(p21);
        d0 = d0 - RED(p22);
        d0 &= (d0 >> 15) - 1;
        res = d0 & 0xff;

        d0 = GREEN(p11) << 3;
        d0 = d0 - GREEN(p00);
        d0 = d0 - GREEN(p01);
        d0 = d0 - GREEN(p02);
        d0 = d0 - GREEN(p10);
        d0 = d0 - GREEN(p12);
        d0 = d0 - GREEN(p20);
        d0 = d0 - GREEN(p21);
```

Figure 11-11. Pipelined image filter process. (*continues*)

```
                d0 = d0 - GREEN(p22);
                d0 &= (d0 >> 15) - 1;
                res = (res << 8) | (d0 & 0xff);

                d0 = BLUE(p11) << 3;
                d0 = d0 - BLUE(p00);
                d0 = d0 - BLUE(p01);
                d0 = d0 - BLUE(p02);
                d0 = d0 - BLUE(p10);
                d0 = d0 - BLUE(p12);
                d0 = d0 - BLUE(p20);
                d0 = d0 - BLUE(p21);
                d0 = d0 - BLUE(p22);
                d0 &= (d0 >> 15) - 1;
                res = (res << 8) | (d0 & 0xff);

                co_stream_write(output_stream, &res, sizeof(int32));
        }
        co_stream_close(r0);
        co_stream_close(r1);
        co_stream_close(r2);
        co_stream_close(output_stream);
    }
```

Figure 11-11. *continued*

The Configuration Function

These four processes and the corresponding stream, memory, and signal declarations are described using Impulse C and are interconnected using the configuration function shown in Figure 11-13.

This configuration function includes the following:

- Signal declarations for **startsig** and **donesig**. These signals are used to
  coordinate the use of shared memories (the image buffers) between the
  software test bench and the image filter hardware. **startsig** indicates that
  the image buffer is ready for processing, while **donesig** indicates that the
  filtering of the image is complete.

- Stream declarations for the input and output pixel streams and for the
  three streams that connect **prep_run** to **filter_run**.

- A memory declaration for **shrmem**. This memory represents the image
  buffer.

```
        void from_stream(co_stream input_stream, co_memory imgmem,
                                                    co_signal done) {
            uint8 err;
            int16 i;
            int32 offset, low, data, d0;
            int32 rowout[IMG_WIDTH / 2];

            co_stream_open(input_stream, O_RDONLY, INT_TYPE(32));
            offset = 0;
            do {
                for ( i = 0; i < (IMG_WIDTH / 2); i++ ) {
#pragma CO PIPELINE
                    err = co_stream_read(input_stream, &d0, sizeof(d0));
                    if ( err != co_err_none ) break;
                    low = (d0 >> 19) & 0x1f;
                    low = (low << 5) | ((d0 >> 11) & 0x1f);
                    low = (low << 6) | ((d0 >>  2) & 0x3f);
                    err = co_stream_read(input_stream, &d0, sizeof(d0));
                    if ( err != co_err_none) break;
                    data = d0 >> 19;
                    data = (data << 5) | ((d0 >> 11) & 0x1f);
                    data = (data << 6) | ((d0 >>  2) & 0x3f);
                    rowout[i] = (data << 16) | low;
                }
                if ( err != co_err_none) break;
                co_memory_writeblock(imgmem, offset, rowout,
                                    IMG_WIDTH * sizeof(int16));
                offset += IMG_WIDTH * sizeof(int16);
            } while ( 1 );
            co_stream_close(input_stream);
            co_signal_post(done, 0);
        }
```

Figure 11-12. Stream to shared memory process, pipelined image filter.

- Process declarations for the four required hardware processes, plus one additional software test bench process, cpu_proc. This software test bench process is listed in Appendix E.

- Calls to co_signal_create for the two signals startsig and donesig.

- A call to co_memory_create for memory shrmem. The memory is created at location heap0, which is a location specific to the Altera platform being targeted for this example. heap0 represents an on-chip memory accessible to both the processor (where the software test bench will reside) and to the image filter running on the FPGA as dedicated hardware.

```
void config_img(void *arg) {
  co_signal startsig, donesig;
  co_memory shrmem;
  co_stream istream, row0, row1, row2, ostream;
  co_process reader, writer;
  co_process cpu_proc, prep_proc, filter_proc;

  startsig = co_signal_create("start");
  donesig = co_signal_create("done");
  shrmem = co_memory_create("image", "heap0",
                      IMG_WIDTH * IMG_HEIGHT * sizeof(uint16));
  istream = co_stream_create("istream", INT_TYPE(32), IMG_HEIGHT/2);
  row0 = co_stream_create("row0", INT_TYPE(32), 4);
  row1 = co_stream_create("row1", INT_TYPE(32), 4);
  row2 = co_stream_create("row2", INT_TYPE(32), 4);
  ostream = co_stream_create("ostream", INT_TYPE(32), IMG_HEIGHT/2);

  cpu_proc = co_process_create("cpu_proc", (co_function)call_fpga,
                      3, shrmem,   startsig, donesig);
  reader = co_process_create("reader",   (co_function)to_stream,
                      3, startsig, shrmem,   istream);
  prep_proc = co_process_create("prep_proc", (co_function)prep_run,
                      4, istream, row0,   row1,   row2);
  filter_proc = co_process_create("filter",   (co_function)filter_run,
                      4, row0,   row1,   row2,   ostream);
  writer = co_process_create("writer",   (co_function)from_stream,
                      3, ostream, shrmem,   donesig);

  co_process_config(reader, co_loc, "PE0");
  co_process_config(prep_proc, co_loc, "PE0");
  co_process_config(filter_proc, co_loc, "PE0");
  co_process_config(writer, co_loc, "PE0");
}
```

Figure 11-13. Pipelined edge detector configuration function.

- Calls to co_stream_create for the five streams (the input stream, the output stream, and the three intermediate streams). Notice that the three intermediate streams are given stream depths of four, while the input and output streams are given stream depths of one-half the number of scan lines in the image. This is done in part to mitigate stalls that may result from longer-than-expected memory read and write times. (Bear in mind that deep stream buffers such as this can incur substantial penalties in hardware resources, however.)

- Calls to **co_process_create** for the four hardware processes and the one software test bench process.

- Calls to **co_process_config** for the four hardware processes, indicating that these four processes are to be compiled to the FPGA as hardware blocks.

11.5 MOVING THE APPLICATION TO HARDWARE

In this section we will detail the steps required to move the completed application (the image filter and its corresponding embedded test bench process) onto an Altera-based FPGA reference board. The board that we will use is included with the Altera's Nios Development Kit, Stratix Edition, and includes a single Stratix EP1S10 device and a variety of peripherals useful for creating and testing system prototypes.

Note: The FPGA design software used in this example is Altera's Quartus II, version 4.0 and Altera's SOPC Builder. The specific steps shown may differ in later versions of the Altera tools.

The required steps using the Altera software are similar to those described in Chapter 9, in which the triple-DES example was implemented on a Xilinx platform. As with that earlier example, our goal is to create an embedded software test bench operating on the Nios II processor that sends test data (in this case using shared memory) to the FPGA-based image filter.

Generating the FPGA Hardware

The first step, after validating the application in desktop simulation, is to compile and generate the HDL and related output files. We'll do this by first selecting a platform target, which will be an Altera FPGA used in conjunction with an embedded Nios II soft processor core. To specify a platform target, we open the Generate Options dialog in the Impulse tools (as shown in Figure 11-14) and specify Altera Nios II. We also specify hw and sw for the hardware and software directories, as shown, and we specify Quartus for the hardware and software export directories.

When we select the Generate HDL item from the Impulse tools, as shown in Figure 11-15, a series of processing steps are performed, as described in earlier chapters. These steps include C preprocessing, analysis, optimization, and HDL code generation, as well as the generation of other files that are specific to the Altera design tool flow. When processing is complete, a

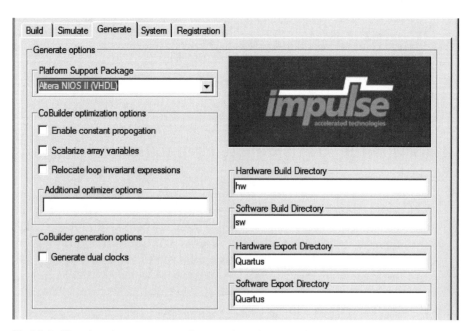

Figure 11-14. Setting hardware generation options for the Altera Nios II target.

number of files are created in the hw and sw subdirectories of our project directory. These files include the following:

- Generated VHDL source files (img_comp.vhd, img_top.vhd, and subsystem.vhd) representing the hardware process and the generated hardware stream and memory interfaces. These files are generated in the hw subdirectory.

- An hw/lib subdirectory containing required VHDL library elements.

- An hw/class subdirectory containing generated files required by the Altera SOPC Builder tools.

- C source and header files extracted from the project that are required for compilation to the embedded processor (in this case, img_sw.c, testfpga.c, testcpu.c, test.h, and img.h). These files are generated in the sw subdirectory.

- A generated C-language file (co_init.c) representing the hardware initialization function. This file will also be compiled to the embedded processor.

- An sw/class subdirectory containing additional software libraries to be compiled as part of the embedded software application. These libraries implement the software side of the hardware/software interface.

Figure 11-15. Generating the FPGA hardware.

Exporting the Generated Files

As you saw in the previous step, the Impulse tools create a number of hardware and software-related output files that must all be used to create a complete hardware/software application on the target platform. You can, if you wish, copy these files manually and integrate them into an existing Altera project. You can also export the files into the Altera tools semi-automatically.

Recall that we specified the directory **Quartus** as the export target for hardware and software. These export directories specify where the generated hardware and software processes are to be copied when the Export Software and Export Hardware functions are invoked in the Impulse tools. Within these target directories (in this case we have specified both directories as **Quartus**), the specific destination for each file is determined from the platform support package architecture library files. It is therefore important that the correct platform support package (in this case Altera Nios II) be selected prior to starting the export process.

To export the files from the build directories (in this case **hw** and **sw**) to the export directories (in this case the **Quartus** directory), we select Export Generated Hardware (HDL) and Export Generated Software, as shown in Figure 11-16.

We have now exported all necessary files from the Impulse tools to the **Quartus** project directory.

Note: The sections that follow make extensive use of the Altera SOPC Builder software. This software is updated frequently and the steps shown may change substantially in future versions.

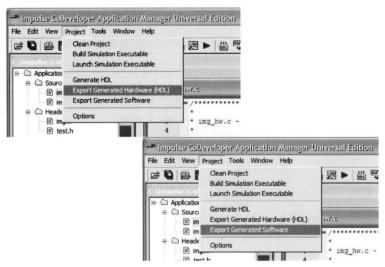

Figure 11-16. Exporting the image filter hardware and software.

Creating a New Quartus and SOPC Builder Project

Now we'll move into the Altera tool environment. We begin by launching Altera Quartus II and opening a new project using the New Project Wizard. In the field prompting us for the new project's working directory, we click the Browse button and find the directory (**Quartus**) to which we exported the hardware and software files in the previous step. We select the **Quartus** directory and click Open. On page one of the New Project Wizard dialog, we enter **ImageFilterDMA** in both the project name and the top-level design entity fields as shown in Figure 11-17. We click Next to move to the next page.

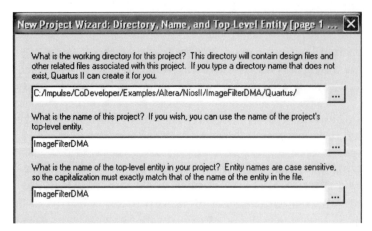

Figure 11-17. Creating and naming a Quartus project.

Now we will import the generated VHDL files generated by the tools, as well as create a block diagram file representing the entire system. (Creating the block diagram is not shown here, but it is not difficult using the Altera tools and following the Altera tutorial examples.) We import the files to Quartus in the following order:

1. Block diagram file: ImageFilterDMA.bdf

2. Bus interface files: user_logic_img_arch_module/subsystem.vhd and impulse_lib/avalon_if.vhd

3. Core logic files in the user_logic_img_arch_module subdirectory: img_comp.vhd and img_top.vhd

4. All .vhd files in the impulse_lib project subdirectory, except for the file avalon_if.vhd

The files are listed in Quartus in the opposite order from which they were added. For example, the impulse_lib files appear at the top of the list, as shown in Figure 11-18.

We click Next to proceed and then skip the EDA Tool Settings page (page 3) by again clicking Next.

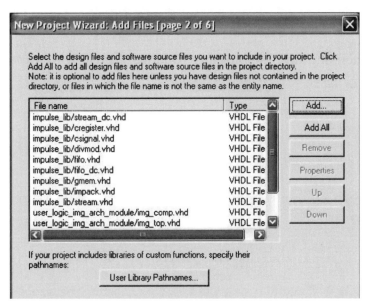

Figure 11-18. Adding files to the Quartus project.

In the Device Family page (page 4), we select the device family we will be targeting—in this case Stratix, as shown in Figure 11-19. We click Next, then in the Wizard page that follows we select the specific device we will be targeting. For this example we have chosen **EP1S10F780C6ES** (see Figure 11-20), which is the device supplied on our prototyping board.

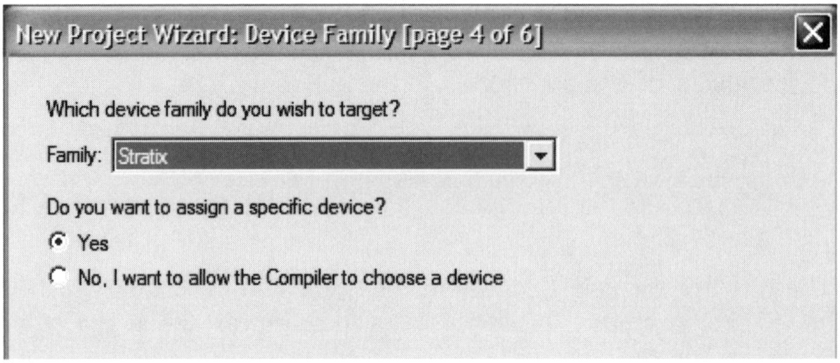

Figure 11-19. Selecting the Altera Stratix device.

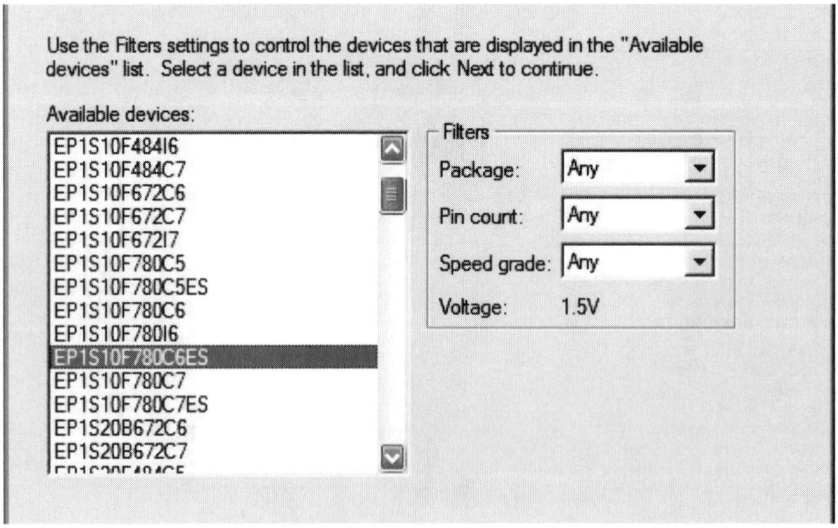

Figure 11-20. Selecting the target device.

We click Next again to see a summary page listing the options we have chosen. We click Finish to exit the Wizard and return to Quartus.

In the next steps we will create and configure a hardware system with a Nios II processor and the necessary I/O interfaces for our sample application. This is a somewhat lengthy procedure but it needs to be done only once, and the resulting platform can be copied and modified for other, similar projects.

Creating the New Platform Using SOPC Builder

Now that we have created a Quartus project using the wizard, we need to specify additional information about our platform to support the requirements of the hardware/software application. These steps include the creation of a hardware system with a Nios II processor and the necessary I/O elements.

We will use SOPC Builder to create a hardware system containing an Altera Nios II embedded processor, the generated FPGA module representing the **ImageFilterDMA** hardware process, and several necessary peripherals. To do this, we start SOPC Builder from the Quartus environment, as shown in Figure 11-21. We give the new SOPC Builder project a name (see Figure 11-22) and specify that it is a VHDL project.

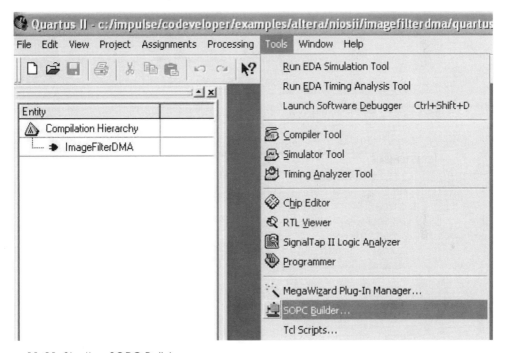

Figure 11-21. Starting SOPC Builder.

Configuring the FPGA Platform

In SOPC Builder, we begin by selecting our specific development board in the SOPC Builder dialog, as shown in Figure 11-23. The SOPC Builder software includes a variety of Altera boards as supported targets, including our Stratix-based board.

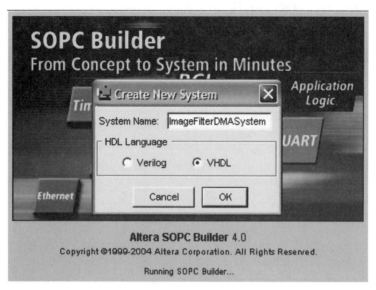

Figure 11-22. Naming the SOPC Builder project.

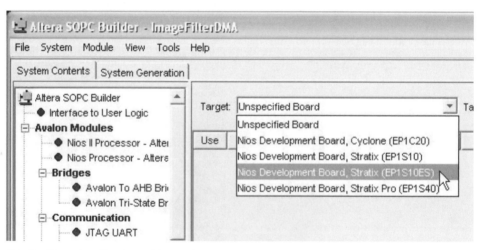

Figure 11-23. Choosing the target board.

After selecting the board, the next task is to add the largest component of the system, the Nios II processor. From the System Contents tab (on the left side of the SOPC Builder window), we select Nios II Processor - Altera Corporation under Avalon Modules, then click Add. The Altera Nios II configuration Wizard appears. We select the Nios II/s core, as shown in Figure 11-24. We click Finish to add the Nios II CPU to the system and return to SOPC Builder.

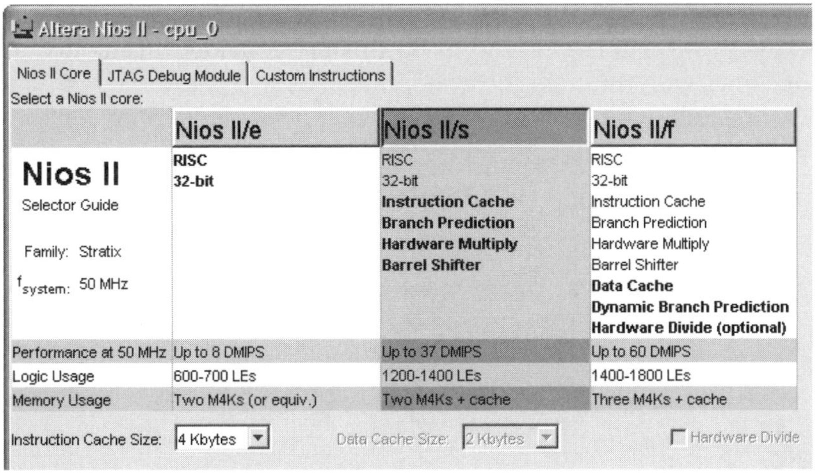

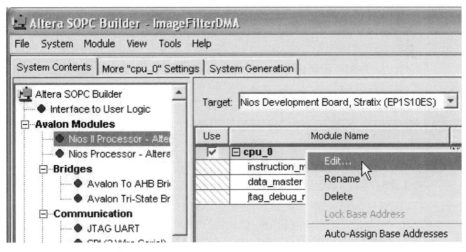

Figure 11-24. Adding the Nios II processor core.

Now we need to rename the Nios II module. We do this by right-clicking cpu_0 in the Module Name column and selecting Rename. We enter the name "cpu" and press the Enter key (see Figure 11-25).

Figure 11-25. Naming the CPU.

Adding Nios II Peripherals

Next we must add the necessary peripherals to the new Nios II system. Detailed instructions on how to do this are provided in the documentation that comes with the Nios II Development Kit and in tutorials provided by Altera. The following steps summarize what needs to be done for each component in the system.

The Timer Peripheral

To add the timer peripheral, **system_timer**, we perform the following steps:

1. We select Interval Timer under Other and click Add. The Avalon Timer - timer_0 wizard appears (see Figure 11-26).

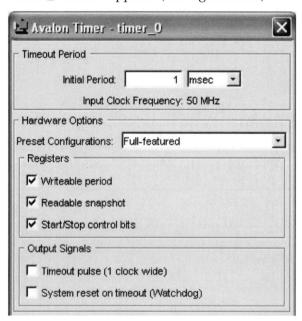

Figure 11-26. Adding a timer IP block.

2. We leave the options at their default settings.

3. We click Finish to add the timer to the system. We are returned to the Altera SOPC Builder window.

4. We right-click on **timer_0** under Module Name and then choose choose Rename from the pop-up menu. We rename **timer_0** to **system_timer** and press the Enter key.

External Flash Memory Interface

To add the external Flash peripheral, **ext_flash**, we perform the following steps:

    1. We select Flash Memory (Common Flash Interface) under Memory and click Add. The Flash Memory (Common Flash Interface) - cfi_flash_0 wizard appears (see Figure 11-27).

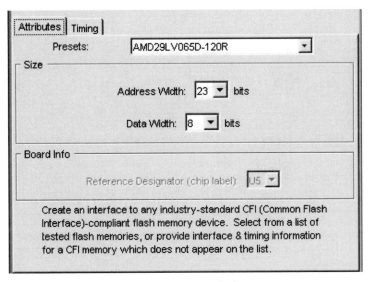

Figure 11-27. Adding the Flash memory controller.

    2. We make sure that AMD29LV065D-120R is selected in the Presets drop-down box. We leave any other options at their default settings.

    3. We click Finish to add the Flash memory interface. We are returned to the Altera SOPC Builder window.

    4. We right-click **cfi_flash_0** under Module Name and then choose Rename from the pop-up menu. We rename **cfi_flash_0** to **ext_flash** and press the Enter key.

External RAM Interface

To add the external RAM peripheral, **ext_ram**, we perform the following steps:

    1. We select IDT71V416 SRAM under Memory and click Add. The SRAM (two IDT71V416 chips) - sram_0 wizard appears (see Figure 11-28).

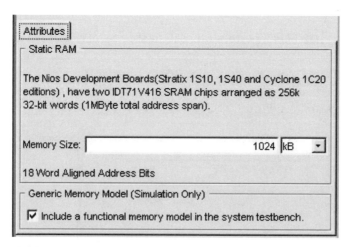

Figure 11-28. Adding the SRAM controller.

2. In the Attributes tab, we make sure the memory size is set at 1024 kB, and we click Finish. We are returned to the Altera SOPC Builder window.

3. We right-click sram_0 under Module Name, and then choose Rename from the pop-up menu. We rename sram_0 as ext_ram and press Enter.

JTAG UART Interface

The JTAG UART is used for communication between the board and the host machine and for debugging software running on the Nios II processor. To add the JTAG UART peripheral, jtag_uart, we perform the following steps:

1. We select JTAG UART under Communication and click Add. The JTAG UART - jtag_uart_0 wizard appears (see Figure 11-29).

2. We leave all options at their default settings.

3. We click Finish and are returned to the Altera SOPC Builder window.

4. We right-click jtag_uart_0 under Module Name, then choose Rename from the pop-up menu. We rename jtag_uart_0 as jtag_uart and press the Enter key.

External RAM Bus (Avalon Tri-State Bridge)

For the Nios II system to communicate with memory external to the FPGA on the development board, we must add a bridge between the Avalon bus and the bus or buses to which the external memory is connected. To add the Avalon tri-state bridge, ext_ram_bus, we perform the following steps:

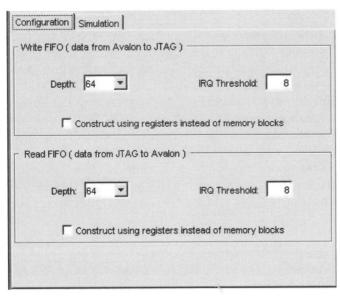

Figure 11-29. Adding the UART device.

1. We select Avalon Tri-State Bridge under Bridges and click Add. The Avalon Tri-State Bridge - tri_state_bridge_0 wizard appears (see Figure 11-30). We see that the Registered option is turned on by default.

2. We click Finish to return to the Altera SOPC Builder window.

3. We right-click tri_state_bridge_0 under Module Name, and then choose Rename from the pop-up menu.

4. We rename tri_state_bridge_0 to ext_ram_bus and then press Enter.

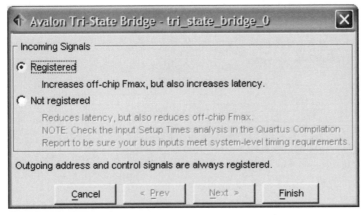

Figure 11-30. Adding the Avalon interface.

Adding the Hardware Process Module (img_arch)

We will now add the img_arch module, which implements the ImageFilter-DMA hardware process. To do this, we double-click User Logic under Avalon Modules in the System Contents pane. We then select the img_arch module and click Add (see Figure 11-31). We make sure the name assigned to the img_arch module item is user_logic_img_arch_module_0.

Next, we connect the new img_arch module to the shared Avalon Bus. To do this, we position our mouse pointer over the open circle at the intersection of the user_logic_img_arch_module_0 (avalon) column and the row named avalon_slave under ext_ram_bus, and then click the mouse button (see Figure 11-32). The open circle turns black to indicate a bus connection.

We have now finished adding the necessary peripherals.

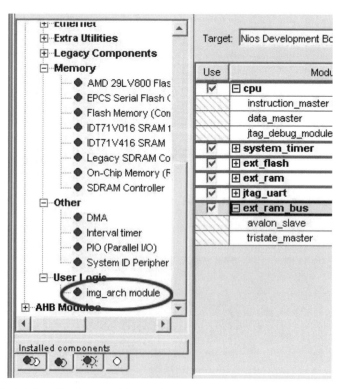

Figure 11-31. Adding the image filter logic.

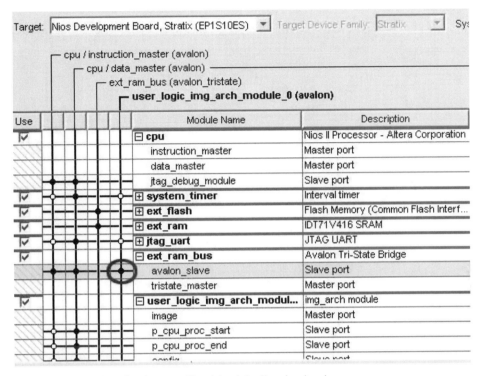

Figure 11-32. Connecting the image filter block to the Avalon bus.

Setting Additional CPU Settings

A few additional settings can be configured using the More "cpu" Settings dialog in SOPC Builder. For our platform, we will need only to change the Memory Module setting of the Reset Address Processor Function to ext_flash, as shown in Figure 11-33. Once this is done, our new Nios II platform is ready for system generation.

Generating the System

At this point we have set up and configured our new Nios II-based platform, including the hardware module generated for the image filter. We can now start the system generation process within SOPC Builder.

To do this, we click the System Generation tab in SOPC Builder and make sure the HDL option is selected. We click Generate to generate the system. This process, which takes several minutes, generates a series of log messages similar to those shown in Figure 11-34. When generation is complete, we exit SOPC Builder and return to Quartus.

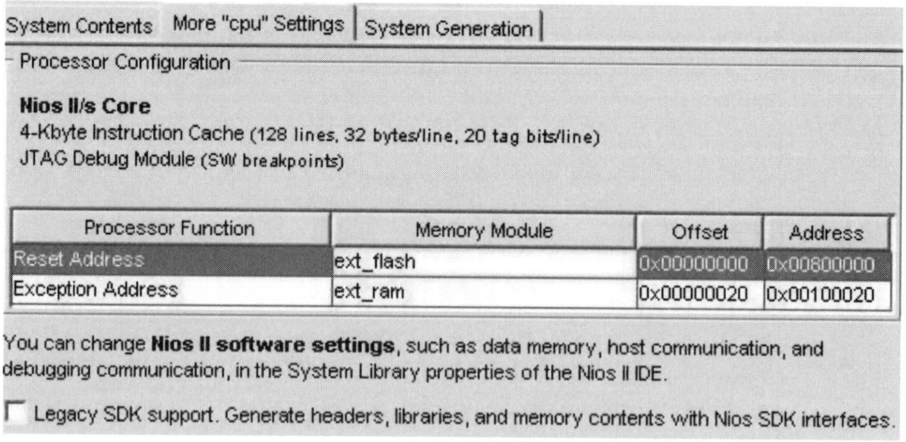

Figure 11-33. Changing the Memory Module setting.

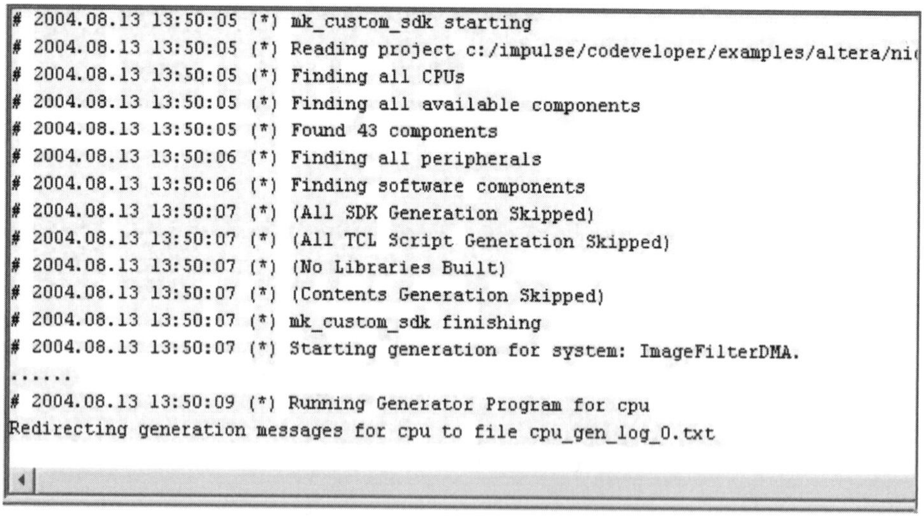

```
# 2004.08.13 13:50:05 (*) mk_custom_sdk starting
# 2004.08.13 13:50:05 (*) Reading project c:/impulse/codeveloper/examples/altera/ni
# 2004.08.13 13:50:05 (*) Finding all CPUs
# 2004.08.13 13:50:05 (*) Finding all available components
# 2004.08.13 13:50:05 (*) Found 43 components
# 2004.08.13 13:50:06 (*) Finding all peripherals
# 2004.08.13 13:50:06 (*) Finding software components
# 2004.08.13 13:50:07 (*) (All SDK Generation Skipped)
# 2004.08.13 13:50:07 (*) (All TCL Script Generation Skipped)
# 2004.08.13 13:50:07 (*) (No Libraries Built)
# 2004.08.13 13:50:07 (*) (Contents Generation Skipped)
# 2004.08.13 13:50:07 (*) mk_custom_sdk finishing
# 2004.08.13 13:50:07 (*) Starting generation for system: ImageFilterDMA.
......
# 2004.08.13 13:50:09 (*) Running Generator Program for cpu
Redirecting generation messages for cpu to file cpu_gen_log_0.txt
```

Figure 11-34. Building the SOPC Builder project.

Connecting the Generated System to FPGA Pins

Now we need to use the Quartus block diagram editor to connect the complete SOPC Builder-generated system (which includes the ImageFilterDMA hardware process module, the Nios II processor, and peripherals) to the pins on the FPGA. To do this, we open the block diagram file by selecting the Files tab in the project explorer window. We double-click on the entry for file ImageFilterDMA.bdf, which is located in the Device Design Files folder. The block diagram file appears, as shown in Figure 11-35.

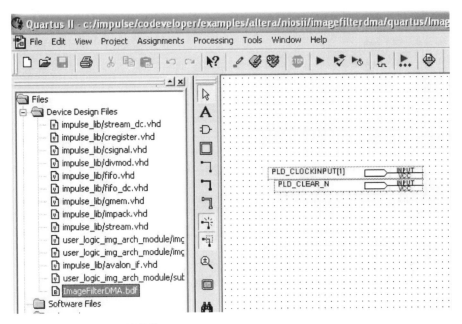

Figure 11-35. Opening the BDF file.

Now we add the block representing the SOPC Builder-generated system. We double-click anywhere in the open block diagram file to bring up the Symbol dialog and then open the Project folder and select the symbol named ImageFilterDMASystem, as shown in Figure 11-36. After we click OK, a symbol outline appears attached to the mouse pointer. We align the outline with the pins on the block diagram and click once to place the symbol, as shown in Figure 11-37.

FPGA Pin Assignment

The next step is to assign pins on the FPGA. Instead of assigning each individual pin (a tedious process), we can use a Tcl script that does the pin assignments for us. To run the Tcl script, we choose Tcl Scripts from the Quartus Tools menu. The dialog of Figure 11-38 appears. We select stratix-pins in the Project folder and click Run to assign the pins in the design.

The project is now (finally!) ready for bitmap generation and subsequent downloading. At this point we will probably want to save the project as a template for use with other Impulse C projects.

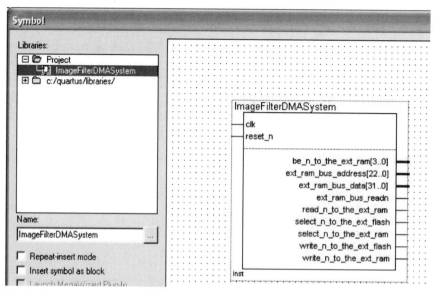

Figure 11-36. Creating a symbol for the image filter block.

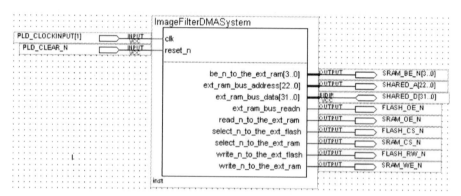

Figure 11-37. Aligning the pins.

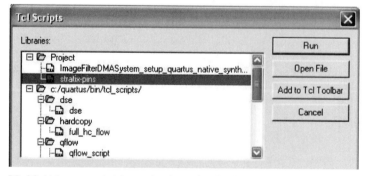

Figure 11-38. Using a script to assigning pins for the image filter block.

Generating the FPGA Bitmap

At this point we have successfully done the following:

- Generated and exported hardware and software files from the Impulse environment.

- Created a new Altera Quartus II project and used SOPC Builder to create a new Nios II-based platform.

- Imported the generated files to the Altera tools environment.

- Completed a block diagram and assigned pins for the selected FPGA device.

We are now ready to generate the bitmap and download the complete application to the target platform. This process is not complicated (at least in terms of your actions at the keyboard), but it can be time-consuming due to the large amount of processing that is required within the Altera tools.

First, we must apply some compiler settings related to pin assignment. We select Assignments -> Device from the Quartus menu to open the Device Settings dialog, as shown in Figure 11-39. We then click the Device & Pin Options button to open the Device & Pin Options dialog, as shown in Figure 11-40.

In the dialog that appears, we select the Unused Pins tab, then set the Reserve all unused pins option to As inputs, tri-stated (see Figure 11-41).

Next, we select the Dual-Purpose Pins tab and specify "Use as regular IO" for all dual-purpose pins listed (Figure 11-42).

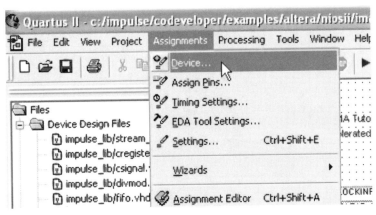

Figure 11-39. Setting device options in Quartus.

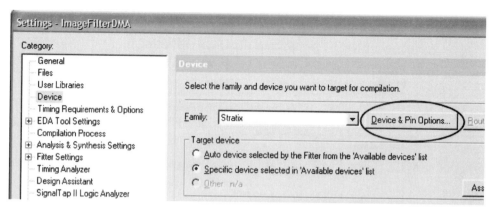

Figure 11-40. Setting device and pin options.

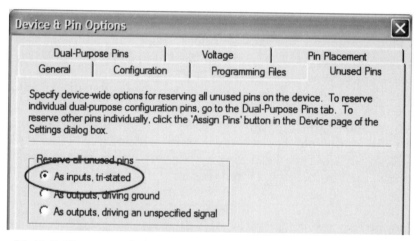

Figure 11-41. Setting unused pins as inputs.

Now we're ready to synthesize, download, and run the application. To generate the bitmap, we select Processing -> Start Compilation, as shown in Figure 11-43. This process requires around 30 minutes to complete on our development system.

During compilation, Quartus analyzes the generated VHDL source files, synthesizes the necessary logic, and creates logic that is subsequently placed and routed into the FPGA along with the Nios II processor and interface elements that were previously specified. The result is a bitmap file (in the appropriate Altera format) ready for downloading to the device.

When the bitstream has been generated, we select Tools > Programmer to open a new programming file. We select File -> Save As and save the chain description file as ImageFilterDMA.cdf, making sure the "Add file to current project" option is selected.

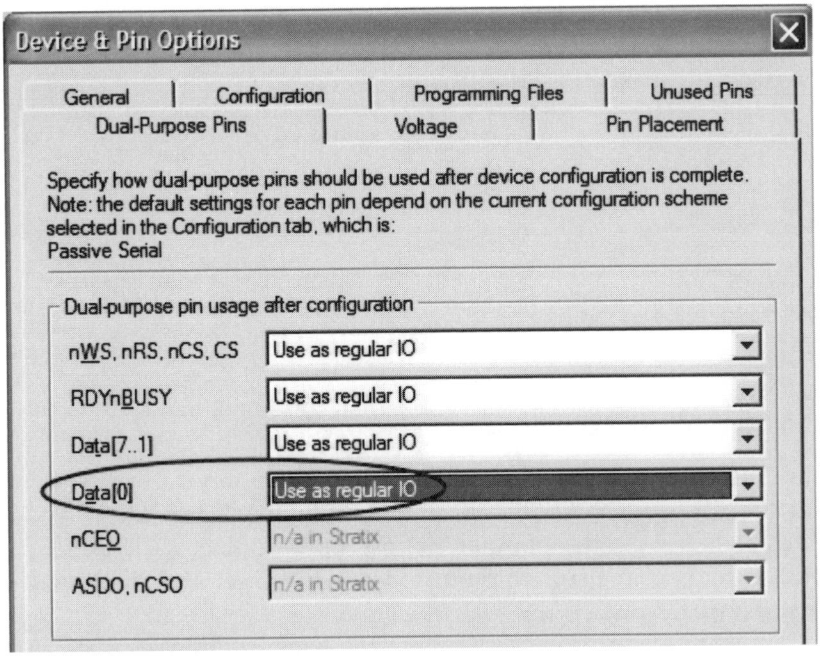

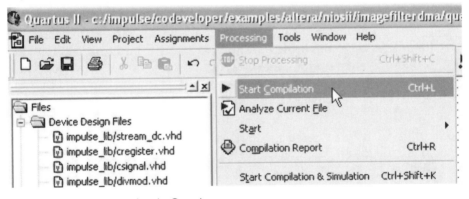

Figure 11-42. Setting dual-purpose pins.

Figure 11-43. Starting processing in Quartus.

The programming file ImageFilterDMA.sof is now visible in the programming window. We enable Program/Configure for the download file named ImageFilterDMA.sof, and make sure our programming hardware (in this case the ByteBlasterMV cable) is configured properly. We click Start (see Figure 11-44) to begin downloading the ImageFilterDMA.sof file to the target device. With the hardware programmed, we are ready to download and run the software application on the Nios II platform.

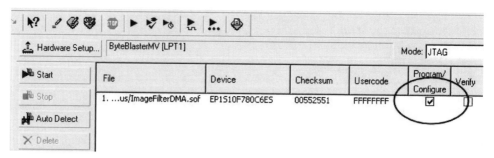

Figure 11-44. Programming the device.

Running the Test Application on the Platform

In the previous step, we programmed the FPGA device with the design we created in Quartus and SOPC Builder. Now we will use Altera's Nios II IDE (which is based on the popular Eclipse open-source development environment) to compile the software portion of the project and run it on the development board.

We begin this process by starting the Nios II IDE. We create a new Nios II project to manage the ImageFilterDMA software files. After we select File -> New -> Project, the wizard shown in Figure 11-45 appears. We select Altera Nios II and C/C++ Application, as shown, and click Next.

On the next page we select the project path, target hardware, and project template, using the Browse buttons to locate the appropriate Path and SOPC Builder System options, as shown in Figure 11-46. The project path is set to the directory containing the software files that were exported by the Impulse tools. We click Finish to create the new project.

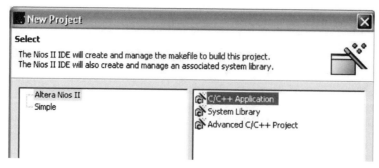

Figure 11-45. Creating a new software project.

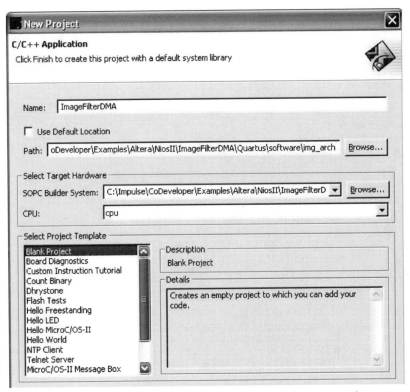

Figure 11-46. Creating the new Nios II project.

Two new projects (ImageFilterDMA and ImageFilterDMA_syslib) now appear in the C/C++ Projects window in the Nios II IDE, as shown in Figure 11-47. We can now build the project by right-clicking the ImageFilterDMA project and selecting Build Project from the menu that appears (see Figure 11-48). The IDE builds the ImageFilterDMA_syslib system library, which includes a driver for the Impulse C hardware module created by the Impulse tools, along with the application software code in the ImageFilterDMA project.

Once the software has finished building, we are ready to run the application in hardware. We right-click the ImageFilterDMA project and select Run As -> Nios II Hardware. We see printed output in the Console window (see Figure 11-49) that lets us verify the output of the hardware against some expected results, which in this test are statically-defined values as shown.

A final note about this example...

If you have been following along with this image filter example, it might have occured to you that the design has a fatal flaw: What happens if the software test application attempts to stream a second set of image pixel values into the process, as would be the case if it was processing a sequence of video frames? In fact, as written, this image filter will accept only one image, after which it will shut down, closing its input and output streams forever.

This is the most useful behavior for performing a simple functional test, but for an actual hardware process you will almost certainly want the process to remain alive, handling inputs until the system is powered down. To allow this, you simply add an outermost infinite loop to each of your processes, as follows:

```
void filter_run(co_stream r0, co_stream r1, co_stream r2, co_stream output_stream) {
    uint32 data,res, p00, p01, p02, p10, p11, p12, p20, p21, p22;
    uint16 d0;

    do {
        co_stream_open(r0, O_RDONLY, INT_TYPE(32));
        co_stream_open(r1, O_RDONLY, INT_TYPE(32));
        co_stream_open(r2, O_RDONLY, INT_TYPE(32));
        co_stream_open(output_stream, O_WRONLY, INT_TYPE(32));
        . . .


        . . .
        co_stream_close(r0);
        co_stream_close(r1);
        co_stream_close(r2);
        co_stream_close(output_stream);
        IF_SIM(break;)
    } while (1);
}
```

Notice that an **IF_SIM** statement has been used to indicate that the infinite **do-while** loop is to be used only when we are generating hardware. During desktop simulation, the **do** loop is terminated, allowing the test application to exit cleanly without looping endlessly.

When specifying an outer loop in this way, be sure that any required initializations (such as setting variables to starting values) are performed within the infinite loop, and verify that you are properly closing and reopening the streams with each loop iteration.

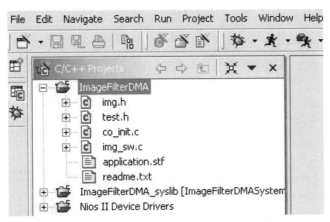

Figure 11-47. Adding software files to the new project.

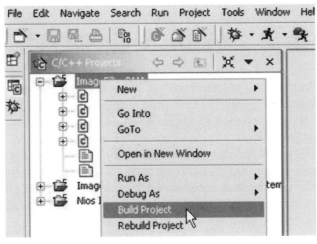

Figure 11-48. Building the project.

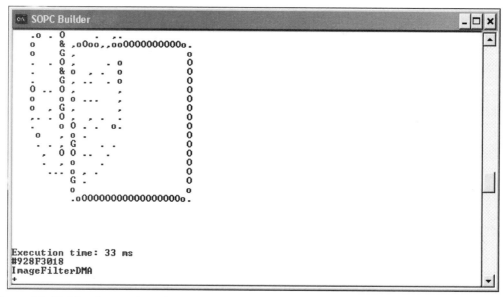

Figure 11-49. Nios II console output.

11.6 SUMMARY

In this chapter we have described how the performance of an application can be improved by applying techniques of system-level parallelism. We have also demonstrated how the complete application (consisting of the four hardware processes and one software test process) can be implemented on an Altera FPGA platform for the purpose of in-system testing.

These techniques for system-level partitioning and parallelism are widely applicable to FPGA-based applications and are a critical part of performance optimization. While it's tempting to think that a large C application (one originally written for a traditional processor) can be adapted with little change to a parallel platform such as an FPGA, the reality is that most algorithms, like the image filter described here, require some new thinking about how to achieve the best balance of parallel hardware to meet both performance and size constraints. Using a streaming programming model for partitioning can be an effective way to create such parallelism.

In the following chapter, we'll continue our exploration of this image filter example. You will see how a software test bench can be used in conjunction with an embedded operating system to create a complete, single-board, FPGA-based computing system.

CHAPTER 12

Combining Impulse C with an Embedded Operating System

Using a small-footprint operating system in conjunction with an embedded FPGA processor can dramatically increase the power and flexibility of FPGA-based computing. An operating system provides access to standard hardware devices (including network interfaces and Flash memory) as well as powerful application services, such as standardized file systems and multitasking capabilities.

By combining software running under the control of an operating system with custom-designed hardware accelerators residing in the FPGA, it's possible to create high-performance computing applications in which critical algorithms reside as dedicated hardware in the FPGA, while non-critical software components (or software test benches) reside in the embedded processor and take advantage of the high-level features provided by the operating system.

This chapter describes how to develop such an application, using the uClinux open-source operating system for demonstration purposes.

12.1 THE uCLINUX OPERATING SYSTEM

uClinux is based on the popular open-source Linux operating system. This operating system has been ported to many different embedded processors, including both the MicroBlaze embedded processor provided by Xilinx, and the Nios II processor from Altera.

The uClinux kernel is designed to be compact and simple and is appropriate for a wide variety of 32-bit, non-MMU processor cores. The tradeoff is that multitasking must be carefully managed to avoid memory conflicts.

The lack of memory management on uClinux target processors means that there is no protection against badly behaved processes, which can write anywhere in memory. There is also no support for virtual memory, which also requires memory management. For most embedded applications, however, these restrictions are not critical. In fact, the lack of memory management can provide performance advantages and opportunities for more direct control over hardware through memory-mapped I/O interfaces.

The port of the uClinux kernel to MicroBlaze was performed by Dr. John Williams, Research Fellow at the University of Queensland in Brisbane, Australia using the Memec V2MB1000 prototyping board, which is coincidentally the same board we used for the triple-DES test application of Chapter 9.

Combining uClinux and Impulse C

The lack of memory management in uClinux greatly simplifies the use of the MicroBlaze processor as an embedded test bench or as the primary processing element in a mixed hardware/software application. Because there is no limitation in uClinux on writing to specific memory-mapped addresses, the same code that implements an Impulse C process on MicroBlaze without an operating system (as was the case in Chapter 9) can be compiled with little or no change to operate in a multitasking environment under uClinux. Additionally, the lack of a driver layer results in very little impact on the potential throughput rate of data moving between the processor and the FPGA. If memory management existed on the processor, as is the case in PowerPC-based Virtex II Pro and Virtex-4 platforms, it would be necessary to create a device driver or other abstraction layer between the Impulse C software process and the FPGA-resident processes, resulting in substantial performance degradation.

There are many benefits to using uClinux (or a commercial operating system of similar capabilities) on an embedded FPGA processor. Because the operating system includes a wide variety of common peripheral interfaces as standard, it is relatively easy to create applications in which test data in the form of actual files, or from other standard interfaces, may be read using standard C functions, such as calls to **fopen**, **fread**, and **getc**, then subsequently streamed to the FPGA using the software-to-hardware streaming macros provided with Impulse C. Using a board such as the Memec V2MB1000 it is possible—even trivial—to create a network-enabled device in which the FPGA board becomes a computing node connected to other such nodes, and to desktop computers via TFTP, via an embedded web server or through the use of other such mechanisms.

Because Impulse C allows hardware processes to be expressed entirely in the C language, and the CoDeveloper tools automatically generate the necessary hardware-to-software interfaces, software engineers using uClinux and Impulse C have everything they need to generate and test complex, hardware-accelerated algorithms—entirely in the context of C-language programming.

12.2 A uClinux Demonstration Project

To demonstrate the feasibility of combining C-to-hardware programming with uClinux, we'll make use of the same type of application presented in Chapter 11, an image filter. As in the previous chapter, this image filter accepts pixel data on a data stream representing all pixels in a 512-by-512 image. It is implemented in Impulse C using two hardware processes that communicate with a software test bench (also written using Impulse C library calls) residing on the embedded processor.

During processing, pixels are sent by the software test bench (via the FSL bus interconnect provided with MicroBlaze) into the input stream of the first hardware process. In this process the pixels are cached, packaged into three streams (representing three streaming scan lines of the image), and fed in turn to the image filter process in a parallel pipeline with an effective maximum throughput of two clock cycles per pixel. The resulting pixels are subsequently streamed to the filter's output and back to the software test bench. (Refer to Chapter 11 for details of this filter's behavior.)

In order to test this image filter with a variety of images, a software test bench has been written that reads image data from an input file (in TIFF format) and streams this data across the FSL interconnect, using Impulse C software macros, to the hardware column generator. After filtering, the processed pixels are read back in by the software test bench using the Impulse C non-blocking stream read function (co_stream_read_nb). The software test bench then writes the processed image to a new TIFF-format file containing the filtered image. The uClinux operating system supports the required file I/O operations, using a RAM disk provided in the uClinux platform.

To complete the test and allow images to be easily moved from a host PC to the Memec V2MB1000 board, a TFTP client, run from the uClinux console, communicates with a TFTP server run from the PC. The complete demonstration is diagrammed in Figure 12-1.

The remainder of this chapter demonstrates step-by-step how this test was accomplished, using the 6.3i release of the Xilinx tools.

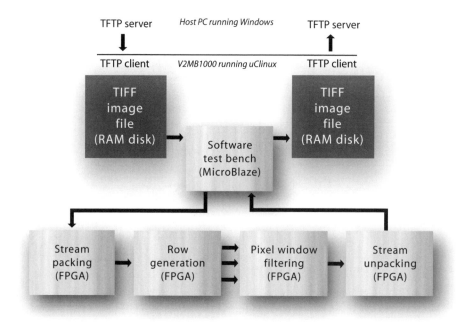

Figure 12-1. uClinux demonstration project system diagram.

The Impulse C Project Files

The files required for this project (see Figure 12-2) are as follows:

- Source files img_sw.c, img_hw.c, and img.h—These source files represent the complete application, including the main() function, consumer and producer software processes, and a single hardware process.

- Source file tiff.c—This source file includes routines for reading and writing TIFF-format data.

- A prebuilt Xilinx Platform Studio project that is properly configured for use with uClinux and includes the necessary uClinux kernel binaries.

The C code contained in img_hw.c describes two hardware processes representing the filter itself (prep_run and filter_run) along with two additional processes that pack and unpack the streaming pixel data. These processes will be generated as hardware (in the form of synthesizable HDL files), while the remaining software elements (in img_sw.c and tiff.c) will be implemented as software running on MicroBlaze.

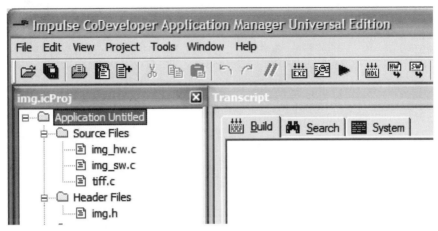

Figure 12-2. The Impulse C project files.

Building the Application for the Target Platform

In this step we will generate HDL for the image filter hardware processes and export that hardware (as well as the software files) into an EDK project directory. Before doing this, we must copy the sample EDK platform files for the prebuilt uClinux platform to a directory within the image filter sample project directory. (These files are provided by Impulse Accelerated Technologies and are available on the Impulse website at **www.ImpulseC.com**.)

Copying the Sample uClinux Platform Files

To set up the sample uClinux Platform files, we create a new subdirectory called uclinux_edk within the project directory. We copy the files supplied for the platform into the new directory, as shown in Figure 12-3.

Specifying the Platform Support Package

Returning to the Impulse tools, we will now set the options needed for generating hardware and exporting the generated files (both hardware and software) to the uClinux platform.

To specify a platform target, we open the Generate Options dialog as shown in Figure 12-4. Notice that we have set the platform name to Xilinx MicroBlaze FSL uClinux and we have set the Hardware and Software Export Directories to uclinux_edk.

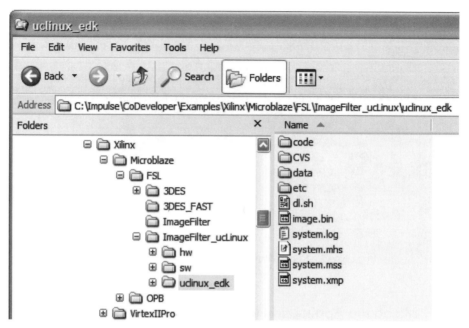

Figure 12-3. Copying the pre-built uClinux MicroBlaze platform.

Building the Image Filter Hardware

To generate the image filter hardware, we select Generate HDL from the Project menu, as shown in Figure 12-5. A transcript window opens, showing us the progress of the hardware generation.

Exporting the Software and Hardware

CoDeveloper generates hardware and software output files in the build subdirectories specified in the Generate Options dialog (see Figure 12-4). We can view the generated files by navigating to the hw and sw subdirectories of the project directory. To move the generated files into the EDK project directory (which in this case is the uclinux_edk subdirectory created earlier), we will use the Software Export and Hardware Export feature of the Impulse tools.

We export the software files to the uclinux_edk directory by selecting Export Generated Software from the Project menu, as shown in Figure 12-6. We then export the hardware files to the uclinux_edk directory by selecting Export Generated Hardware, as shown in Figure 12-7.

We now have all the files needed to build the platform, including the image filter hardware and software. The remaining steps make use of the Xilinx EDK (Platform Studio) tools.

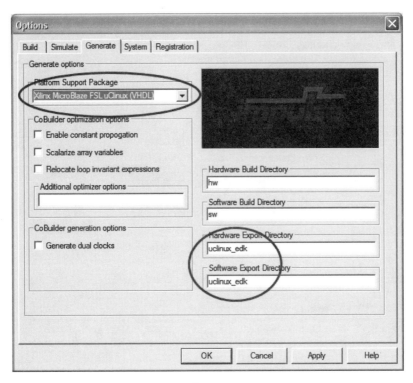

Figure 12-4. Setting HDL generation options.

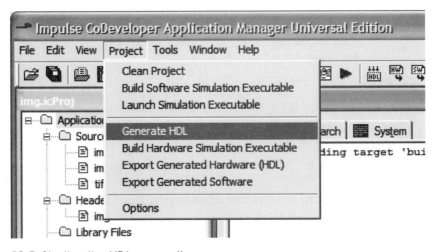

Figure 12-5. Starting the HDL generation process.

Figure 12-6. Exporting the software components.

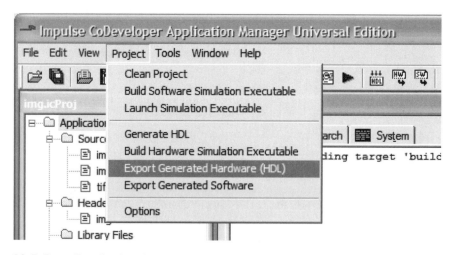

Figure 12-7. Exporting the hardware components.

Modifying the Sample uClinux Platform

We now move into the Xilinx tools environment. We open the prebuilt sample project by starting Xilinx Platform Studio and selecting Open Project from the File menu and navigating to the uclinux_edk directory (see Figure 12-8) to find the project file named system.xmp.

Now we will modify the sample project to add and connect the image filter hardware block that was generated in the previous step.

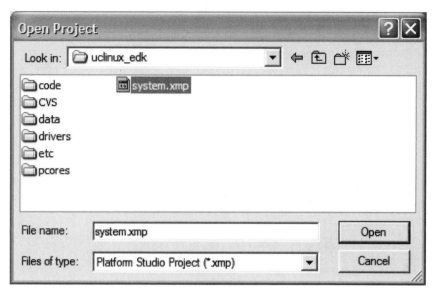

Figure 12-8. Opening the Platform Studio project.

Configuring the Platform with the Image Filter IP Core

The first step in modifying the standard uClinux platform is to add sufficient FSL links for the image filter's stream I/O. The image filter has just two stream interfaces (the pixel filter input and the pixel filter output), so we will need to add two FSL connections to the platform. We do this by selecting the Add/Edit Cores item from the Platform Studio Project menu, selecting the Parameters tab, and then selecting the IP instance named microblaze_0. We find a parameter named C_FSL_LINKS in the list of parameters on the left side of the dialog and we increase the value specified for this parameter by 2 to accomodate the two streams used by the image filter. For example, we change the C_FSL_LINKS parameter from a value of 1 to a value of 3, as shown in Figure 12-9.

The next step is to add the image filter as a peripheral. We do this by selecting the Peripherals tab and finding the entry for fsl_img in the list of IP blocks on the right side of the dialog. This is the hardware block that was exported to the EDK project by the Impulse tools when we invoked the Export Hardware menu item.

We click the Add button to add one instance of the fsl_img block to the platform, as shown in Figure 12-10.

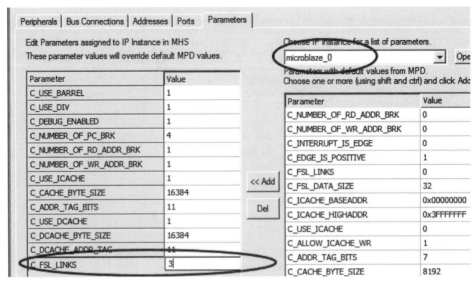

Figure 12-9. Adding FSL links for the image filter data streams.

Making the FSL Connections

Now we need to connect the new peripheral to the MicroBlaze processor via the FSL links. To do this, we select the Bus Connections tab and find the entry for fsl_v20_v1_00_b in the bus selection. We click the Add button twice to add two FSL connections to the platform. We then use the connection matrix (on the left side of the Bus Connections dialog) to connect the microblaze_0 mfsl1 master to the fsl_img_0 mfsl0 slave, and the microblaze_0 sfsl2 slave to the fsl_img_0 mfsl1 master, as shown in Figure 12-11. This establishes the FSL stream connections between the MicroBlaze processor and the image filter hardware block.

Connecting the Image Filter Clock and Reset Lines

Now that we have the image filter peripheral connected to the MicroBlaze processor via the FSL bus, we need to add clock and reset connections for the FSL links and for our new peripheral. In this example, we are using a single clock for both the MicroBlaze core and the FPGA logic.

To connect the required clock and reset lines, we select the Ports tab of the Add/Edit Cores dialog and add the fsl_img_0 clock port (clk), as well as the clock and reset ports for the fsl_v20_0 and fsl_v20_1 blocks. We verify that the clock ports are all connected to Net Name sys_clk, and we change the SYS_Rst ports so they connect to net_gnd as shown in Figure 12-12.

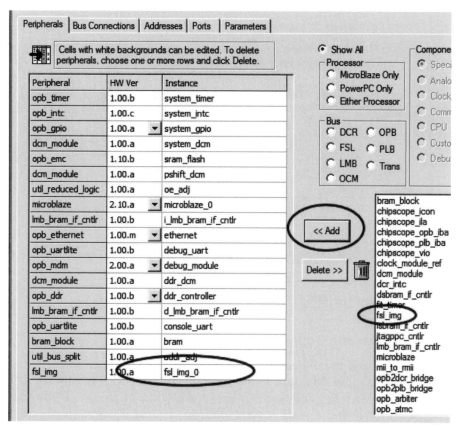

Figure 12-10. Adding the generated image filter peripheral.

We have now finished connecting the image filter block to the Micro-Blaze core and configuring the platform. We click the OK button to save the changes and exit the Add/Edit Cores dialog.

Generating the FPGA Bitmap

We are now ready to generate the bitstream and download it to the FPGA. To generate the bitstream, we select Generate Bitstream from the Tools menu (see Figure 12-13). The Xilinx software performs a number of steps, including logic synthesis and place-and-route, resulting in an FPGA bitmap file.

To download the bitmap file to the board, we connect the JTAG cable and select Download from the Tools menu (see Figure 12-14).

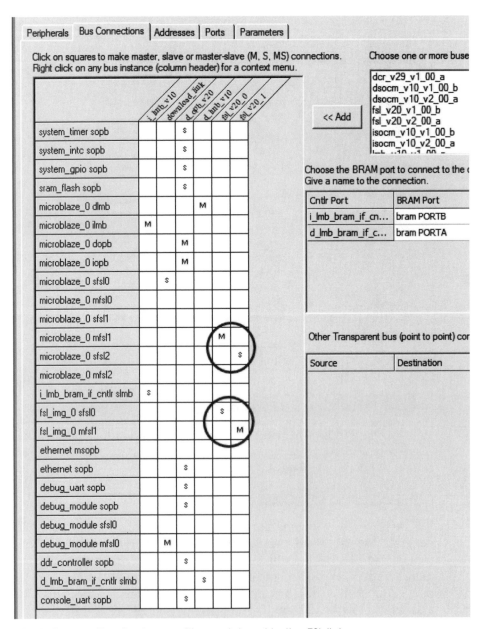

Figure 12-11. Connecting the image filter peripheral to the FSL links.

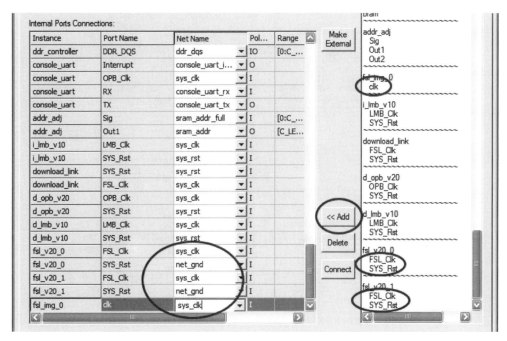

Figure 12-12. Connecting the image filter clock and reset ports.

Figure 12-13. Generating the FPGA bitmap.

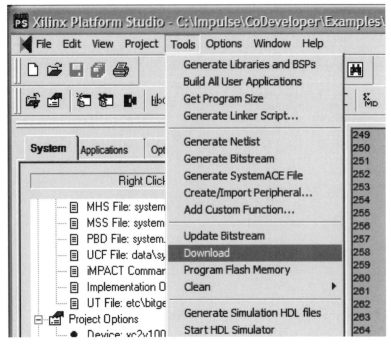

Figure 12-14. Downloading the FPGA bitmap.

After downloading is complete, the FPGA is configured with all the necessary components for use with uClinux, including the MicroBlaze processor core, peripheral devices, and the image filter IP core. The next steps are to download the uClinux kernel image and the image filter test application.

Downloading the Kernel Image

The FPGA is now programmed and ready to accept a MicroBlaze software image, which in the case of this example will be the uClinux kernel image. The uClinux kernel will be downloaded using the serial port located on the V2MB1000 main board, while a second serial cable will be attached to the P160 expansion module and used for a console. To open the console connection, we start a Hyperterminal session on a second available COM port of our development machine and connect to the V2MB1000 board via the P160 expansion module. We give the Hyperterminal session the following properties:

Speed: 115200

Bits: 8

Parity: None

Stop bits: 1

Flow control: None

These settings are shown in Figure 12-15. Note that this is not the COM port that will be used to download the uClinux kernel. On our development system, COM1 is used for the kernel download and is configured automatically by the kernel download script.

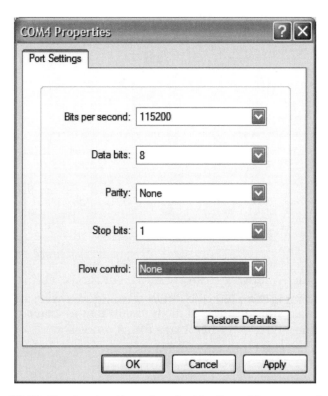

Figure 12-15. Configuring Hyperterminal for the uClinux console.

To download the kernel image, we open a Xilinx Xygwin shell window and use the cd command to change directories to the uclinux_edk subdirectory created earlier. We execute the following command to download the uClinux kernel image over COM1:

./dl.sh

This command script, supplied with the uClinux for MicroBlaze distribution, causes the binary image of the operating system to be downloaded to the MicroBlaze processor via the serial port. This takes a few moments to complete and generates messages similar to those shown in Figure 12-16.

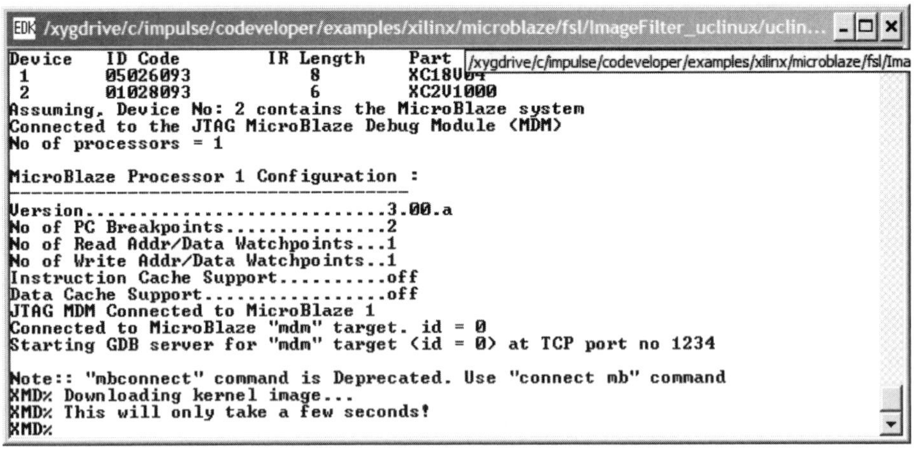

Figure 12-16. Downloading the uClinux kernel to the V2MB1000 board.

When the kernel download process is complete, we see a login prompt on the uClinux console window. We can log into the uClinux session with the "root" username (no password is required), as shown in Figure 12-17. We are now communicating directly with the uClinux operating system running on the embedded MicroBlaze FPGA processor.

Building and Testing the Image Filter Software

At this point, we have successfully configured our FPGA board as a single-board, network-enabled computer running the uClinux operating system. The next step is to download and run the image filter sample application. This test application reads data from TIFF format files, processes the data pixel-by-pixel using the hardware image filter, and writes the resulting image data to a TIFF file using the RAM disk accessible from uClinux. Performing this test, of course, requires that we move sample image files from our host development machine to the uClinux file system running on the MicroBlaze. We also need to move the necessary software application (the software side of our complete Impulse C application) to MicroBlaze. Both of these operations are accomplished using TFTP (Trivial File Transfer Protocol). First, however, we need to compile the Impulse C software application, which is represented by the source file img_sw.c and the automatically generated co_init.c.

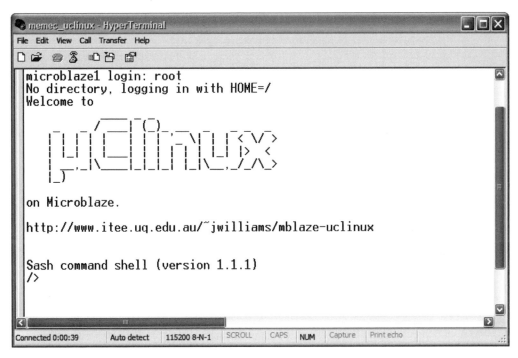

Figure 12-17. The uClinux login screen.

To compile the sample application, we return to the Xygwin shell window and type the following commands:

```
cd code
make FSL=1
```

The **make** command builds the Impulse C software application, named **img**, that will run on the MicroBlaze processor (see Figure 12-18). The makefile used in this build process, along with other supporting files, is automatically generated by the Impulse C compiler because we specified the MicroBlaze uClinux platform support package when setting the Impulse hardware generation options for the project (see Figure 12-4).

As described previously, the resulting test application reads data from a TIFF format file, processes that file using the FPGA hardware previously generated, and writes the results back to another TIFF format file.

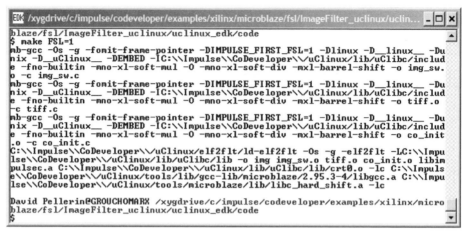

Figure 12-18. Building the Impulse C software application.

Using TFTP to Transfer Files to the Board

We will now copy the compiled img program and a test image file to the platform using TFTP. To do this, we must first start a TFTP server on our host machine, or copy the files to any operational TFTP server. (Freeware TFTP servers are widely available.) For this example we will start a TFTP server on the host PC, which has a network address of 192.168.0.3.

For simplicity, we set the base directory of the TFTP server to the directory containing the img executable file and the TIFF file to be copied.

Returning to the uClinux console (a Hyperterminal session connected to the P160 COM port of the V2MB1000 board via COM4), we type the following commands:

```
cd /tmp
tftp -g -r peppers.tiff 192.168.0.3
tftp -g -r img 192.168.0.3
```

Note that the /tmp directory on uClinux is a writable RAM disk. We now have the img program on our uClinux system, along with a sample image file.

Running the Image Filter Program

To run the img program on the board, we enter the following commands at the uClinux console:

chmod 777 img
./img peppers.tiff results.tiff

The complete set of commands is shown in Figure 12-19. Note that the chmod command is necessary to set the img file to an executable mode. When executed, the img program prints a message and processes the input file (peppers.tiff) to create an output file named results.tiff. The processing runs quickly, in just a few seconds. (If the process does not finish running, this indicates that an error has been made somewhere in the hardware generation process or—more likely—we have made an error in the configuration of the platform or FSL bus connections.)

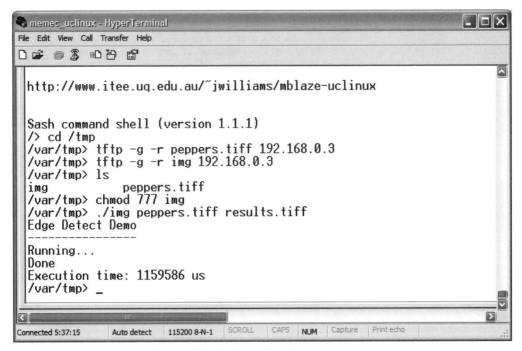

Figure 12-19. Running the image filter test application.

We have now successfully generated a MicroBlaze-based uClinux platform on the V2MB1000 board, combined this platform with the image filter hardware accelerator, and downloaded and run a test application. To view the results of the image filtering, we use TFTP to upload the results.tiff file to our host PC.

The original and filtered images are shown in Figures 12-20 and 12-21.

Figure 12-20. The test image before processing.

Figure 12-21. The test image after processing.

12.3 SUMMARY

This chapter has described how to use the uClinux operating system, running on a Xilinx MicroBlaze processor, to support highly capable software test benches for custom hardware modules developed using Impulse C.

There are many other potential applications for uClinux and Impulse C, including the ability to describe multiple (threaded) Impulse C software processes and to create highly reconfigurable systems consisting of multiple soft processor cores, each running uClinux in combination with Impulse C hardware processes. Research is continuing into such multiprocessor, mixed software/hardware applications, which appear to be limited only by our imaginations.

CHAPTER 13

Mandelbrot Image Generation

The preceding chapters have shown how FPGAs can be combined with traditional processors to create highly parallel, high-performance computing platforms for direct hardware acceleration of key algorithms. You have seen as well how a streaming programming model can be used to exploit parallelism at the system level while at the same time allowing modern compilers to handle the automatic generation of parallelism at the level of individual C statements and for inner code loops.

The combination of automated generation of process-level hardware and system-level programmability provides system designers with a powerful, efficient way to create applications for programmable hardware. However, it does introduce new complexities to the programming problem—complexities that grow in proportion to the size of the application and the number of component processes.

This chapter presents a final example that allows us to demonstrate some key ideas related to larger parallel applications. The example we will present is a Mandelbrot set generator, which is one type of fractal model generator. Fractal models are used in many supercomputing applications. They are useful for predicting systems that demonstrate chaotic behavior (such as might be observed when modeling the weather, genetic mutations, or the formation of planets). We use the Mandelbrot set here because the underlying calculation is easy to understand and yet still demonstrates the characteristics of other, possibly more complex fractal objects. More importantly, this example demonstrates a system in which a potentially large number of parallel calculations must be performed, but for which the calculations themselves vary enormously in terms of the time required to complete. This presents us with a *load balancing* challenge, which is the primary topic of this chapter.

13.1 Design Overview

Simply stated, fractals are infinitely complex objects with dimensions that are expressed as complex numbers. These types of objects were discovered by mathematicians around 1870, but it wasn't until Benoit Mandelbrot found (in the mid-1970s) a common mathematical relationship between the fractal objects discovered to date that the field of fractal geometry was born. One of Mandelbrot's most important contributions was to discover that fractal geometry could describe many real-world objects (which consist of both order and chaos) far better than could traditional mathematical models. Mandelbrot found a way to describe many of the structures and phenomena found in nature as an infinite set of fractal models. Such models now form the basis of weather forecasting as well as providing a basis for more mundane uses such as data compression.

Fractals are generated by performing repeated calculations on complex numbers (those with both real and imaginary components). The same formulae are repeated over and over (perhaps thousands or millions of times), with the results of one step being fed back into the next step.

The Mandelbrot Set

The Mandelbrot set is the best-known fractal object and was first described by Benoit Mandelbrot in 1977. This object is created by iterating a simple formula over an X-Y plane of complex numbers. For each complex number C in the plane, the algorithm iterates on the following basic calculation:

$$Z_0 = 0$$
$$Z_{n+1} = Z_n^2 + C$$

This formula iteratively takes the value Z_n, squares it, and adds the value of C (from an X-Y plane of complex values) to arrive at a new value for Z_{n+1}. This point is again squared, C is added, and so on for some maximum number of iterations or until Z diverges toward infinity, whichever comes first.

The key to fractals is that the value of Z (through the course of these many iterative calculations) demonstrates dramatically different behavior depending on the value of C, which is of course a complex number. If the iteration formula is applied to every point on the X-Y plane and colored according to the behavior of Z (one color, such as white, if it diverges to infinity, and another color, such as black, if it converges), you obtain a graph such as that shown in Figure 13-1.

Figure 13-1. A subregion of the Mandelbrot set.

This is a plot of the Mandelbrot set for a specific subregion. As this plot demonstrates, by zooming in to specific regions at the edges of the fractal objects that appear, we can find fascinating patterns. These patterns represent an infinite number of variations on the same basic theme and are similar to many of the complex (and yet seemingly random) patterns found in nature.

Accuracy Versus Processing Requirements

The image shown in Figure 13-1 was generated using double-precision floating-point calculations and a maximum number of iterations equal to 2000. This is the number of iterations at which we have decided—although not proven—that a given point will not diverge to infinity. Although a maximum iteration value of 2000 is adequate for drawing reasonably attractive pictures of fractal objects, it is probably insufficient for generating a high-quality result. This is because many of the points in a given X-Y plane diverge only after many more iterations have been performed. For a high level of accuracy, in fact, we may want to have the maximum iteration value set substantially higher, perhaps even in the millions.

Why such a high iteration value? While most points in a given plane (which in Figure 13-1 range from a value of negative .3965625 to negative .380470703125 in the X axis and positive .588052734375 to positive .604144531250 in the Y axis) diverge rather quickly, other points may not diverge until a very large number of iterations have been completed. This fact becomes more apparent as the image is magnified. If the number of iterations is not sufficiently high, the edges of the diagram will not be clearly defined, and some points appear (incorrectly) to converge, perhaps showing up as black or white points of "noise" in the generated image.

The higher the iteration count, the higher the accuracy of the fractal image. From this fact it should be obvious that generating high-quality fractal images is computationally expensive. The more detail you want to observe (and the more interesting the region of the plane you are examining), the more calculations that are required. This means that if the fractal image is generated just one pixel at a time in a serial fashion on a traditional processor (such as a PC), a high-quality image may require many hours—or potentially many days—to complete.

13.2 EXPRESSING THE ALGORITHM IN C

As with any application destined for FPGA acceleration, the first step is to model the application at a high level, using C language. In the case of the Mandelbrot set generation algorithm, we could write the relatively simple program shown in Figure 13-2. This program generates an image (in the form of a Windows bitmap format file) representing one region of the Mandelbrot set, as defined by the following declarations:

```
#define XL -0.396562500000
#define XH -0.380470703125
#define YL  0.588052734375
#define YH  0.604144531250
```

```
// Mandelbrot generation program, floating point version.

#include <windows.h>
#include <sys/types.h>
#include <sys/stat.h>
#include <stdio.h>
#include "mand.h"

#define XL -0.396562500000
#define XH -0.380470703125
#define YL  0.588052734375
#define YH  0.604144531250

void mandelbrot(double xmax, double xmin, double ymax, double ymin,
          double dx, double dy, double hdx, double hdy) {
  int max_iterations = MAX_ITERATIONS;
  double c_imag,c_real;
  int j, i, t, k, R, G, B;
  double result, tmp, z_real, z_imag;

  // 1. Open the BMP file and write the header information
  BITMAPFILEHEADER header;  /* Bitmap header */
  BITMAPINFO info;        /* Bitmap information */
  unsigned int pixelValue byteValue;
  const char * FileName = "Mandelbrot_float.bmp";
  FILE * outfile;
  outfile = fopen(FileName, "wb");
  if (outfile==NULL) {
      fprintf(stderr, "Error opening BMP file %s for writing\n", FileName);
      exit(-1);
  }
  header.bfType = 0x4d42; // BMP file
  header.bfSize = XSIZE * YSIZE + 54;
  header.bfOffBits = 54;
  info.bmiHeader.biSize = 40;
  info.bmiHeader.biWidth = XSIZE;
  info.bmiHeader.biHeight = YSIZE;
  info.bmiHeader.biPlanes = 1;
  info.bmiHeader.biBitCount = 24;
  info.bmiHeader.biCompression = 0;
  info.bmiHeader.biSizeImage = XSIZE * YSIZE;
  info.bmiHeader.biXPelsPerMeter = 11811;
  info.bmiHeader.biYPelsPerMeter = 11811;
  info.bmiHeader.biClrUsed = 0;
```

Figure 13-2. Mandelbrot generation, initial floating-point version. (*continues*)

```
      info.bmiHeader.biClrImportant = 0;
      if (fwrite(&header, 1, sizeof(BITMAPFILEHEADER), outfile) < 1) {
        fprintf(stderr, "Error writing BMP header for %s\n", FileName);
        exit(-1);
      }
      if (fwrite(&info, 1, sizeof(BITMAPINFO), outfile) < 1) {
        fprintf(stderr, "Error writing BMP info for %s\n", FileName);
        exit(-1);
      }
      // 2. Calculate the value at each point
      c_imag = ymax;
      for (j = 0; j < YSIZE; j++) {
        c_real = xmin;
        for (i = 0; i < XSIZE; i++) {
          z_real = z_imag = 0;
          // Calculate z0, z1, .... until divergence or maximum iterations
          k = 0;
          do {
            tmp = z_real*z_real - z_imag*z_imag + c_real;
            z_imag = 2.0*z_real*z_imag + c_imag;
            z_real = tmp;
            result = z_real*z_real + z_imag*z_imag;
            k++;
          } while (result < 4.0 && k < max_iterations);

          // 3. Map points to gray scale: change to suit your preferences
          B = G = R = 0;
          if (k != MAX_ITERATIONS) {
            R = G = G = k > 255 ? 255 : k;
          }
          putc(B, outfile); putc(G, outfile); putc(R, outfile);
          c_real += dx;
        }
        c_imag -= dy;
      }
      fclose(outfile);
    }

    int main(int argc, char *argv[]) {
      double xmin,xmax,ymin,ymax,dx,dy;

      xmin = XL; xmax = XH; ymin = YL; ymax = YH;
      dx = (XH-XL)/XSIZE; dy = (YH-YL)/YSIZE;
      mandelbrot(xmax,xmin,ymax,ymin,dx,dy,dx/2.0,dy/2.0);
      return(0);
    }
```

Figure 13-2. *continued*

These values could be passed into the program on the command line, but for simplicity we have hard-coded them as shown.

As written, the program consists of just two functions, main and mandelbrot. The main function simply calls the mandelbrot function with the predefined arguments representing the X and Y bounding values, which are assumed to be double-precision floating-point values in the range of positive 2 to negative 2. The main function also calculates scaling values for X and Y, based on the height and width (in pixels) of the desired image.

The mandelbrot function has three important sections that correspond to the following operations:

1. The output file (a BMP format file) is opened, and appropriate header information is generated. The program makes use of the Windows types **BITMAPFILEHEADER** and **BITMAPINFO**, which are further defined (for non-Windows platforms) in the file **mand.h**.

2. For each pixel in the desired output (which has a size of **HSIZE**-by-**YSIZE** pixels), the program calculates the value of the point. It does this by iterating on the same point until the value at that point either diverges to infinity (has a value greater than or equal to 4) or has exceeded the maximum number of iterations without yet diverging (indicating a likely convergence).

3. For each value generated in this way, the program assigns a color to the pixel based on the iteration value reached before the value diverged to infinity. For the sake of brevity, we have simply chosen a grayscale output, and have accomplished this by setting all three color values to the iteration value, up to a maximum of 255.

If you compile this program and run it, you obtain an image file very much like the image shown in Figure 13-1. This version of the application can now serve as our benchmark for the modifications to come.

13.3 CREATING A FIXED-POINT EQUIVALENT

The next step is to create a fixed-point equivalent of the same floating-point algorithm. This is important because floating-point math operations are inefficient when rendered in hardware, and also because the C-to-hardware compiler used to create this example does not support floating-point datatypes and corresponding operations.

Converting a floating-point algorithm to a fixed-point equivalent can be a time-consuming process, and the accuracy of the calculations will nearly always result in subtle (and in some cases extreme) differences in behavior. It's critical, then, to perform adequate software simulations to ensure that the fixed-point rendering of the results (including the results of intermediate calculations as appropriate) is of adequate precision for the algorithm in question. The general case of floating- to fixed-point conversion is beyond the scope of this book, but may often involve considerations of rounding (toward infinity and toward 0), saturation, and other factors. The amount of precision required by a given algorithm must also be balanced against the range of numbers to be processed.

For this algorithm we have chosen to use 32-bit fixed-point values, 24 bits of which represent the fractional part. This reflects the fact that the values we will be calculating fall within a narrow range in terms of whole numbers.

The Impulse C libraries include limited support for fixed-point operations, allowing the mandelbrot function to be rewritten as shown in Figure 13-3. (Note that the BMP file header generation section of the fix_mandelbrot function has been removed for brevity.)

13.4 CREATING A STREAMING VERSION

At this point we have created a fixed-point version of the algorithm that might be appropriate for compiling to an embedded processor. If we want to move this algorithm to an FPGA, we first need to do some additional work to allow the configuration data to be transferred from the test routine (an equivalent of the existing main function) to the image generator and to allow the resulting pixel data to be transferred back out to the test routine. To do this, we will make some minor changes to the fix_mandelbrot routine and more substantial changes to the test routine by introducing a producer/consumer module, as demonstrated in previous chapters.

The Impulse C Process

The modified Mandelbrot image generator function is shown in Figure 13-4. Notice that this version of the generator (which is now an Impulse C process) does not include the file I/O operations. These operations will instead be performed on the software side of the application.

Examining this process in more detail, you see the following:

- The process run function declaration, including the two streams config_stream and pixel_stream.

- Declarations for all local variables in the form of unsigned 32-bit integers.

```
void fix_mandelbrot(uint32 xmax, uint32 xmin, uint32 ymax, uint32 ymin,
                    uint32 dx, uint32 dy, uint32 hdx, uint32 hdy) {
  uint32 max_iterations = MAX_ITERATIONS;
  uint32 c_imag,c_real,two;
  int j, i, t, R, G, B;
  uint32 k, result, tmp, z_real, z_imag;

  // BMP file opened here, etc...

  two = 2*(1<<FRACBITS);   // FRACBITS defined as 24
  // Calculate points
  c_imag = ymax;
  for (j = 0; j < YSIZE; j++) {
    c_real = xmin;
    for (i = 0; i < XSIZE; i++) {
      z_real = z_imag = 0;
      // Calculate z0, z1, .... until divergence or max_iterations
      k = 0;
      do {
        tmp = FXMUL(z_real,z_real);
        tmp = FXSUB(tmp,FXMUL(z_imag,z_imag));
        tmp = FXADD(tmp,c_real);
        z_imag = FXMUL(two,FXMUL(z_real,z_imag));
        z_imag = FXADD(z_imag,c_imag);
        z_real = tmp;
        tmp = FXMUL(z_real,z_real);
        result = FXADD(tmp,FXMUL(z_imag,z_imag));
        k++;
      } while (result < (4*(1<<FRACBITS)) && k < max_iterations);

      // Map points to gray scale: change to suit your preferences
      B = G = R = 0;
      if (k != MAX_ITERATIONS) {
        R = G = G = k > 255 ? 255 : k;
      }

      putc(B, outfile); putc(G, outfile); putc(R, outfile);

      c_real = FXADD(c_real,dx);
    }
    c_imag = FXSUB(c_imag,dy);
  }
  fclose(outfile);
}
```

Figure 13-3. Mandelbrot generation, fixed-point version.

```
void mandelbrot(co_stream config_stream, co_stream pixel_stream) {
  co_uint32 xmax,xmin,ymax,ymin,dx,dy,hdx,hdy;
  co_uint32 B,G,R,BGR;
  co_uint32 i,j,k,t;
  co_uint32 c_imag,c_real,two,four;
  co_uint32 result,tmp;
  co_uint32 z_real,z_imag;

  two = FXCONST(2);
  four = FXCONST(4);

  co_stream_open(config_stream, O_RDONLY, UINT_TYPE(32));

  // Read in parameters
  while (co_stream_read(config_stream, &xmax, sizeof(co_uint32)) ==
                                                    co_err_none) {
    co_stream_read(config_stream,&xmin,sizeof(co_uint32));
    co_stream_read(config_stream,&ymax,sizeof(co_uint32));
    co_stream_read(config_stream,&ymin,sizeof(co_uint32));
    co_stream_read(config_stream,&dx,sizeof(co_uint32));
    co_stream_read(config_stream,&dy,sizeof(co_uint32));
    co_stream_read(config_stream,&hdx,sizeof(co_uint32));
    co_stream_read(config_stream,&hdy,sizeof(co_uint32));

    IF_SIM(printf("x: %f - %f\n",FX2REAL(xmin),FX2REAL(xmax));)
    IF_SIM(printf("y: %f - %f\n",FX2REAL(ymin),FX2REAL(ymax));)

    // Loop over region
    co_stream_open(pixel_stream, O_WRONLY, UINT_TYPE(24));
    c_imag = ymax;
    for (j=0; j<YSIZE; j++) {
      c_real = xmin;
      for (i=0; i<XSIZE; i++) {
        z_real = z_imag = 0;
        // Calculate point
        k = 0;
        do {
          tmp = FXMUL(z_real,z_real);
          tmp = FXSUB(tmp,FXMUL(z_imag,z_imag));
          tmp = FXADD(tmp,c_real);
          z_imag = FXMUL(two,FXMUL(z_real,z_imag));
          z_imag = FXADD(z_imag,c_imag);
          z_real = tmp;
          tmp = FXMUL(z_real,z_real);
          result = FXADD(tmp,FXMUL(z_imag,z_imag));
```

Figure 13-4. Mandelbrot generation, Impulse C version. (*continues*)

```
            k++;
        } while ((result < four) && (k < MAX_ITERATIONS));

        // Map points to gray scale: change to suit your preferences
        B = G = R = 0;
        if (k != MAX_ITERATIONS) {
          R = G = G = k > 255 ? 255 : k;
        }
        BGR = ((B<<16) & BLUEMASK) |
              ((G<<8) & GREENMASK) |
              (R & REDMASK);
        co_stream_write(pixel_stream,&BGR,sizeof(co_uint32));
        c_real = FXADD(c_real,dx);
      }
      c_imag = FXSUB(c_imag,dy);
    }
    co_stream_close(pixel_stream);
  }

  co_stream_close(config_stream);
}
```

Figure 13-4. *continued*

- Assignments (to variables named **two** and **four**) of constant fixed-point representations for the values 2 and 4, respectively. These constants are defined using the **FXCONST** macro, which is declared elsewhere (in mand.h) as FXCONST32(a,FRACBITS). Recall that FRACBITS has been defined for this application as 24, indicating the number of bits associated with the fractional part of a number.

- A call to **co_stream_open**, which opens the configuration stream. The configuration stream accepts seven input values that collectively define the region of the X-Y plane to be processed.

- A **co_stream_open** function call corresponding to the **pixel_stream** output stream. Notice that this stream has a width of 24 bits. This corresponds to the three 8-bit color values that will be calculated for each pixel in the generated image.

- A nested loop of size (**YSIZE** times **XSIZE**) that processes the image line-by-line and pixel-by-pixel to produce the desired pattern.

- An inner code loop (the **do** loop) performs the iterative calculations for each pixel using fixed-point versions of the add, subtract, multiply, and divide operations.

- After this **do** loop, color values are assigned based on the number of iterations required for that pixel.

- At the completion of the nested loops (the processing of the entire image), the output stream is closed, and a new set of configuration data is read, continuing the operation repeatedly until it has been detected that the configuration stream is closed.

The result of this process is a stream of pixel values that begin to appear on the process output stream (**pixel_stream**) after the seven input values have been read into the process input stream (**config_stream**).

We now have a process that can be compiled to hardware; if we also provide a software test bench, we have a complete application that may be simulated within a standard C environment or tested in-system using an embedded processor, as described in previous chapters. Before doing so, however, let's look at how we might improve this process by introducing more parallelism.

13.5 PARALLELIZING THE ALGORITHM

As we have described in previous chapters, there are two fundamental ways in which parallelism can be added to a C application that will be targeting FPGA hardware. We have seen how statement-level parallelism and system-level parallelism can together provide enormous speedups. Let's apply both statement-level and system-level parallism to this application. We'll begin by considering an effective partitioning strategy.

Partitioning the Problem

Because the Mandelbrot image generation algorithm performs iterative calculations on the same point, with no need to share data between points, it's natural to consider a partitioning strategy in which all of the computations for a given point are performed within a single process, which is then replicated some number of times to create a complete application. At its most extreme, we could imagine a system in which there is one dedicated processor for each point.

Although this approach might provide the highest possible performance, it is clearly impractical due to the huge amount of hardware required (640,000 processing elements, or PEs, for a relatively modest 800-by-800 image). In order to balance the desire for performance with the practical limits of the design size, we need to consider a few important points:

- For any given processor performing calculations on a given point, it is not known in advance how many iterations will be required to obtain a result for that point.

- Given the preceding point, we need to design a system in which there is minimal wasted/idle time. All the processing elements need to be kept busy, which implies some kind of master, controlling process.

- If we are generating a sorted output (such as would be required when streaming the outputs to a file), we will need some kind of consumer process that can properly arrange the results or store them intelligently in a random-access memory for later use.

The fact that the different point processors have no need to interact with one another gives us a great deal of flexibility in how points are allocated to processors. We might choose a scheme, for example, in which the X-Y plane is sliced into horizontal sections. Each of these sections is streamed to a different process, which then performs its calculations on each point in the section in sequence to return an ordered result. This seems at first to be a good choice, until you consider that in many regions of the Mandelbrot set entire sections of the plane consist entirely of points requiring a greater-than-average number of iterations, while other horizontal slices (consisting of a few or perhaps a hundred or more lines) may consist of points that diverge rapidly and therefore require a minimum amount of computation. Because of this irregular characteristic of the data, the potential exists for an imbalance in the processing requirements; we will waste computing resources if we do not choose a partitioning strategy that smooths out such irregularities.

A better method is to interlace the lines such that each processor handles every nth line, where n is the number of processors. This method is most efficient, of course, when n divides evenly into the number of lines in the region. For the sake of demonstration we'll choose a relatively small value for n (the number of processors), just three processing elements, or *worker processes*, to perform the operations. The method we will demonstrate to synchronize the activities of these three processes is easily scalable to much larger arrays of processors.

Output Synchronization

The method we will choose, that of creating three worker processes that handle the processing of alternate lines, will allow the interlaced regions of the image to be processed and their outputs ordered through the use of a randomly accessed image buffer—a shared memory. Alternate methods of synchronization are also possible, including the use of tokens (implemented as either signals or streams) indicating the correct order of output.

The partitioned and load-balanced version of the Mandelbrot image generator, consisting of two processes (a master controller and a worker process that will be instantiated three times), is shown in Figures 13-5 and 13-6. The process shown in Figure 13-5 represents the worker process and is named **genline**. This process includes the following:

- The process declaration for **genline**, including pointers to the image buffer shared memory (**img**) and two streams, **req** and **ack**. These two streams are used to receive configuration information (the pixel value range) and to communicate to the master process that a given scan line has been completed. (Note that a signal could have been used in place of a stream for the **ack** communication channel. As you'll see, however, the existence of a nonblocking stream read function makes the use of a stream somewhat more convenient for this particular application.)

- Local variable declarations (of type **co_uint32**) for the pixel value calculations, loop indices, and so on.

- A local array of type **co_uint8** (and size **XSIZE X 3**) that will be used to store the pixel color values prior to writing them to the shared memory.

- Assignments for the two static values **two** and **four**, as described previously.

- A **co_stream_open** function call for the **ack** stream. Note that this is an output stream, as indicated by the **O_WRONLY** parameter.

- A **co_stream_open** function call for the **req** stream. Note that this is an input stream, as indicated by the **O_RDONLY** parameter.

- An outer **while** loop that reads configuration information (sent by the controller process) from stream **req** for the line to be processed.

- A **for** loop that iterates **XSIZE** times to process the line, one pixel position at a time.

- A inner **do** loop that iteratively performs the required calculations for each pixel position. (Note the use of the **PIPELINE** and **StageDelay** pragmas in this inner code loop. They accelerate the innermost loop by parallelizing and pipelining the critical calculations.)

- After the **do** loop, a set of **if** statements are used to determine the intensities of the three colors red, green, and blue for the pixel just processed. These values are assigned to **R**, **G**, and **B**, respectively.

- After the colors have been determined, the R, G, and B values are stored in the temporary **line** buffer.

```
void genline(co_memory img, co_stream req, co_stream ack) {
    co_uint32 xmax,xmin,dx;
    co_uint32 B,G,R,BGR;
    co_uint32 i,k,t;
    co_uint32 c_imag,c_real,two,four;
    co_uint32 result,tmp;
    co_uint32 z_real,z_imag;
    co_uint32 offset,outpos;
    co_uint8 line[XSIZE*3];

    two = FXCONST(2);
    four = FXCONST(4);

    co_stream_open(ack,O_WRONLY,UINT_TYPE(8));

    // read in config
    co_stream_open(req, O_RDONLY, UINT_TYPE(32));
    while (co_stream_read(req,&xmax,sizeof(co_uint32))==co_err_none) {
        co_stream_read(req,&xmin,sizeof(co_uint32));
        co_stream_read(req,&dx,sizeof(co_uint32));
        // signal ready
        co_stream_write(ack,&tmp,sizeof(co_uint8));

        // read in parameters
        while (co_stream_read(req,&c_imag,sizeof(co_uint32)) == co_err_none) {
            co_stream_read(req,&offset,sizeof(co_uint32));

            c_real = xmin;
            outpos = 0;
            for (i=0; i<XSIZE; i++) {
                z_real=z_imag=0;
                // Calculate point
                k = 0;
                do {
#pragma CO PIPELINE
#pragma CO set stageDelay 96
                    tmp = FXMUL(z_real,z_real);
                    tmp = FXSUB(tmp,FXMUL(z_imag,z_imag));
                    tmp = FXADD(tmp,c_real);
                    z_imag = FXMUL(two,FXMUL(z_real,z_imag));
                    z_imag = FXADD(z_imag,c_imag);
                    z_real = tmp;
                    tmp = FXMUL(z_real,z_real);
```

Figure 13-5. Worker processing for a single line of the Mandelbrot image. (*continues*)

```
            result = FXADD(tmp,FXMUL(z_imag,z_imag));
            k++;
        } while ((result < four) && (k < MAX_ITERATIONS));

        // Map points to gray scale: change to suit your preferences
        B = G = R = 0;
        if (k != MAX_ITERATIONS) {
            R = G = G = k > 255 ? 255 : k;
        }
        line[outpos++] = B;
        line[outpos++] = G;
        line[outpos++] = R;
        c_real=FXADD(c_real,dx);
    }
    co_memory_writeblock(img,offset,line,XSIZE*3);
    co_stream_write(ack,&tmp,sizeof(co_uint8));
    }
    co_stream_close(req);
    co_stream_open(req, O_RDONLY, UINT_TYPE(32));
  }
  co_stream_close(ack);
}
```

Figure 13-5. *continued*

- After an entire line has been processed, the result (in the line array) is written to the shared memory through the use of the co_memory_writeblock function.

- After the line has been written to shared memory, a message is sent to the controller via co_stream_write.

This process is replicated (instantiated) three times and operates under the control of a second hardware called ctrl, which is shown in Figure 13-6.

Examining the ctrl function in detail, you find the following:

- A declaration of the process that includes references to the config_stream input, a done signal, three req streams (one for each worker process), and three ack streams (again, one for each worker process).

- An outer code loop that reads configuration data from the config_stream input stream. This configuration data defines the area of interest (the subregion of the Mandelbrot set) as a set of eight 32-bit fixed-point values.

```
void ctrl(co_stream config_stream, co_signal done,
      co_stream req0, co_stream req1, co_stream req2,
      co_stream ack0, co_stream ack1, co_stream ack2) {
  co_uint32 xmax,xmin,ymax,ymin,dx,dy,hdx,hdy;
  co_uint32 i,j;
  co_uint32 c_imag,c_real;
  co_uint32 offset,outpos;

  co_stream_open(config_stream,O_RDONLY,UINT_TYPE(32));
  co_stream_open(ack0,O_RDONLY,UINT_TYPE(8));
  co_stream_open(ack1,O_RDONLY,UINT_TYPE(8));
  co_stream_open(ack2,O_RDONLY,UINT_TYPE(8));

  // Read in parameters
  while (co_stream_read(config_stream,&xmax,sizeof(co_uint32))
                                      == co_err_none) {
    co_stream_read(config_stream,&xmin,sizeof(co_uint32));
    co_stream_read(config_stream,&ymax,sizeof(co_uint32));
    co_stream_read(config_stream,&ymin,sizeof(co_uint32));
    co_stream_read(config_stream,&dx,sizeof(co_uint32));
    co_stream_read(config_stream,&dy,sizeof(co_uint32));
    co_stream_read(config_stream,&hdx,sizeof(co_uint32));
    co_stream_read(config_stream,&hdy,sizeof(co_uint32));

    co_stream_open(req0,O_WRONLY,UINT_TYPE(32));
    co_stream_write(req0,&xmax,sizeof(co_uint32));
    co_stream_write(req0,&xmin,sizeof(co_uint32));
    co_stream_write(req0,&dx,sizeof(co_uint32));

    co_stream_open(req1,O_WRONLY,UINT_TYPE(32));
    co_stream_write(req1,&xmax,sizeof(co_uint32));
    co_stream_write(req1,&xmin,sizeof(co_uint32));
    co_stream_write(req1,&dx,sizeof(co_uint32));

    co_stream_open(req2,O_WRONLY,UINT_TYPE(32));
    co_stream_write(req2,&xmax,sizeof(co_uint32));
    co_stream_write(req2,&xmin,sizeof(co_uint32));
    co_stream_write(req2,&dx,sizeof(co_uint32));

    // Loop over region
    offset=0;
    c_imag=ymax;
```

Figure 13-6. Controller process for the Mandelbrot image generator. (*continues*)

```
for (j=0; j<YSIZE; ) {
    if (co_stream_read_nb(ack0,&tmp,sizeof(co_uint8))) {
        co_stream_write(req0,&c_imag,sizeof(co_uint32));
        co_stream_write(req0,&offset,sizeof(co_uint32));
        offset += XSIZE*3;
        c_imag = FXSUB(c_imag,dy);
        j++;
    } else if (co_stream_read_nb(ack1,&tmp,sizeof(co_uint8))) {
        co_stream_write(req1,&c_imag,sizeof(co_uint32));
        co_stream_write(req1,&offset,sizeof(co_uint32));
        offset += XSIZE*3;
        c_imag = FXSUB(c_imag,dy);
        j++;
    } else if (co_stream_read_nb(ack2,&tmp,sizeof(co_uint8))) {
        co_stream_write(req2,&c_imag,sizeof(co_uint32));
        co_stream_write(req2,&offset,sizeof(co_uint32));
        offset += XSIZE*3;
        c_imag = FXSUB(c_imag,dy);
        j++;
    }
}
co_signal_post(done,1);
co_stream_close(req0);
co_stream_close(req1);
co_stream_close(req2);
}

co_stream_open(req0,O_WRONLY,UINT_TYPE(32));
co_stream_close(req0);
co_stream_open(req1,O_WRONLY,UINT_TYPE(32));
co_stream_close(req1);
co_stream_open(req2,O_WRONLY,UINT_TYPE(32));
co_stream_close(req2);
co_stream_close(config_stream);
}
```

Figure 13-6. *continued*

- Three sets of stream writes, each of which passes the necessary configuration data to the individual worker processes via their **req** stream inputs. The passing of data via the **req** stream begins processing for each of the workers.

- An inner loop iterates through each line of the image (the Y dimension). Using nonblocking read functions (**co_stream_read_nb**), it polls each of the three worker processes in sequence to determine if they have a completed value waiting in their output streams. If a value is detected, it is read from the **ack** stream. A new value and X offset are written to the process via the **req** stream, starting a new calculation for the current X-Y point. Note that as each point is processed, its value is written by the worker process to a shared memory buffer, as described earlier. This is important because the design of the application and the use of **co_stream_read_nb** means that the lines will not necessarily be processed in order. By using a shared memory to store the image, the lines may be written (using **co_memory_writeblock** from within the **genline** process) in any order.

- After all points have been processed, the loop terminates and a message is posted via the **done** signal. This message indicates to the test bench (the controlling software process, which might be driven from a user interface of some kind) that the image generation is complete.

The Configuration Function

Figure 13-7 shows the configuration function that connects the three instances of the worker processes to the controller process and connects these four processes to a software test bench. The software test bench (named **test_proc**) accepts the results of the fractal image through the use of a shared memory buffer. The results are then written to a Windows format image file for viewing.

13.6 FUTURE REFINEMENTS

There are a number of ways that this application could be improved for even higher levels of performance, above and beyond simply adding more worker processes and modifying the configuration function accordingly.

For example, depending on the nature of the target hardware platform, it might be desirable to eliminate the need for shared memory, which must (as currently written) hold the entire image. If the goal is instead to stream the pixel results in a sorted manner, scan line by scan line, an alternative means of worker process synchronization would be required. One possible approach would be to implement a token-passing mechanism, in which each worker processor passes a message (using a signal) to subsequent processes, such that each process streams its output (representing a single scan line) to the controlling process only after the preceding process has completed its operations. This optimization and other potential means of acceleration are left for you, the reader, to contemplate.

```
#define BUFSIZE 2        /* buffer size for FIFO in hardware */
void config_mand(void *arg)
{
   int i;
   co_memory img;
   co_process test_proc, ctrl_proc;
   co_process genline_proc[NUM_UNITS];
   co_stream req[NUM_UNITS],ack[NUM_UNITS];
   co_stream config_stream;
   co_signal done;

   img = co_memory_create("img","ext0",YSIZE*XSIZE*3);
   config_stream = co_stream_create("config_stream", UINT_TYPE(32),
                                    BUFSIZE);
   for (i=0; i < NUM_UNITS; i++)
      req[i] = co_stream_create("req",UINT_TYPE(32),BUFSIZE);
   for (i=0; i < NUM_UNITS; i++)
      ack[i] = co_stream_create("ack",UINT_TYPE(8),BUFSIZE);

   done=co_signal_create("done");

   test_proc=co_process_create("test_proc",(co_function)test_mandelbrot,
                               3, config_stream,img,done);
   for (i=0; i<NUM_UNITS; i++)
      genline_proc[i] = co_process_create("gen",(co_function)genline,
                                          3, img,req[i],ack[i]);

   ctrl_proc = co_process_create("ctrl_proc",(co_function)ctrl,
                                 8, config_stream,done,
                                 req[0],req[1],req[2],
                                 ack[0],ack[1],ack[2]);

   co_process_config(ctrl_proc,co_loc,"PE0");
   for (i=0; i < NUM_UNITS; i++)
      co_process_config(genline_proc[i],co_loc,"PE0");
}

co_architecture co_initialize()
{
   return(co_architecture_create("mand","generic_vhdl",config_mand, NULL));
}
```

Figure 13-7. Configuration function, Mandelbrot image generator.

13.7 SUMMARY

Fractal models are used today to predict many kinds of systems that show chaotic behavior, such as the weather, population growth, or brain waves. We used the Mandelbrot set here because it is simple to generate, yet nicely demonstrates all characteristics of a fractal object. The program could be easily adapted to generate other fractal objects by changing the iteration formula.

Fractal geometry and its many applications were not, of course, the primary goal of this chapter. (Entire books have been written on the subject; we have only scratched the surface here.) More important are the insights you have gained into methods for achieving increased performance through the creation of multiple communicating processes—more specifically, for processes that may have widely differing processing requirements and irregular output timing.

CHAPTER 14

The Future of FPGA Computing

Throughout most of this book we've presented FPGAs as relatively small computing elements capable of accelerating key algorithms for embedded systems and, to some extent, for more esoteric scientific applications (the fractal image generator being one such example). FPGAs have become increasingly useful for such applications as the cost of the devices has fallen and as their densities have increased. You have also seen how the use of software programming methods, and C programming in particular, can simplify the creation of hardware-accelerated systems—systems in which the FPGA serves as a computational resource alongside other, more traditional processing resources.

Looking forward, it seems increasingly likely that applications requiring even higher levels of performance—supercomputing applications—will come to rely on FPGAs or FPGA-like devices to provide greater levels of raw computing power. This trend toward FPGA-based supercomputing is increasing in practicality to the point where major supercomputing platform vendors and research teams are investing substantially in FPGA-related research. As of this writing, two major vendors of high-performance computing platforms (Cray and Silicon Graphics) have begun offering FPGA-based computers. This trend will only intensify as the benefits of FPGA computing become more apparent and more widely accepted.

14.1 THE FPGA AS A HIGH-PERFORMANCE COMPUTER

Low-cost, high-performance computing using highly parallel platforms is not a new concept. It has long been recognized that many, perhaps most, of the problems requiring supercomputer solutions are well-suited to parallel processing. By partitioning a problem into smaller parts that can be solved simultaneously, it's possible to gain massive amounts of raw processing power, at the expense of increased hardware size, system cost, and power consumption. It is quite common, for example, for supercomputing researchers to create clusters of traditional desktop PC (often running the Linux operating system) to create what can be thought of as a "coarse-grained" parallel processing system. To create applications for such a system, software programmers may make use of various available programming tools, including PVM (Parallel Virtual Machine) and MPI (Message Passing Interface). Both of these libraries provide the ability to express parallelism through message passing and/or shared memory resources using the C language.

Using such an approach, traditional supercomputers can be replaced for many applications by clusters of readily available, low-cost computers. The computers used in such clusters can be of standard design and are programmed to act as one large, parallel computing platform. Higher-performance (and correspondingly higher-cost) clusters often employ high-bandwidth, low-latency interconnects to reduce the overhead of communication between nodes. Another path to reducing communication overhead is to increase the functionality of each node through the introduction of low-level custom hardware to create high-speed datapaths. More recently, supercomputer cluster designers have pursued this approach further and have exploited node-level parallelism by introducing programmable hardware into the mix. In this environment, an FPGA can be thought of as one node in a larger, highly parallel computer cluster.

Taking Parallelism to Extreme Levels

When FPGAs are introduced into a supercomputing platform strategy, opportunities exist for improving both coarse-grained (application-level) and fine-grained (instruction-level) parallelism. Because FPGAs provide a large amount of highly parallel, configurable hardware resources, it is possible to create structures (such as parallel multiply/add instructions) that can greatly accelerate individual operations or higher-level statements such as loops. Through the use of instruction scheduling, instruction pipelining, and other techniques, inner code loops can be further accelerated. And at a somewhat higher level these parallel structures can themselves be replicated to create further levels of parallelism, up to the limit of the target device capacity. You have seen some of these techniques used on a more modest scale in earlier chapters.

As you learned in Chapter 2, FPGA devices were introduced in the mid-1980s to serve as desktop-programmable containers for random ("glue") logic and as alternatives to custom ASICs. These devices were almost immediately discovered by researchers seeking low-cost, hardware-based computing solutions. Thus, the field of FPGA-based reconfigurable computing (RCC) was born, or at least came of age. The applications for which FPGA-based reconfigurable computing platforms are appropriate include computationally intensive real-time image processing and pattern recognition, data encryption and cryptography, genomics, biomedical, signal processing, scientific computing (including algorithms in physics, astronomy, and geophysics), data communications, and many others.

It should be noted that today most such uses of FPGAs are relatively static, meaning that the algorithm(s) being implemented on the FPGA are created at system startup and rarely, if ever, are changed during system operation. There has been extensive research into dynamic reconfiguration of FPGAs (changing the hardware-based algorithm "on the fly" to allow the FPGA to perform multiple tasks). However, to date there has not been widespread use of FPGAs for dynamically reconfigurable computing, due in large part to the system costs, measured in performance and power, of performing dynamic FPGA reprogramming. In most cases, then, the goal of FPGA-based computing is to increase the raw power and computational throughput for a specific algorithm.

Through a combination of design techniques that exploit *spatial parallelism*, key algorithms and algorithmic "hot spots" can be accelerated in FPGAs by multiple orders of magnitude. Until recently, however, it has been difficult or impossible for software algorithm developers to take advantage of these potential speedups because the software development tools (in particular, the compiler technologies) required to generate low-level structures from higher-level software descriptions have not been widely available. Instead, software algorithm developers have had to learn low-level hardware design methods or have turned to more experienced FPGA designers to take on the daunting task of manually optimizing an algorithm and expressing it as low-level hardware. With the advent of software-to-hardware compilers, however, this process is becoming easier.

Many Platforms to Choose From

The smallest FPGA-based computing platforms combine one or more widely available, reasonably priced FPGA devices with standard I/O devices (such as PCI or network interfaces) on a prototyping board. There are many examples of such boards, which typically make use of FPGA devices produced by Altera or Xilinx. These board-level solutions may or may not include adjacent microprocessors, but in most cases it is possible to make use of embedded processors to create mixed hardware/software applications, as you've seen in earlier chapters. Using such boards (or a collection of such boards) in combination with existing tools, it is possible to create hardware implementations of computationally intensive software algorithms using software-based methods.

When combined with other peripherals on a board, an FPGA—or a collection of FPGAs—can become an excellent platform for algorithm experimentation, given appropriate design expertise and/or design tools.

By making appropriate use of one or more embedded ("soft") processor cores, a savvy application developer or design team can construct a computing environment roughly analogous to the multiprocessor cluster systems described earlier. Soft processors are appropriate for this because they are generally configurable (you can select only those peripheral interfaces you need) and they generally support high-performance connectivity with the adjacent FPGA logic.

Using such a mixed processor/FPGA platform and carefully evaluating the application's processing and bandwidth requirements, it is possible to create mixed hardware/software solutions in which the FPGA logic serves as a hardware coprocessor for one or more on-chip embedded processors. And by using the multiprocess, streaming programming model presented in this book, it's possible to achieve truly astonishing levels of performance for many types of algorithms.

In terms of cost, a board-level prototyping system consisting of a mainstream FPGA device (capable of hosting one or more embedded processors along with other FPGA-based computations) and various input/output peripherals will range in price from a few hundred to a few thousand dollars, depending on the capacity of the onboard FPGA device(s). In fact, because the streaming programming model (with its network of connected processing nodes) requires relatively little node-to-node connectivity (typically a small number of streams, which may be implemented externally as high-speed serial I/O), it is becoming practical to use simple, low-cost FPGA prototyping boards arranged in a grid computing matrix.

The next evolutionary step is to combine larger, multiple-FPGA processing arrays with custom interconnect schemes and dynamic reprogramming to create large-scale, reconfigurable supercomputing applications. All of the major supercomputing vendors today are working on such platforms and have either announced, or will soon announce, related supercomputing products. Smaller FPGA-based supercomputing (or supercomputing-capable) platforms are available from Nallatech, SBS Technologies, Annapolis Microsystems, Gidel, the Dini Group, and many others.

Taking a Software Approach

Much of the research activity in FPGA-based computing has been oriented around the problem of software programming for massively parallel targets. As described in earlier chapters, parallelism in an application can be considered at two distinct levels: at the system level and at the level of specific instruction within a computational process or loop. The ideal software development tools for such targets would exploit both levels of parallelism with a high degree of automation. For now, however, the best approach seems to be to focus the efforts of automation (represented by compiler and optimizer technologies) on the lower-level aspects of the problem. At the same time, the software programmer is provided with an appropriate programming model and related tools and libraries allowing higher-level, coarse-grained parallelism to be expressed in a natural way. (This has, of course, been a major theme of this book.)

In this way the programmer, who has knowledge of the application's external requirements that are not necessarily reflected in a given set of source files, can make decisions and experiment with alternative algorithmic approaches while leaving the task of low-level optimization to the compiler tools. A number of programming models can be applied to FPGA-based programmable platforms, but the most successful of these models share a common attribute: they support modularity and parallelism through a dataflow (or dataflow-like) method of design partitioning and abstraction. The key to allocating processing power within such a system, and using such a programming model, is to use FPGA to implement one or more processes that handle the heavy computation, and provide other processes running on embedded or external microprocessors to handle file I/O, memory management, system setup, and other non-performance-critical tasks.

14.2 THE FUTURE OF FPGA COMPUTING

This book has introduced the concept of FPGAs as high-performance computing platforms, using the C language as a vehicle. While the use of a software programming language for FPGA programming may seem exotic, it is our belief that FPGAs and the factors that drive them into new applications are at an important point in their 20-year history—a tipping point if you will—beyond which they can and will be applied to an increasing number of computational problems.

There are important signals that a fundamental shift in programmable platform architectures is underway as well. The first important signal is the existence of multiple, well-funded startups that have worked (with varying levels of success) on entirely new computing platforms that make use of what has variously been called dynamically reconfigurable logic (DRL), reconfigurable computing, or adaptive computing. What this new technology does is merge, almost completely, the idea of programmable hardware as represented in something like an FPGA with a software application as represented by embedded software operating in an embedded processor.

Creating such a platform—one that balances the desire for general-purpose programmability with the need for application-specific hardware optimizations—has proven to be a challenge. A number of companies have tried and failed to produce a broadly applicable reconfigurable platform. Some of these companies have already folded their tents after grappling for years with the difficulty of providing both a hardware platform (the reconfigurable computer itself) and the design tools and application libraries needed to promote and support it.

The volatility of this technology and of reconfigurable computing in general has kept mainstream system designers from embracing these new reconfigurable computing systems. A major part of the problem is the lack of a universal, portable method of design. This is the same factor that limited the acceptance of earlier programmable hardware technologies including programmable logic devices and FPGAs in their early years. Until such time as a reasonably portable method of design becomes available, reconfigurable computing will remain an exotic, high-risk technology.

Bigger FPGAs and Increased System Integration

The most exciting FPGA developments in recent years have been in the area of systems-on-programmable-chips. We have explored some aspects of these developments in this book as we made use of embedded "soft" processors and other IP components to create mixed software/hardware applications. As FPGAs grow in density, in lockstep with Moore's Law, we will see a dramatic increase in the use of such predefined cores, which will be provided by FPGA vendors, by third-party IP core vendors, and through open-source efforts.

What will make these cores more compelling, however, will be improved software tools for hardware integration. The EDK tools from Xilinx and the SOPC Builder tools from Altera are powerful, but they remain challenging for users who are not already intimately familiar with FPGA devices. In the future, then, we can expect FPGA vendors or third-party tool suppliers to invest heavily in "system builder"-type products, making it easier for system designers to assemble complete, custom platforms within one or more FPGA devices. Such tools should make it easier to select and assemble a wide variety of system components, emulating at the level of an FPGA what is possible today in the domain of board-level (or even rack-level) system design.

As the effective prices of FPGAs continue to decline and the range of potential applications increases as a result, we can also expect to see FPGAs appearing in products with much higher volumes than were historically practical. FPGAs are already appearing—and not as simple placeholders for ASICs—in such domains as automotive telematics and network data communication equipment. Given the demonstrated practicality of programming these devices from software, it is not unreasonable to expect FPGAs to appear as reconfigurable software coprocessing elements in higher-end workstations and PCs. As a concrete example, it would be entirely possible—using one of today's PCI-equipped FPGA-based development cards—to create an FPGA hardware accelerator for applications such as Adobe Photoshop. As the need for more complex image processing tasks grows, the demand for such specialized (but perhaps infrequently used) hardware accelerators will drive the use of FPGAs for more general types of computing problems, and into more widely available computing platforms. It is at that point, when you will be able to order the FPGA accelerator option with your new notebook PC, that the FPGA will have truly arrived as a computing resource.

14.3 SUMMARY

In this chapter we have taken a brief look at the future of FPGA-based computing. We have described how current efforts in the area of supercomputing show great promise, and we have touched on their more-recent relatives, reconfigurable computing devices. We have avoided making predictions regarding the actual architectures of tomorrow's devices (we are not qualified to make such predictions), but we have stated what we believe to be true: that the future of FPGA and FPGA-like devices lies in the domain of software applications. It is software that drives the need for increased computing performance, and FPGAs have demonstrated conclusively that they are viable alternatives to more traditional computing platforms.

FPGAs are no longer just the "poor man's ASIC"; instead they have evolved into computing platforms of the highest order. The history of their next 20 years will be one in which software applications—and software design methods—play a dominant role.

APPENDIX A

Getting the Most Out of Embedded FPGA Processors

Bryan H. Fletcher
Technical Program Manager
Memec
San Diego, California
www.memec.com
bryan_fletcher@mei.memec.com

Note: This appendix is adapted from a paper titled "FPGA Embedded Processors: Revealing True System Performance," which was presented by Bryan Fletcher at the Embedded Systems Conference in San Francisco in March of 2005.

This book has included multiple examples that combine embedded FPGA "soft" processors with custom-designed FPGA logic to create a mixed hardware/software system, primarily for the creation of embedded, software-driven test benches. For larger-scale systems design, embedding a processor inside an FPGA has many advantages. Specific peripherals can be chosen based on the application, with unique user-designed peripherals being easily attached. A variety of available memory controllers enhance the FPGA embedded processor system's interface capabilities.

FPGA embedded processors use general-purpose FPGA logic to construct internal memory, processor buses, internal peripherals, and external peripheral controllers, including external memory controllers. Soft processors are built from general-purpose FPGA logic as well.

As more pieces (buses, memory, memory controllers, peripherals, and peripheral controllers) are added to the embedded processor system, the system becomes increasingly powerful and useful. However, these additions reduce performance and increase the embedded system cost, consuming FPGA resources.

Likewise, large banks of external memory can be connected to the FPGA and accessed by the embedded processor system using included memory controllers. Unfortunately, the latency to access this external memory can have a significant, negative impact on performance. It is important to consider these various aspects of embedded processor performance in order to get the best use of these powerful components.

A.1 FPGA Embedded Processor Overview

As you've learned in the early chapters of this book, the Field-Programmable Gate Array (FPGA) is a general-purpose device filled with digital logic building blocks. The two market leaders in the FPGA industry, Altera and Xilinx, are the focus of this discussion, although other programmable logic companies do exist, including Actel, Quicklogic, and Lattice Semiconductor.

The most primitive FPGA building block is called either a logic cell (LC) by Xilinx or a logic element (LE) by Altera. In either case, this building block consists of a lookup table (LUT) for logical functions and a flip-flop for storage. In addition to the LC/LE block, FPGAs also contain memory, clock management, input/output (I/O), and multiplication blocks. For the purposes of estimation and comparison, LC/LE consumption is most often used in determining system cost.

Soft Versus Hard Processor

Both Xilinx and Altera produce FPGA families that embed a physical processor core into the FPGA silicon. A processor built from dedicated silicon is referred to as a "hard" processor. Such is the case for the ARM922T in the Altera Excalibur family and the PowerPC 405 in the Xilinx Virtex-II Pro and Virtex-4 families.

A "soft" processor, on the other hand, is built using the FPGA's general-purpose logic. The soft processor is typically described in a Hardware Description Language (HDL) or in a lower-level netlist. Unlike the hard processor, a soft processor must be synthesized and fit into the FPGA fabric.

In both soft and hard processor systems, the local memory, processor buses, internal peripherals, peripheral controllers, and memory controllers must be built from the FPGA's general-purpose logic.

Advantages of an FPGA Embedded Processor

An FPGA embedded processor system offers many exceptional advantages compared to typical microprocessors:

- Customization
- Obsolescence mitigation
- Component and cost reduction
- Hardware acceleration

Customization

The designer of an FPGA embedded processor system has complete flexibility to select any combination of peripherals and controllers. In fact, the designer can invent new, unique peripherals that can be connected directly to the processor's bus. If a designer has a nonstandard requirement for a peripheral set, this can be met easily with an FPGA embedded processor system. For example, a designer would not easily find an off-the-shelf processor with 10 UARTs. However, in an FPGA, this configuration is very easily accomplished.

Obsolescence Mitigation

Some companies, in particular those supporting military contracts, have a design requirement to ensure a product lifespan that is much longer than the lifespan of a standard electronics product. Component obsolescence mitigation is a difficult issue. FPGA soft processors are an excellent solution in this case since the source HDL for the soft processor can be purchased. Ownership of the processor's HDL code may fulfill the requirement for product lifespan guarantee.

Component and Cost Reduction

With the versatility of the FPGA, previous systems that required multiple components can be replaced with a single FPGA. Certainly this is the case when an auxiliary I/O chip or coprocessor is required next to an off-the-shelf processor. By reducing the component count in a design, a company can reduce board size and inventory management, both of which save design time and cost.

Hardware Acceleration

Perhaps the most compelling reason to choose an FPGA embedded processor is the ability to make trade-offs between hardware and software to maximize efficiency and performance. If an algorithm is identified as a software bottleneck, a custom coprocessing engine can be designed in the FPGA specifically for that algorithm. This coprocessor can be attached to the FPGA embedded processor through special, low-latency channels, and custom instructions can be defined to exercise the coprocessor. With modern FPGA hardware design tools, transitioning software bottlenecks from software to hardware is much easier since the software C code can be readily adapted into hardware with only minor changes to the C code.

Disadvantages

The FPGA embedded processor system is not without disadvantages. Unlike an off-the-shelf processor, the hardware platform for the FPGA embedded processor must be designed. The embedded designer becomes the hardware processor system designer when an FPGA solution is selected.

Because of the integration of the hardware and software platform design, the design tools are more complex. The increased tool complexity and design methodology requires more attention from the embedded designer.

Since FPGA embedded processor software design is relatively new compared to software design for standard processors, the software design tools are likewise relatively immature, although workable. Significant progress in this area has been made by both Altera and Xilinx. Within the next year, this disadvantage should be further diminished, if not eliminated.

Device cost is another aspect to consider. If a standard, off-the-shelf processor can do the job, that processor will be less expensive in a head-to-head comparison with the FPGA capable of an equivalent processor design. However, if a large FPGA is already in the system, consuming unused gates or a hard processor in the FPGA essentially makes the embedded processor system cost inconsequential.

A.2 PERIPHERALS AND MEMORY CONTROLLERS

To facilitate FPGA embedded processor design, both Xilinx and Altera offer extensive libraries of intellectual property (IP) in the form of peripherals and memory controllers. This IP is included in the embedded processor toolsets provided by these manufacturers. To emphasize the versatility and flexibility afforded the embedded designer using an FPGA, the following is a partial list of IP included with the embedded processor design tools from Altera and Xilinx:

Peripheral Types

- General-purpose I/O
- UART
- Timer
- Debug
- SPI
- DMA controller
- Ethernet (interface to external MAC/PHY chip)

Memory Controllers

- SRAM
- Flash
- SDRAM
- DDR SDRAM (Xilinx only)
- CompactFlash

A.3 INCREASING PROCESSOR PERFORMANCE

Some embedded designers get frustrated when the performance of their selected FPGA embedded processor is half or less of what was expected based on published performance numbers. In some cases, a designer is unable to duplicate the manufacturer's published results for a benchmark. The reason often lies with the designer's lack of understanding of how these benchmarks are produced, but it can also reflect a lack of experience with the specific processor being used and the available techniques for enhancing processor performance.

Manufacturers' Benchmarks

The industry standard benchmark for FPGA embedded processors is Dhrystone millions of instructions per second (DMIPs). Both Altera and Xilinx quote DMIPs for most, if not all, of the available embedded processors.

The achieved DMIPs reported by the manufacturers are based on several things that maximize the benchmark results:

- Optimal compiler optimization level
- Fastest available device family (unless otherwise noted)
- Fastest speed grade in that device family
- Executing from the fastest, lowest-latency memory, typically on-chip
- Optimization of the processor's parameterizable features

Because of all these interrelated factors, meeting the published benchmark numbers can be a challenge for designers.

Performance-Enhancing Techniques

Achieved performance might be lacking because the designer has not taken advantage of all of the performance-enhancing techniques available to FPGA embedded processors. The manufacturers obviously know what must be done to get the most out of their chips, and they take full advantage of every possible enhancement when benchmarking. Embedded designers, who are familiar with standard microprocessor performance optimization, need to learn which software optimization techniques apply to FPGA embedded processors. Designers must also learn performance-enhancing techniques that apply specifically to FPGAs.

The design landscape is certainly more complicated with an FPGA embedded processor. The incredible advantages gained with this type of design are not without trade-offs. Specifically, the increased design complexity is overwhelming to many, including experienced embedded or FPGA designers. Manufacturers and their partners put significant effort into training and providing support to designers experimenting with this technology. Taking advantage of a local field applications engineer is therefore essential to FPGA embedded processor design success.

As an introduction to this type of design, a few performance-enhancing techniques are highlighted in the following sections. Specific references are based on research of Xilinx FPGAs and tools, although users of Altera processors are likely to have access to similar features and techniques.

A.4 OPTIMIZATION TECHNIQUES THAT ARE NOT FPGA-SPECIFIC

An experienced embedded system designer is familiar with many of the techniques discussed in this section. For that reason, significant detail is not given here. The main objective of this section is to emphasize that many standard microprocessor design optimization techniques apply to FPGA embedded processor design and can have excellent benefits.

Code Manipulation

Many optimizations are available to affect the application code. Some techniques apply to how the code is written. Other techniques affect how the compiler handles the code.

Optimization Level

Compiler optimizations are available in Xilinx Platform Studio (XPS) based on the Gnu C compiler (GCC). The current version of the MicroBlaze and PowerPC GCC-based compilers in EDK 6.3 is 2.95.3-4. These compilers have several levels of optimization, including Levels 0, 1, 2, and 3, and also a size reduction optimization. An explanation of the strategy for the different optimization levels briefly follows:

- Level 0—No optimization.
- Level 1—First-level optimization. Performs jump and pop optimizations.
- Level 2—Second-level optimization. This level activates nearly all optimizations that do not involve a speed-space trade-off, so the executable should not increase in size. The compiler does not perform loop unrolling, function inlining, or strict aliasing optimizations. This is the standard optimization level used for program deployment.
- Level 3—Highest optimization level. This level adds more expensive options, including those that increase code size. In some cases this optimization level actually produces code that is less efficient than the O2 level, so it should be used with caution.
- Size—Size optimization. The objective is to produce the smallest possible code size.

Use of Manufacturer-Optimized Instructions

Xilinx provides several customized instructions that have been streamlined for Xilinx embedded processors. One example is xil_printf. This function is nearly identical to the standard printf, with the following exceptions: support for type real numbers is removed, the function is not reentrant, and no long-long (64-bit) types are supported. For these differences, the xil_printf function is 2,953 bytes, making it much smaller than printf, which is 51,788 bytes.

Assembly

Assembly, including inline assembly, is supported by GCC. As with any microprocessor, assembly becomes very useful in fully optimizing time-critical functions. Be aware, however, that some compilers do not optimize the remaining C code in a file if inline assembly is also used in that file. Also, assembly code does not enjoy the code portability advantages of C.

Miscellaneous

Many other code-related optimizations can and should be considered when optimizing an FPGA embedded processor:

- Locality of reference
- Code profiling
- Careful definition of variables (Xilinx provides a Basic Types definition)
- Strategic use of small data sections, with accesses that can be twice as fast as large data sections
- Judicious use of function calls to minimize pushing/popping of stack frames
- Loop length (especially where cache is involved)

Memory Usage

Many processors provide access to fast, local memory, as well as an interface to slower, secondary memory. The same is true with FPGA embedded processors. The way this memory is used has a significant affect on performance. Like other processors, the memory usage in an FPGA embedded processor can be manipulated with a linker script.

Local Memory Only

The fastest possible memory option is to put everything in local memory. Xilinx local memory is made up of large FPGA memory blocks called BlockRAM (BRAM). Embedded processor accesses to BRAM happen in a single bus cycle. Since the processor and bus run at the same frequency in MicroBlaze, instructions stored in BRAM are executed at the full MicroBlaze processor frequency. In a MicroBlaze system, BRAM is essentially equivalent in performance to a Level 1 (L1) cache. The PowerPC can run at frequencies greater than the bus and has true, built-in L1 cache. Therefore, BRAM in a PowerPC system is equivalent in performance to a Level 2 (L2) cache.

Xilinx FPGA BRAM quantities differ by device. For example, the 1.5 million gate Spartan-3 device (XC3S1500) has a total capacity of 64KB, whereas the 400,000 gate Spartan-3 device (XC3S400) has half as much at 32KB. An embedded designer using FPGAs should refer to the device family datasheet to review a specific chip's BRAM capacity.

If the designer's program fits entirely within local memory, the designer achieves optimal memory performance. However, many embedded programs exceed this capacity.

External Memory Only

Xilinx provides several memory controllers that interface with a variety of external memory devices. These memory controllers are connected to the processor's peripheral bus. The three types of volatile memory supported by Xilinx are static RAM (SRAM), single-data-rate synchronous dynamic RAM (SDRAM), and double data rate (DDR) SDRAM. The SRAM controller is the smallest and simplest inside the FPGA, but SRAM is the most expensive of the three memory types. The DDR controller is the largest and most complex inside the FPGA, but fewer FPGA pins are required, and DDR is the least expensive per megabyte.

In addition to the memory access time, the peripheral bus also incurs some latency. In MicroBlaze, the memory controllers are attached to the On-chip Peripheral Bus (OPB). For example, the OPB SDRAM controller requires a four- to six-cycle latency for a write and an eight- to ten-cycle latency for a read, depending on bus clock frequency. The worst possible program performance is achieved by having the entire program reside in external memory. Since optimizing execution speed is a typical goal, an entire program should rarely, if ever, be targeted solely at external memory.

Cache External Memory

The PowerPC in Xilinx FPGAs has instruction and data cache built into the silicon of the hard processor. Enabling the cache is almost always a performance advantage for the PowerPC.

The MicroBlaze cache architecture is different from the PowerPC because the cache memory is not dedicated silicon. The instruction and data cache controllers are selectable parameters in the MicroBlaze configuration. When these controllers are included, the cache memory is built from BRAM. Therefore, enabling the cache consumes BRAM that otherwise could have been used for local memory. Cache consumes more BRAM than local memory for the same storage size because the cache architecture requires address line tag storage. Additionally, enabling the cache consumes general-purpose logic to build the cache controllers.

For example, an experiment in Spartan-3 enables 8KB of data cache and designates 32MB of external memory to be cached. This cache requires 12 address tag bits. This configuration consumes 124 logic cells and six BRAMs. Only four BRAMs are required in Spartan-3 to achieve 8KB of local memory. In this case, cache is 50% more expensive in terms of BRAM usage than local memory. The two extra BRAMs are used to store address tag bits.

If 1MB of external memory is cached with an 8KB cache, the address tag bits can be reduced to seven. This configuration then requires only five BRAMs rather than six (four BRAMs for the cache and one BRAM for the tags). This is still 25% greater than if the BRAMs are used as local memory.

Additionally, the achievable system frequency may be reduced when the cache is enabled. In one example, the system without any cache is capable of running at 75MHz; the system with cache is capable of running at only 60MHz. Enabling the cache controller adds logic and complexity to the design, decreasing the achieved system frequency during FPGA place and route. Therefore, in addition to consuming FPGA BRAM resources that may have otherwise been used to increase local memory, the cache implementation may also cause the overall system frequency to decrease.

Considering these cautions, enabling the MicroBlaze cache, especially the instruction cache, may improve performance, even when the system must run at a lower frequency. Testing has shown that a 60MHz system with instruction cache enabled can have a 150% advantage over a 75MHz system without instruction cache (both systems store the entire program in external memory). When both instruction and data caches are enabled, the 60MHz outperforms the 75MHz system by 308%.

Note that this particular test example is not the most practical since the entire test program fits in the cache. A more realistic experiment would be to use an application that is larger than the cache. Another precaution concerns applications that frequently jump beyond the size of the cache. Multiple cache misses degrade the performance, sometimes making a cached external memory worse than the external memory without cache.

Enabling the cache is always worth an experiment to determine if it improves the performance for your particular application.

Partitioning Code into Internal, External, and Cached Memory

The memory architecture that provides the best performance is one that has only local memory. However, this architecture is not always practical since many useful programs exceed the available capacity of the local memory. On the other hand, running from external memory exclusively may have more than an eight times performance disadvantage due to the peripheral bus latency. Caching the external memory is an excellent choice for PowerPC. Caching the external memory in MicroBlaze definitely improves results, but an alternative method is presented that may provide more optimal results.

For MicroBlaze, perhaps the optimal memory configuration is to wisely partition the program code, maximizing the system frequency and local memory size. Critical data, instructions, and stack are placed in local memory. Data cache is not used, allowing for a larger local memory bank. If the local memory is not large enough to contain all instructions, the designer should consider enabling the instruction cache for the address range in external memory used for instructions.

By not consuming BRAM in data cache, the local memory can be increased to contain more space. An instruction cache for the instructions assigned to external memory can be very effective. Experimentation or profiling shows which code items are most heavily accessed; assigning these items to local memory provides a greater performance improvement than caching.

For example, Express Logic's Thread-Metric test suite has been used to demonstrate how partitioning a small piece of code in local memory can result in a significant performance improvement. One function in the Thread-Metric Basic Processing test is identified as time-critical. The function's data section (consisting of 19% of the total code size) is allocated to local memory, and the instruction cache is enabled. The 60MHz cached and partitioned-program system achieves performance that is 560% better than running a noncached, nonpartitioned 75MHz system using only external memory.

However, the 75MHz system shows even more improvement with code partitioning. If the time-critical function's data and text sections (22% of the total code size) are assigned to local memory on the 75MHz system, a 710% improvement is realized, even with no instruction cache for the remainder of the code assigned to external memory.

In this one case, the optimal memory configuration is one that maximizes local memory and system frequency without cache. In other systems where the critical code is not so easily pinpointed, a cached system may perform better. Designers should experiment with both methods to determine what is optimal for their design.

A.5 FPGA-SPECIFIC OPTIMIZATION TECHNIQUES

Because the designer is actually building and creating the embedded processor system hardware in an FPGA, much can be done to improve the performance of the hardware itself. Additionally, with an FPGA embedded processor residing next to additional FPGA hardware resources, a designer can consider custom coprocessor designs specifically targeted at a design's core algorithm.

Increasing the FPGA's Operating Frequency

Employing FPGA design techniques to increase the operating frequency of the FPGA embedded processor system increases performance. Several methods are considered.

Logic Optimization and Reduction

Connect only the peripherals and buses that will be used. Here are a few examples:

- If a design does not store and run any instructions using external memory, do not connect the instruction side of the peripheral bus. Connecting both the instruction and data side of the processor to a single bus creates a multimaster system, which requires an arbiter. Optimal bus performance is achieved when a single master resides on the bus.

- Debug logic requires resources in the FPGA and may be the hardware bottleneck. When a design is completely debugged, the debug logic can be removed from the production system, potentially improving the system's performance. For example, removing a MicroBlaze Debug Module (MDM) with an FSL acceleration channel saves 950 LCs. In MicroBlaze systems with the cache enabled, the debug logic is typically the critical path that slows down the entire design.

- The Xilinx OPB External Memory Controller (EMC) used to connect SRAM and Flash memories creates a 32-bit address bus even if 32 bits are not required to address the memory. Xilinx also provides a bus-trimming peripheral that removes the unused address bits. When using this memory controller, the bus trimmer should always be used to eliminate the unused addresses. This frees up routing and pins that would have otherwise been used. The Xilinx Base System Builder (BSB) now does this automatically.

- Xilinx provides several general-purpose I/O (GPIO) peripherals. The latest GPIO peripheral version (v3.01.a) has excellent capabilities, including dual-channel support, bidirectionality, and interrupt capability. However, these features also require more resources, which affect timing. If a simple GPIO is all that the design requires, the designer should use a more primitive version of the GPIO, or at least ensure that the unused features in the enhanced GPIO are turned off. In the optimized examples in this study, GPIO v1.00.a is used, which is much less sophisticated, much faster, and approximately half the size (304 LCs for seven GPIO v1.00.a peripherals as compared to 602 LCs for v3.01.a).

Area and Timing Constraints

Xilinx FPGA place and route tools perform much better when given guidelines as to what is most important to the designer. In the Xilinx tools, a designer can specify the desired clock frequency, pin location, and logic element location. By providing these details, the tools can make smarter trade-offs during the hardware design implementation.

Some peripherals require additional constraints to ensure proper operation. For example, both the DDR SDRAM controller and the 10/100 Ethernet MAC require additional constraints to guarantee that the tools create correct and optimized logic. The designer must read the datasheet for each peripheral and follow the recommended design guidelines.

Hardware Acceleration

Dedicated hardware outperforms software. The embedded designer who is serious about increasing performance must consider the FPGA's ability to accelerate the processor performance with dedicated hardware. Although this technique consumes FPGA resources, the performance improvements can be extraordinary.

Turn on the Hardware Divider and Barrel-Shifter

MicroBlaze can be customized to use a hardware divider and a hardware barrel-shifter rather than performing these functions in software. Enabling these processor capabilities consumes more logic but improves performance. In one example, enabling the hardware divider and barrel-shifter adds 414 LCs, but the performance is improved by 18.1%.

Software Bottlenecks Converted to Coprocessing Hardware

Custom hardware logic can be designed to offload an FPGA embedded processor. When a software bottleneck is identified, a designer can choose to convert the bottleneck algorithm into custom hardware. Custom software instructions can then be defined to operate the hardware coprocessor.

Both MicroBlaze and Virtex-4 PowerPC include very low-latency access points into the processor, which are ideal for connecting custom coprocessing hardware. Virtex-4 introduces the Auxiliary Processing Unit (APU) for the PowerPC. The APU provides a direct connection from the PowerPC to coprocessing hardware. In MicroBlaze, the low-latency interface is called the Fast Simplex Link (FSL) bus. The FSL bus contains multiple channels of dedicated, unidirectional, 32-bit interfaces. Because the FSL channels are dedicated, no arbitration or bus mastering is required. This allows an extremely fast interface to the processor.

Converting a software bottleneck into hardware may seem like a very difficult task. Traditionally, a software designer identifies the bottleneck, after which the algorithm is transitioned to an FPGA designer who writes VHDL or Verilog code to create the hardware coprocessor. Fortunately, this process has been greatly simplified by tools that can generate FPGA hardware from C code. One such tool is CoDeveloper from Impulse Accelerated Technologies. This tool allows one designer who is familiar with C to port a software bottleneck into a custom piece of coprocessing FPGA hardware using CoDeveloper's Impulse C libraries.

Here are some examples of algorithms that could be targeted for hardware-based coprocessors:

- Inverse Discrete Cosine Transformation, used in JPEG 2000

- Fast Fourier Transform

- MP3 decode
- Triple-DES and AES encryption
- Matrix manipulation

Any operation that is algorithmic, mathematical, or parallel is a good candidate for a hardware coprocessor. FPGA logic consumption is traded for performance. The advantages can be enormous, improving performance by tens or hundreds of times.

A.6 SUMMARY

Based on experiments performed at Memec, several conclusions have been reached.

Have Reasonable Expectations

All manufacturers want to publish the best possible benchmark. FPGA manufacturers take full advantage of system flexibility and FPGA design techniques to achieve high benchmarks. The embedded designer must understand how these benchmarks are achieved and realize that an actual FPGA embedded processor system cannot achieve such high marks. However, the benchmarks are still useful as a means of comparing one manufacturer to another.

Regarding the design process, remember that the hardware platform is part of the FPGA embedded processor design. Unlike off-the-shelf processors where the hardware is predefined and fixed, FPGAs have the flexibility and added complexity to create a multitude of different systems. A member of the design team must have knowledge of hardware development and optimization of the FPGA embedded processor.

If an application does not require high performance or any of the other advantages that an FPGA can provide, an off-the-shelf processor is most likely a less complicated, less expensive, and better solution.

Optimization Through Experimentation Yields the Best Results

Much can be done to optimize an FPGA embedded processor system. In addition to standard software optimization, the embedded designer can perform many hardware optimizations. Careful use of memory has a large impact on the system's performance. Experimenting with different memory strategies is necessary to achieve the best performance.

Understand that the addition or removal of each peripheral, peripheral controller, or bus alters the design size, cost, and speed. Use only what is necessary and no more!

Take Advantage of Superior Flexibility in FPGAs

An FPGA embedded processor has the power and ability to provide previously unachievable flexibility and performance. With an FPGA, a designer can specify exactly the peripherals required in a system. With an FPGA soft processor, a designer can purchase and own the source code for the processor. With an FPGA, a designer can adapt C code from a software bottleneck to create a custom hardware coprocessing unit. These are excellent advantages that can be realized only in programmable hardware.

If the designer chooses not to take advantage of these capabilities, the realized FPGA embedded processor system performance may be a huge disappointment. However, for those who take full advantage of the FPGA embedded processor, specific application performance greatly exceeding typical microprocessor expectations is possible!

Acknowledgments

Thanks to Shalin Sheth of Xilinx and Ron Wright of Memec for their advice and assistance.

APPENDIX B

Creating a Custom Stream Interface

Note: This appendix is based on an application note and related VHDL source code titled "An Impulse C Compatible Stream Interface for the National DS92LV16 Serializer/Deserializer," developed by Scott Bekker and Dr. Ross Snider of the Signal Processing and Neural Instrumentation Laboratory at Montana State University in Bozeman, Montana.

While most of this book has focused on applications that communicate with other parts of a hardware/software system via predefined stream, signal, and memory interfaces, for many real-world applications it is necessary for performance reasons to create direct connections between hardware processes written in C and other hardware elements such as high-speed I/O devices. The creation of such interfaces requires a good grasp of FPGA hardware design fundamentals, but for those with the right expertise it is not a difficult problem, given the standarized nature of the stream and signal interfaces.

This chapter describes one such application, presenting the VHDL code written to implement an Impulse C-compatible stream over a standard serial device, the National DS92LV16 serializer/deserializer (SERDES). The VHDL code (which appears at the end of this chapter, in Figure B-5) describes a handshaking protocol that establishes a stream connection and provides a fault-tolerent means of data transfer. Once the connection is established, data can be transferred in either direction with the standard Impulse C stream interface.

B.1 APPLICATION OVERVIEW

A team at Montana State University led by Dr. Ross Snider has been funded by the National Science Foundation to develop a computational platform that will help further the understanding of how the nervous system encodes information. The system being developed, which is pictured in Figure B-1, includes an array of 27 Xilinx Virtex-II FPGAs connected via high-speed serial interfaces (with larger future scale-ups likely) that will be used to analyze and build a model of an insect brain.

Dr. Snider is collaborating with scientists at the Center for Computational Biology at MSU, where the aim is to connect this hardware directly to an insect (a cricket), analyze neural data in real time, and insert new models that will interact with the recorded data. To do so, the data needs to be processed in real time and the models run as fast as practical within the limits of the computational resources.

Dr. Snider's team evaluated a conventional DSP approach, but the benchmarks indicated that DSPs would not provide the needed performance. An array of FPGAs, however, does provide the computational horsepower required at a relatively low cost. Benchmarks conducted using a single Virtex-II 3000 device indicate speed enhancements of up to 200 times are practical for certain neural simulations when compared to conventional processors.

The create the system, performance-critical algorithms describing how the sodium or potassium channels work inside a neuron are developed as FPGA-resident processes. These processes are mapped onto the FPGA-based platform, creating a grid computing system for biological simulations.

The end goal of the project is to create a platform, supported with high-level tools, that can be an enabling force in the discovery of how to decode neural information in real time. This research will eventually enable individuals to control prosthetic limbs just by thinking, in the same way healthy limbs are controlled by the brain. Xilinx has helped with this project by donating the necessary FPGA devices. The National Institutes of Health (NIH) has

Figure B-1. An array of 27 Xilinx FPGAs connected via high-speed serial links (photo courtesy of Dr. Ross Snider, Montana State University).

expressed interest in the project and is acting as a funding source. More about this project can be found at www.hylitech.com.

B.2 THE DS92LV16 SERIAL LINK FOR DATA STREAMING

As described in earlier chapters, the Impulse C stream is a means by which two Impulse C processes can communicate through a standard interface. For this project, which involves a relatively large array of processing elements (the FPGAs) communicating between different FPGA boards, it was necessary to implement an Impulse C-compatible stream that can communicate over DS92LV16 serial links. The hardware stream that was created for this application is not as general as the standard Impulse C stream, because limitations such as data width are imposed by the hardware. Nonetheless, for this application, the Impulse C stream provides a good abstraction for the kind of data movement required by the algorithm.

The serializer hardware allows a maximum of 16 bits to be sent during each clock cycle. Since it is also necessary to send control flags such as FIFO status, write enable, and end-of-stream signals in addition to the raw data, not all of the 16 transceiver bits are available for data transmission. Of the 16 bits, 3 are used for control flags, and the rest are available for data, with data widths of up to 13 bits. Another difference between this stream definition and the Impulse C stream is the need to initialize the serial link before using it. A handshaking scheme is implemented to ensure that both transceivers are locked to the serial data stream before allowing data transmission. The following sections discuss the details of the handshaking process as well as the specifics of data transmission.

Initializing the Serial Connection

The implementation uses a simple handshaking process to ensure that both the remote and local hardware transceivers are locked onto the serial data stream prior to proceeding with data transfer. Figure B-2 shows the control flow for the handshaking. The first step is sending out a synchronization pattern while waiting for the local transceiver to lock to the serial data stream. After being locked on for a period of a few tens of milliseconds (to allow for switch bounce effects—for example, if a cable was pulled off and replaced), the state machine alternately sends out flags and the synchronization pattern for 240 clock cycles each. The synchronization pattern allows the remote DS92LV16 to lock onto the data stream if it has not done so already.

The SERDES must receive the synchronization pattern for 150 clock cycles to ensure synchronization with the serial data stream. Therefore, the slightly larger value of 240 clock cycles was chosen for the synchronization period to make sure the remote receiver can lock to the serial stream.

The transmitted flags indicate the status of the local state machine, allowing the remote state machine to know its state. If a valid flag response is received, the first acknowledge state is entered, which simply waits for the remote transceiver to reach the same state. When the remote state machine reaches the first acknowledge state, the local state machine progresses to the second acknowledge state. In the second acknowledge state, a sequence of flags is sent out for a period of time until it is determined that the remote state machine is also in this state, or until a timeout occurs. Upon leaving the second acknowledge state, the connected state is entered, signifying that both the remote and local transceivers are locked and data transmission can begin. If the local lock signal is ever lost, such as because the serial cable got disconnected, the state machine reverts to the initial state, and the handshaking process is repeated.

B.3 STREAM INTERFACE STATE MACHINE DESCRIPTION

While reading the following descriptions, refer to the state diagram shown in Figure B-2. The VHDL source code listings corresponding to these descriptions are found in Figure B-5 at the end of this chapter.

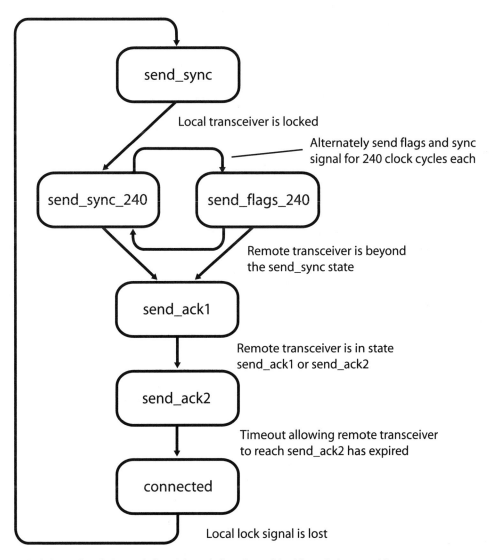

Figure B-2. Impulse C-to-serializer/deserializer handshaking state machine.

State: send_sync

In state send_sync, the sync_out bit is set high. This tells the transceiver to send a synchronization pattern allowing the remote transceiver to lock to the serial stream. The state machine remains in this state until the counter variable reaches 0x200000, signifying that the local transceiver has been locked to the serial stream for 26ms. This value was arbitrarily chosen, but it works with this particular hardware. After being locked for 26ms, state send_sync_240 is entered.

State: send_sync_240
State: send_flags_240

In state send_sync_240, the sync_out bit is held high to enable the sending of the synchronization pattern. The variable flag_cnt is incremented each clock cycle. When flag_cnt reaches 240, the state changes to send_flags_240 and flag_cnt is set to 0. In send_flags_240, the sync_out bit is cleared and a flag sequence is sent out instead of the synchronization pattern. In both of these states, the data received from the local transceiver is monitored. The **counter** variable is used to keep track of for how many clock cycles specific flags were received. If the counter reaches 0x000FFF in either of these two states, the local state machine advances to state send_ack1. However, if the local transceiver loses synchronization with the serial data stream in either state, the state machine reverts back to state send_sync.

State: send_ack1

State send_ack1 is used to let the remote state machine know that it has advanced beyond the previous two states. This is done by sending a new, different flag sequence corresponding to this state. The local state machine remains in this state until the remote state machine has reached either this state or an advanced state for 240 clock cycles, as determined by the value in flag_cnt. Upon receiving the appropriate flags for 240 clock cycles, state send_ack2 is entered.

State: send_ack2

Once this state is reached, both ends of the serial connection know that the opposite end has established a connection. This state is simply to allow the remote state machine to exit the send_ack1 state. Flags are sent out until it has been determined that the remote state machine has been in this same state for 240 clock cycles or until a timeout occurs. When either scenario happens, the connected state is entered.

State: connected

In the connected state, the link_established bit is held high, indicating to the other processes that the serial connection is available for their use. The data sent to the transceiver is set to data_to_transceiver, which is controlled by the data transmission logic. The state machine remains in this state unless the local transceiver loses synchronization with the serial data stream, in which case the send_sync state is entered and the handshaking process is repeated.

B.4 DATA TRANSMISSION

After initializing the serial connection, it is possible to send data. As mentioned previously, up to 13 of the 16 bits can be used for data transmission. The lower bits were arbitrarily chosen for this purpose, while the high bits are used for flags. The data width can be specified during instantiation by means of a VHDL generic. The diagram shown in Figure B-3 illustrates the packet sent to the serializer.

15	14	13	12	11	10	9	8	7	6	5	4	3	2	1	0
FIFO too full	write enable	end of stream	data	data	data	data	data	data	data	data	data	data	data	data	data

Figure B-3. Serializer/deserializer data packet.

A feedback mechanism is implemented to stop the remote sender from sending data if the local FIFO is too full. If the local FIFO is too full, the "FIFO too full" flag is set in the outgoing packet. The remote transceiver then reads this flag and stops sending data. This process works in both directions because the serial links are bidirectional. The "Write Enable" and "End of Stream" flags are associated with the data in the corresponding packet. These flags let the receiver know if the data is valid and if the stream is finished.

The interface to this VHDL stream is identical to the standard Impulse C stream with the addition of a hardware transceiver interface. The VHDL stream must be instantiated for each hardware transceiver that is to be used for an Impulse C stream. Each stream has an interface for a producer and for a consumer. The data sent to the producer port appears on the consumer port of the remote stream instantiation, and vice versa. Figure B-4 illustrates a typical stream setup.

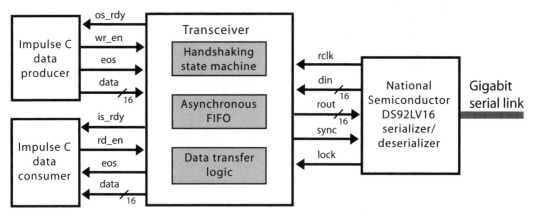

Figure B-4. Serializer/deserializer-to-Impulse C stream connections.

Future Refinements

Note that the stream implementation described here can only be used for data widths up to 13 bits. If it is necessary to send data consisting of more than 13 bits, the data can first be broken down into smaller portions directly in the Impulse C program, and the pieces can be sent sequentially. A drawback of this implementation is that, depending on the word size, several of the bits sent across the serial link may never contain any data, which reduces potential bandwidth. If a more sophisticated method were used to send the data, such as consecutively sending several 16-bit data words followed by 16 bits of flags, higher data throughput could be achieved.

B.5 SUMMARY

This example has demonstrated how two Impulse C software processes (the producer and consumer processes in Figure B-4) can be connected via a standard, readily available serial interface device for creating an FPGA-based grid computing system. When you combine this kind of high-speed serial communication with embedded software processors and related embedded operating systems (as demonstrated in earlier chapters) you can create extremely powerful, FPGA-based grid computing platforms for relatively little cost.

```
--transceiver.vhdl
--This VHDL code describes the implementation of an eight-bit ImpulseC
--compatible stream for use over the DS92LV16 serializer/deserializer.
--
--Scott Bekker
--6/24/04

library IEEE;
use IEEE.STD_LOGIC_1164.ALL;
use IEEE.STD_LOGIC_ARITH.ALL;
use IEEE.STD_LOGIC_UNSIGNED.ALL;

entity transceiver is
    generic(data_width : natural := 8); -- Max data width: 13 due to
                                        -- 16 bit serdes bus width - 3 flag bits used
    port ( clk : in std_logic;
        -- hardware transceiver ports
        rclk : in std_logic;
        din : out std_logic_vector(15 downto 0);
        rout : in std_logic_vector(15 downto 0);
        sync_out : out std_logic;
        local_lock_active_low : in std_logic;

        --producer ports
        producer_os_rdy : out std_logic;
        producer_wr_en : in std_logic;
        producer_eos : in std_logic;
        producer_data_in : in std_ulogic_vector(data_width - 1 downto 0);

        --consumer ports
        consumer_os_rdy : out std_logic;
        consumer_rd_en : in std_logic;
        consumer_eos : out std_logic;
        consumer_data_out : out std_ulogic_vector(data_width - 1 downto 0));

end transceiver;

architecture Behavioral of transceiver is

component async_fifo IS
    port ( din: IN std_logic_VECTOR(15 downto 0);
        wr_en: IN std_logic;
        wr_clk: IN std_logic;
        rd_en: IN std_logic;
```

Figure B-5. Transceiver VHDL source code. (*continues*)

```
        rd_clk: IN std_logic;
        ainit: IN std_logic;
        dout: OUT std_logic_VECTOR(15 downto 0);
        full: OUT std_logic;
        empty: OUT std_logic;
        rd_count: OUT std_logic_VECTOR(6 downto 0));
END component;

TYPE state_type is (send_sync, send_sync_240, send_flags_240,
                    send_ack1, send_ack2, connected);
signal present_state : state_type := send_sync;
signal counter : std_logic_vector(23 downto 0);
signal din_sig : std_logic_vector(15 downto 0);
signal data_to_transceiver : std_logic_vector(15 downto 0);
signal link_established : std_ulogic;
signal rout_sig : std_logic_vector(15 downto 0);
signal flag_cnt : std_logic_vector(7 downto 0);
signal fifo_out_sig : std_logic_vector(15 downto 0);
signal wr_en_sig : std_logic;
signal rd_en_sig : std_logic;
signal empty_sig : std_logic;
signal rd_count_sig : std_logic_vector(6 downto 0);
signal remote_fifo_too_full : std_logic;
signal local_fifo_too_full : std_logic;
signal producer_data_in_sig : std_logic_vector(data_width - 1 downto 0);
signal consumer_data_out_sig : std_ulogic_vector(data_width - 1 downto 0);

begin
  async_fifo0: async_fifo
    port map( din => rout_sig,
          wr_en => wr_en_sig,
          wr_clk => rclk,
          rd_en => rd_en_sig,
          rd_clk => clk,
          ainit => '0',--reset_sig,
          dout => fifo_out_sig,
          full => open,
          empty => empty_sig,
          rd_count => rd_count_sig);

--state machine process to perform handshaking and establish a serial connection
process(clk)
begin
    if rising_edge(clk) then
```

Figure B-5. *continued*

```
        --allow producer to write if remote fifo is not too
        --full and link is established
        producer_os_rdy <= (not remote_fifo_too_full) and link_established;
                                 --rdy signal must be synchronous to clk

        case present_state is
        when send_sync =>   --send sync pattern until local transceiver
                                 --has been locked for a period of time
           sync_out <= '1';
           link_established <= '0';
           flag_cnt <= X"00";
           if local_lock_active_low = '0' then
              counter <= counter + 1;
              if counter > X"200000" then   --26 ms at 80 MHz (delay for switch bounce)
                 counter <= X"000000";
                 present_state <= send_sync_240;
              else
                 present_state <= send_sync;
              end if;
           else
              present_state <= send_sync;
              counter <= X"000000";
           end if;

        when send_sync_240 =>    --send sync pattern until a valid flag
                                 --pattern is received or until timeout occurs
           sync_out <= '1';
           link_established <= '0';
           if local_lock_active_low = '0' then
              if counter > X"000FFF" then
                 counter <= X"000000";
                 present_state <= send_flags_240;
              else
                 counter <= counter + 1;
                 if (rout = X"00BB") or (rout = X"00CC")
                       or (rout = X"00DD") then
                            --remote transceiver is in send_xxxx_160 or send_ackx
                    flag_cnt <= flag_cnt + 1;
                    if flag_cnt > X"F0" then
                       counter <= X"000000";
                       flag_cnt <= X"00";
                       present_state <= send_ack1;
                    end if;
                 else
```

Figure B-5. *continued*

```
              flag_cnt <= X"00";
              present_state <= send_sync_240;
         end if;
       end if;
    else
       present_state <= send_sync;
    end if;

when send_flags_240 =>  --send flag pattern until a valid flag pattern
                        --is received or until timeout occurs
    sync_out <= '0';
    din_sig <= X"00BB";
    link_established <= '0';
    if local_lock_active_low = '0' then
       if counter > X"000FFF" then
          counter <= X"000000";
          present_state <= send_sync_240;
       else
          counter <= counter + 1;
          if (rout = X"00BB") or (rout = X"00CC")
             or (rout = X"00DD") then  --remote transceiver is in
                                       --send_xxxx_240 or send_ackx
             flag_cnt <= flag_cnt + 1;
             if flag_cnt > X"F0" then
                counter <= X"000000";
                flag_cnt <= X"00";
                present_state <= send_ack1;
             end if;
          else
             flag_cnt <= X"00";
             present_state <= send_flags_240;
          end if;
       end if;
    else
       present_state <= send_sync;
    end if;

when send_ack1 =>   --wait here for remote transceiver to reach this state
                    --or the next state
    sync_out <= '0';
    link_established <= '0';
    din_sig <= X"00CC";
    if local_lock_active_low = '0' then
       if (rout_sig = X"00CC") or (rout_sig = X"00DD") then
```

Figure B-5. *continued*

```
                    --remote transceiver is in send_ack1 or send_ack2
            flag_cnt <= flag_cnt + 1;
            if flag_cnt > X"F0" then
               counter <= X"000000";
               present_state <= send_ack2;
            end if;
         else
            flag_cnt <= X"00";
            present_state <= send_ack1;
         end if;
      else
         present_state <= send_sync;
      end if;

   when send_ack2 => --when this state is reached both transceivers are
                        --locked, send flags for a period then go to connected state
      sync_out <= '0';
      link_established <= '0';
      din_sig <= X"00DD";
      counter <= counter + 1;
      if local_lock_active_low = '0' then
         if counter > X"0000FFF" then
            present_state <= connected;
         else
            if rout_sig = X"00DD" then
               flag_cnt <= flag_cnt + 1;
               if flag_cnt > X"F0" or counter > X"0000FFF" then
                  present_state <= connected;
               end if;
            else
               flag_cnt <= X"00";
               present_state <= send_ack2;
            end if;
         end if;
      else
         present_state <= send_sync;
      end if;

   when connected =>   --both transceivers are connected, assert
                        --link_established signal and allow data transmission
      sync_out <= '0';
      din_sig <= data_to_transceiver;
      link_established <= '1';
      if local_lock_active_low = '0' then
```

Figure B-5. *continued*

```
                present_state <= connected;
            else
              counter <= X"000000";
              present_state <= send_sync;
            end if;

        when others =>
          present_state <= send_sync;

      end case;
    end if;
    end process;

    --retrieve remote fifo too full signal off of transceiver output when link is established
    with link_established select
      remote_fifo_too_full <= rout_sig(15) when '1',
                              '1' when others;

    --retrieve end of stream signal off of transceiver output when link is established,
    --consumer enables read, and data is valid
    with std_logic_vector'(link_established, consumer_rd_en, fifo_out_sig(14)) select
        consumer_eos <= fifo_out_sig(13) when "111",
                '0' when others;

    --retrieve data from fifo and output to the consumer
    consumer_data_out <= consumer_data_out_sig;
    with std_logic_vector'(link_established, consumer_rd_en, fifo_out_sig(14)) select
        consumer_data_out_sig <= std_ulogic_vector(fifo_out_sig(data_width - 1 downto 0))
                        when "111",
                        consumer_data_out_sig when others;

    --if there is data in the fifo, allow consumer to read it
    consumer_os_rdy <= not empty_sig;

    --send flags and data to the transceiver for sending
    data_to_transceiver(15 downto 13) <= local_fifo_too_full & producer_wr_en &
    producer_eos;
    data_to_transceiver(data_width - 1 downto 0) <= producer_data_in_sig;

    --latch data from producer when producer write enable is asserted
    with producer_wr_en select
        producer_data_in_sig <= std_logic_vector(producer_data_in) when '1',
                    producer_data_in_sig when others;

    --assert fifo write enable when link is established and write flag is set
    with std_logic_vector'(link_established, rout_sig(14)) select
        wr_en_sig <= '1' when "11",
```

Figure B-5. *continued*

```vhdl
            '0' when others;

--assert fifo read enable when consumer asserts read enable or when fifo data
--is not valid
rd_en_sig <= consumer_rd_en or not fifo_out_sig(14);


--latch data from DS92LV16 on rising edge of rclk when locked onto serial data stream
process(rclk)
begin
if rising_edge(rclk) then
   if local_lock_active_low = '0' then
      rout_sig <= rout;
   end if;
end if;
end process;

--update transceiver input data on falling edge of the clock
-- hardware transceiver reads input on rising clock edge so data must be
-- stable during rising edge
process(clk)
begin
if falling_edge(clk) then
   din <= din_sig;
end if;
end process;

--set local_fifo_too_full_sig based on rd_count_sig flag is
-- set if fifo contains more than 64 items and cleared if fifo
-- contains fewer than 31 items
process(clk)
begin
if rising_edge(clk) then
   if rd_count_sig > "1000000" and rd_count_sig < "1111111" then --fifo occasionally
                                                    --reports a count of
                                                    --127 when it is empty

      local_fifo_too_full <= '1';
   elsif rd_count_sig < "0011111" then
      local_fifo_too_full <= '0';
   else
      local_fifo_too_full <= local_fifo_too_full;
   end if;
end if;
end process;
end Behavioral;
```

Figure B-5. *continued*

APPENDIX C

Impulse C Function Reference

Impulse C is a library of predefined functions that extend the ANSI standard C language in support of parallel programming and programming for hardware and reconfigurable software/hardware processing targets. As such, it includes functions related to processes, streams, signals, and memories, as well as functions related to bit manipulations and bit-accurate arithmetic. Additional functions allow you to set up and monitor your application for simulation. The Impulse C library includes the following:

- Stream-related functions (prefixed with **co_stream_**) that allow you to open, close, read, and write streams and check for the end of a stream or an error indicator.

- Process- and memory-related functions (prefixed with **co_process_** and **co_memory_**, respectively) that allow you to create the major processing elements of your application, to start and stop processing, and to store and access data, either locally or shared.

- Signal-related functions (**co_signal_post** and **co_signal_wait**) that allow you to synchronize processes through the use of messages. (One process posts an event, and another process waits for that event to be posted.)

- Bit manipulation functions (such as **co_bit_insert**) that allow you to insert and extract bits and perform other bitwise operations.

- Simulation-related functions (prefixed with **cosim_**) that allow you to instrument your application and gain quick access to internal application data during desktop simulation.

CO_ARCHITECTURE_CREATE

```
co_architecture co_architecture_create(const char *name,
                                       const char *arch,
                                       co_function configure,
                                       void *arg);
```

Header File

co.h

Description

This function associates your application with a specific architecture definition. Architecture definitions are supplied with CoDeveloper and/or with optional Platform Support Packages. This function must be called from within
the co_initialize function of your application.

Arguments

The arguments for co_architecture_create are as follows:

const char *name	A programmer-defined name. This name is used to identify this particular application's use (or instance) of the selected architecture.
const char *arch	This text string specifies which platform architecture (as supplied with CoDeveloper or with an optional Platform Support Package) is to be targeted when compiling this application.
co_function configure	The configuration function associated with this architecture.
void *arg	This argument is passed to your configuration function.

Return Value

A pointer to an architecture. This return value should subsequently be returned by co_initialize to the main function, where it should be passed to the
co_execute function to begin simulation.

Notes

The name argument is used as the basis for the name of the top-level HDL module generated by CoBuilder for the application. The name specified must therefore be compatible with any downstream HDL synthesis and simulation tools.

In simulation, the function prints an error message and terminates the application if the name argument is an empty string ("" or NULL) or the function is called from outside the co_initialize function.

CO_BIT_EXTRACT

```
int32 co_bit_extract(int32 value,
                     uint8 start_bit,
                     uint8 num_bits);
```

Header File

co.h

Description

This function provides a concise and efficient method of extracting bits from a 32-bit signed integer.

Arguments

The arguments for co_bit_extract are as follows:

int32 value	The 32-bit signed integer from which the bits are taken.
uint8 start_bit	The starting bit (LSB) of the extracted bit sequence.
uint8 num_bits	The number of bits to extract.

Return Value

Returns a 32-bit signed integer (int32) representing the value of the extracted bits.

Notes

The co_bit_insert, co_bit_insert_u, co_bit_extract, and co_bit_extract_u functions operate on 32-bit data. If required, you can truncate the input and results to suit your needs. (Types greater than 32 bits are currently not supported.) For example, if you are operating on 16-bit numbers, you can use the functions as follows:

```
co_int16 src = 0xffff;
co_int16 dest;
dest = co_bit_extract(src, 8, 16);
// dest == 0x00ff

dest = co_bit_insert(src, 8, 8, dest);
// dest == 0xffff
```

In this example, src and dest automatically get promoted (and sign-extended because they are signed types) to 32-bit integers when used as arguments. The return value gets cast invisibly to a co_int16, which just truncates 32 to 16 bits.

The difference between signed and unsigned bit extractions (co_bit_extract and co_bit_extract_u, respectively) has to do with what's beyond the most-significant bit. If you try to extract 16 bits from signed value 0xffff, starting (halfway) at bit 8, you get 0xffff. If you extract the same bits from unsigned 0xffffU, you get 0x00ffU, since the sign is not extended beyond the existing 16 bits. Otherwise, the extract functions do the same thing.

CO_BIT_EXTRACT_U

```
int32 co_bit_extract_u(uint32 value,
                       uint8 start_bit,
                       uint8 num_bits);
```

Header File

co.h

Description

This function provides a concise and efficient method of extracting bits from a (32-bit) unsigned integer.

Arguments

The arguments for co_bit_extract_u are as follows:

uint32 value	The 32-bit unsigned integer from which the bits are taken.
uint8 start_bit	The starting bit (LSB) of the extracted bit sequence.
uint8 num_bits	The number of bits to extract.

Return Value

Returns a 32-bit unsigned integer (uint32) value representing the value of the extracted bits.

Notes

The co_bit_insert, co_bit_insert_u, co_bit_extract, and co_bit_extract_u functions operate on 32-bit data. If required, you can truncate the input and results to suit your needs. (Types larger than 32 bits are currently not supported.) For example, if you are operating on 16-bit numbers, you can use the functions as follows:

```
co_int16 src = 0xffff;
co_int16 dest;
dest = co_bit_extract(src, 8, 16);
// dest == 0x00ff

dest = co_bit_insert(src, 8, 8, dest);
// dest == 0xffff
```

In this example, src and dest are automatically promoted (and sign-extended, since they are signed types) to 32-bit integers when used as arguments, and the return value gets cast invisibly to a co_int16, which just truncates 32 to 16 bits.

The difference between signed and unsigned bit extractions (co_bit_extract and co_bit_extract_u, respectively) has to do with what's beyond the most-significant bit. If you try to extract 16 bits from signed value 0xffff, starting (halfway) at bit 8, you get 0xffff. If you extract the same bits from unsigned 0xffffU, you get 0x00ffU, since the sign is not extended beyond the existing 16 bits. Otherwise, the extract functions do the same thing.

CO_BIT_INSERT

```
int32 co_bit_insert(int32 destination,
                    uint8 dest_start_bit,
                    uint8 num_bits,
                    int32 source);
```

Header File

co.h

Description

This function provides a concise and efficient method of inserting bits into a 32-bit signed integer.

Arguments

The arguments for co_bit_insert are as follows:

int32 destination	The 32-bit signed integer into which the bits are inserted.
uint8 dest_start_bit	The bit position in the destination value where the bits will begin to be inserted, from dest_start_bit toward the most-significant bit.
uint8 num_bits	The number of bits to extract from the source, starting at the least-significant bit.
int32 source	The source of the bits to be inserted.

Return Value

Returns a 32-bit signed integer (int32) representing the original destination value, but with num_bits bits, starting at bit location dest_start_bit, replaced with the first num_bits bits of the source integer.

Notes

The co_bit_insert, co_bit_insert_u, co_bit_extract, and co_bit_extract_u functions operate on 32-bit data. If required, you can truncate the input and results to suit your needs. (Values greater than 32 bits are currently not supported.) For example, if you are operating on 16-bit numbers, you can use the functions as follows:

```
co_int16 src = 0xffff;
co_int16 dest;
dest = co_bit_extract(src, 8, 16);
// dest == 0x00ff

dest = co_bit_insert(src, 8, 8, dest);
// dest == 0xffff
```

In this example, src and dest automatically get promoted (and sign-extended, since they are signed values) to 32-bit integers when used as arguments. The return value gets cast invisibly to a co_int16, which just truncates 32 to 16 bits.

CO_BIT_INSERT_U

```
uint32 co_bit_insert_u(uint32 destination,
                       uint8 dest_start_bit,
                       uint8 num_bits,
                       uint32 source);
```

Header File

co.h

Description

This function provides a concise and efficient method of inserting bits into a 32-bit unsigned integer.

Arguments

The arguments for co_bit_insert_u are as follows:

uint32 destination	The 32-bit unsigned integer into which the bits are inserted.
uint8 dest_start_bit	The starting bit in the destination value where the bits will begin to be inserted, from dest_start_bit toward the most-significant bit.

uint8 num_bits   The number of bits to extract from the
                               source.

uint32 source   A 32-bit unsigned value representing the
                               source of the bit values to be inserted.

Return Value

Returns a 32-bit unsigned integer (uint32) representing the original destination value, but with num_bits bits, starting at bit location dest_start_bit, replaced with the first num_bits bits of the source.

Notes

The co_bit_insert, co_bit_insert_u, co_bit_extract, and co_bit_extract_u functions operate on 32-bit data. If required, you can truncate the input and results to suit your needs. (Values greater than 32 bits are currently not
supported.) For example, if you are operating on 16-bit numbers, you can use
the functions as follows:

```
co_int16 src = 0xffff;
co_int16 dest;
dest = co_bit_extract(src, 8, 16);
// dest == 0x00ff

dest = co_bit_insert(src, 8, 8, dest);
// dest == 0xffff
```

In this example, src and dest automatically get promoted (and sign-extended,
since they are signed values) to 32-bit integers when used as arguments. The
return value gets cast invisibly to a co_int16, which just truncates 32 to 16
bits.

CO_EXECUTE

```
void co_execute(co_architecture architecture);
```

Header File

co.h

Description

This function starts execution of an Impulse C application for desktop simulation.

Arguments

The argument for **co_execute** is as follows:

co_architecture architecture   A pointer to an architecture previously
                                 created using the **co_architecture_create**
                                 function.

Return Value

None.

Notes

This function must be called from the application's **main** function.

CO_INITIALIZE

co_architecture co_initialize(void *arg);

Header File

None.

Description

This function, which must be defined as part of every Impulse C application, initializes the application for the purpose of desktop simulation. The co_initialize function also defines the relationship between your application's main function and its configuration function. Finally, co_initialize provides a defined starting point for hardware/software compilation using the Impulse C compiler tools.

Arguments

The argument for **co_initialize** is application-specific. The value of the argument, if it can be determined at compile time, is cast to an integer and substituted for the **co_param** parameter of every hardware process. In software processes and desktop simulation, arguments of type **co_param** are treated normally (that is, they may be used like a **void ***).

Return Value

A pointer to the architecture pointer returned from the function co_architecture_create.

Notes

The co_initialize function is not a predefined Impulse C function. Instead, you define this function in your Impulse C application. The co_initialize function must include at least one call to co_architecture_create and must be called from main.

CO_MEMORY_CREATE

```
co_memory co_memory_create(const char *name,
                           const char *loc,
                           size_t size);
```

Header File

co.h

Description

This function creates a shared memory. It must be called from within the configuration function of your application.

Arguments

The arguments for co_memory_create are as follows:

const char *name	A programmer-defined name. This name may be any text string. It is used to identify the memory externally, such as when using the Application Monitor.
const char *loc	Specifies the architecture-specific physical location associated with this memory. The set of valid values for this argument is platform-dependent. (See your Platform Support Package documentation.)
size_t size	The size of the memory in bytes.

Return Value

A pointer to the created memory. This return value may subsequently be used as an argument to the function co_process_create.

Notes

Memory names, as specified in the first argument to co_memory_create, must be unique across the application when using the Application Monitor.

In simulation, this function prints an error message and terminates the application if it is called outside the configuration function.

CO_MEMORY_PTR


```
void *co_memory_ptr(co_memory mem);
```

Header File

co.h

Description

This function returns a pointer to the data buffer of a shared memory. This function may be called from within a software process to obtain the pointer to a data buffer (memory location), allowing that memory to be directly accessed.

Arguments

The arguments for co_memory_ptr are as follows:

co_memory mem   A memory pointer as passed to the process
                                     on the process argument list.

Return Value

A pointer to the memory buffer. Returns NULL if the mem argument is NULL, in which case co_errno is set to CO_ENULL_PTR.

Notes

This function is normally used in software processes to provide a direct means of accessing memory contents for reading and writing. In hardware processes, co_memory_ptr is not supported; instead, use the functions co_memory_readblock and co_memory_writeblock.

CO_MEMORY_READBLOCK

```
void co_memory_readblock(co_memory mem,
                         unsigned int offset,
                         void *buf,
                         size_t buffersize);
```

Header File

co.h

Description

This function reads a block of data from a shared memory. This function must be called from within a process run function.

Arguments

The arguments for co_memory_readblock are as follows:

co_memory mem	A pointer to a memory as passed on the process argument list.
unsigned int offset	The memory offset from which to begin reading, in bytes.
void *buf	A pointer to the local memory that will receive the shared memory data. In hardware processes, this argument must be an array identifier.
size_t buffersize	The number of bytes to transfer in the block read operation.

Return Value

None.

Notes

co_memory_readblock performs a block DMA transfer between shared memories and local memories. The third argument, buf, is an array that represents a block of local RAM. Note that co_memory_readblock is not designed for efficient random access of individual memory locations. Note also that stream interfaces may actually provide better performance than memory block reads and writes if the system contains a CPU.

The co_errno variable is set to CO_EMEM_OUT_OF_BOUNDS if the offset is outside the bounds of the memory. co_errno is set to CO_ENULL_PTR if the argument mem is NULL.

CO_MEMORY_WRITEBLOCK

```
void co_memory_writeblock(co_memory mem,
                          unsigned int offset,
                          void *buf,
                          size_t buffersize);
```

Header File

co.h

Description

This function writes a block of data to a shared memory. This function must be called from within a process run function.

Arguments

The arguments for co_memory_writeblock are as follows:

co_memory mem	A pointer to a memory as passed on the process argument list.
unsigned int offset	Specifies the memory offset, in bytes.
void *buf	A pointer to the local memory containing the data to be written to shared memory. In hardware processes, this argument must be an array identifier.
size_t buffersize	The number of bytes to transfer in the block write operation.

Return Value

None.

Notes

The co_memory_writeblock function performs a block DMA transfer between shared memories and local memories. The third argument, buf, is an array that represents a block of shared memory. The type of memory being accessed depends on the targeted platform (see your Platform Support Package documentation). Note that co_memory_writeblock is not designed for efficient random access of individual memory locations. Note also that stream interfaces may actually provide better performance than memory block reads and writes if the system contains a CPU.

The co_errno variable is set to CO_EMEM_OUT_OF_BOUNDS if the offset is outside the bounds of the memory. co_errno is set to CO_ENULL_PTR if the argument mem is NULL.

CO_PAR_BREAK

```
void co_par_break();
```

Header File

co.h

Description

This function is used to explicitly specify each cycle boundary in a block of C code. When invoked in your C program, this function instructs the optimizer to insert a cycle boundary and suppress the generation of parallel logic.

Arguments

None.

Return Value

None.

Notes

This function is used to more precisely control the generation of parallelism in your C code. For example, if you want the optimizer to generate four clock cycles for a loop, simply add three **co_par_break** statements, as in the following example:

```
do {
    // Cycle 1 operations
    co_par_break();
    // Cycle 2 operations
    co_par_break();
    // Cycle 3 operations
    co_par_break();
    // Cycle 4 operations
} while (...);
```

If the optimizer generates more than four cycles for this block of code, that means that some operations in one of the cycles cannot be performed in parallel.

CO_PROCESS_CONFIG

```
co_error co_process_config(co_process process,
                           co_attribute attribute,
                           const char *val);
```

Header File

co.h

Description

This function configures attributes of a process. The **co_attribute** argument specifies which attribute of the process to configure. This function must be called from within the configuration function of your application.

Arguments

The arguments for **co_process_config** are as follows:

co_process process	A pointer to a process created using co_process_create.
co_attribute attribute	The attribute whose value is being set.
const char *val	The value to assign to the attribute.

Return Value

Currently, this function always returns co_err_none.

Notes

The only possible value of **attribute** at this time is co_loc (hardware location). The **val** argument specifies the value of the given attribute. Currently, since there is only one attribute to configure, **val** always specifies the name of the process hardware location.

In simulation, this function prints an error message and terminates the application if it is called outside the configuration function.

CO_PROCESS_CREATE

```
co_process co_process_create(const char *name,
                    co_function run,
                    int argc, ...);
```

Header File

co.h

Description

This function creates a process. This function must be called from within the configuration function of your application.

Arguments

The arguments for **co_process_create** are as follows:

const char *name	A programmer-defined name. This name may be any text string. It is used to identify the process externally, such as when using the Application Monitor.
co_function run	The run function associated with this process.
int argc	The number of input/output ports (streams, signals, registers, memories, and parameters) associated with the run func-

tion. The **argc** variable is followed by a corresponding list of port arguments as specified in the run function declaration.

Return Value

A pointer to the created process. This return value may subsequently be used as an argument to the function **co_process_config**.

Notes

Processes created using **co_process_create** represent specific instances of the specified run function. It is possible (and common) for **co_process_create** to be called repeatedly for the same process in order to create multiple instances of a run function.

In simulation, this function prints an error message and terminates the application if it is called from outside the configuration function, or if the Application Monitor is being used and a **name** argument is not supplied (== NULL).

CO_REGISTER_CREATE

```
co_register co_register_create(const char *name,
                               co_type type);
```

Header File

co.h

Description

This function creates a register object. This function must be called from within the configuration function of your application.

Arguments

The arguments for **co_register_create** are as follows:

| const char *name | A programmer-defined name. This name may be any text string. It is used to identify the register externally, such as when using the Application Monitor. |

co_type type   The type of the register. The type is nor-
                                     mally expressed as a signed or unsigned in-
                                     teger of a specific width using the
                                     INT_TYPE(n), UINT_TYPE(n), or
                                     CHAR_TYPE macros.

Return Value

A pointer to the created register. This return value may subsequently be used
as an argument to function co_process_create. It returns NULL and sets
co_errno to CO_ENOERROR if the type argument is NULL.

Notes

Register names, as specified in the first argument to co_register_create, must
be unique across the application when using the Application Monitor.

CO_REGISTER_GET

```
int32 co_register_get(co_register reg);
```

Header File

co.h

Description

This function reads the contents of a register. This function must be called
from within a process run function.

Arguments

The argument for co_register_get is as follows:

co_register reg   A pointer to a register object as passed on
                                     the process argument list.

Return Value

The value of the register is returned as a 32-bit integer. If the register is wider
than 32 bits, the least-significant 32 bits of data are returned as an integer. If
the register is smaller than 32 bits, the integer value of the data in the register
is returned.

CO_REGISTER_PUT

```
void co_register_put(co_register reg,
                     int32 value);
```

Header File

co.h

Description

This function puts a data value to a register object. This function must be called from within a process run function.

Arguments

The arguments for co_register_put are as follows:

co_register reg	A pointer to a register object as passed on the process argument list.
int32 value	Specifies the value to be assigned to the register.

Return Value

None.

Notes

If the register is wider than 32 bits, the data value is written into the least-significant bytes of the register and higher-order bytes are unchanged. If the register is smaller than 32 bits, the least-significant bytes of the value are written to the register until it is full.

CO_REGISTER_READ

```
co_error co_register_read(co_register reg,
                          void *buffer,
                          size_t buffersize);
```

Header File

co.h

Description

This function reads a data value from a register object and copies the value to a local memory. This function must be called from within a process run function.

Arguments

The arguments for co_register_read are as follows:

co_register reg	A pointer to a register object as passed on the process argument list.
void *buffer	A pointer to the destination memory. (The destination is typically an integer variable.)
size_t buffersize	The number of bytes to transfer in the read operation.

Return Value

If the operation is successful, the function returns co_err_none and sets co_errno to CO_ENOERROR. In case of error, it returns co_err_invalid_arg and sets the co_errno variable accordingly; co_errno is set to CO_ENULL_PTR if the reg argument is NULL or to CO_EINVALID_REGIS-TER_WIDTH if the register is smaller than 32 bits.

CO_REGISTER_WRITE

```
co_error co_register_write(co_register reg,
                           const void *buffer,
                           size_t buffersize);
```

Header File

co.h

Description

This function writes a data value from a local variable to a register object. This function must be called from within a process run function.

Arguments

The arguments for co_register_write are as follows:

co_register reg	A pointer to a register object as passed on the process argument list.
const void *buffer	A pointer to the source memory. (The source is typically an integer variable.)
size_t buffersize	The number of bytes to transfer in the write operation.

Return Value

Returns co_err_none on success. Returns co_invalid_arg if the reg argument is NULL (co_errno: CO_ENULL_PTR) or the register is smaller than buffersize bytes (CO_EINVALID_REGISTER_WIDTH).

CO_SIGNAL_CREATE

```
co_signal co_signal_create(const char *name);
```

Header File

co.h

Description

This function creates a signal. Signals created using co_signal_create are subsequently used to connect processes. This function must be called from within the configuration function of your application.

Arguments

The argument for co_signal_create is as follows:

| const char *name | A programmer-defined name. This name may be any text string. It is used to identify the signal externally, such as when using the Application Monitor. |

Return Value

A pointer to the created signal. This return value may subsequently be used as an argument to the function co_process_create.

Notes

Signal names, as specified in the first argument to co_signal_create, must be unique across the application when using the Application Monitor.

   In simulation, the function prints an error and terminates the application if it is called from outside the configuration function.

CO_SIGNAL_POST

```
co_error co_signal_post(co_signal signal,
                        int32 value);
```

Header File

co.h

Description

This function posts a message (a 32-bit value) on a signal. This function must be called from within a process run function.

Arguments

The arguments for co_signal_post are as follows:

| co_signal signal | A pointer to a signal as passed on the process argument list. |
| int32 value | The 32-bit integer value to be posted. |

Return Value

Returns co_err_none on success or co_err_unknown if there was an error during desktop simulation.

Notes

The co_signal_post function is nonblocking and overwrites any message that is already pending on the signal.

CO_SIGNAL_WAIT

```
co_error co_signal_wait(co_signal signal,
                        int32 *ip);
```

Header File

co.h

Description

This function waits for a message on a signal. This function must be called from within a process run function.

Arguments

The arguments for co_signal_wait are as follows:

co_signal signal	A pointer to a signal as passed on the process argument list.
int32 *ip	A pointer to a 32-bit local variable that will receive the message value.

Return Value

Returns co_err_none on success or co_err_unknown if there was an error during desktop simulation.

Notes

The co_signal_wait function blocks (waits) until a message has been posted to the signal by another process using co_signal_post.

Note also that any number of processes can wait on a given signal, but only one process receives and consumes the signal and its value. The first process to receive the signal consumes it, and any other waiting processes continue to wait until another signal is posted.

CO_STREAM_CLOSE

```
co_error co_stream_close(co_stream stream);
```

Header File

co.h

Description

This function closes a previously opened stream. This function must be called from within a process run function.

Arguments

The argument for **co_stream_close** is as follows:

co_stream stream   A pointer to a stream as passed on the process argument list.

Return Value

Possible return values are listed, with values of **co_errno** shown in parentheses:

co_err_none   Success (CO_ENOERROR).

co_err_invalid_arg   Stream argument is NULL (CO_ENULL_PTR).

co_err_not_open   Stream is not open (CO_ENOT_OPEN).

Notes

Closing an input stream flushes any unread data from the stream and consumes one end-of-stream token. An attempt to close an input stream that has not yet received an end-of-stream token blocks until the writer closes the stream.

Closing an output stream sets the end-of-stream condition after the last unread data packet in the buffer.

CO_STREAM_CREATE

```
co_stream co_stream_create(const char *name,
                           co_type type,
                           int numelements);
```

Header File

co.h

Description

This function creates a stream. Streams created using co_stream_create are subsequently used to connect processes. This function must be called from within the configuration function of your application.

Arguments

The arguments for co_stream_create are as follows:

const char *name	A programmer-defined name. This name may be any text string. It is used to identify the stream externally, such as when using the Application Monitor.
co_type type	The type of the stream, including its width. The type is normally expressed as a signed or unsigned integer of a specific width using the INT_TYPE(n), UINT_TYPE(n), or CHAR_TYPE macros.
int numelements	Specifies the size of the stream buffer. Streams are implemented using first-in-first-out (FIFO) buffers. The size of this buffer in bytes is the value of the **numelements** argument multiplied by the size (in bytes) of the type specified in the preceding argument. Note that the size you specify for streams may have a dramatic impact on the size of the resulting logic after compilation to hardware.

Return Value

A pointer to the created stream. This return value may subsequently be used as an argument to function **co_process_create**.

Notes

Stream names, as specified in the first argument to **co_stream_create**, must be unique across the application when using the Application Monitor.

In simulation, this function prints an error message and terminates the application if it is called from outside the configuration function, or if the Application Monitor is being used and a **name** argument is not supplied (== NULL), or if a type wider than 64 bits is specified.

CO_STREAM_EOS

```
int co_stream_eos(co_stream stream);
```

Header File

co.h

Description

This function checks if a stream is at an end-of-stream condition. This function must be called from within a process run function.

Arguments

The argument for **co_stream_eos** is as follows:

co_stream stream   A pointer to a stream as passed on the process argument list.

Return Value

Returns an integer value: 0 if no end-of-stream is detected, and 1 if there is an end-of-stream.

Notes

When **co_stream_eos** returns 1 (true), subsequent calls to **co_stream_eos** continue to return 1 until the reader closes the stream.

CO_STREAM_OPEN

```
co_error co_stream_open(co_stream stream,
                        mode_t mode,
                        co_type type);
```

Header File

co.h

Description

This function opens a stream. It must be called from within a process run function.

Arguments

The arguments for co_stream_open are as follows:

co_stream stream	A pointer to a stream as passed on the process argument list.
mode_t mode	The stream's read/write mode. Valid modes are O_RDONLY and O_WRONLY. (There are no bidirectional streams.)
co_type type	The stream's data type. The type is normally expressed as a signed or unsigned integer of a specific width using the INT_TYPE(n), UINT_TYPE(n), or CHAR_TYPE macros.

Return Value

Possible return values are listed, with the resulting value of co_errno shown in parentheses:

co_err_none	Success (CO_ENOERROR).
co_err_invalid_arg	The stream argument is NULL (CO_ENULL_PTR) or the mode argument is neither O_RDONLY nor O_WRONLY (CO_EINVALID_MODE).
co_err_already_open	The stream is already open in this mode (CO_EALREADY_OPEN).

CO_STREAM_READ

```
co_error co_stream_read(co_stream stream,
                        void *buffer,
                        size_t buffersize);
```

Header File

co.h

Description

This function reads a data packet from a previously opened stream. This function must be called from within a process run function.

Arguments

The arguments for co_stream_read are as follows:

co_stream stream	A pointer to a stream as passed on the process argument list.
void *buffer	A pointer to the destination variable, where read data will be stored. The destination is typically a local variable or array element.
size_t buffersize	The size of the destination buffer in bytes. This number must be at least as large as the stream's packet width (in bytes), as specified by co_stream_create.

Return Value

Possible return values are listed, with the resulting value of co_errno shown in parentheses:

co_err_none	Success (CO_ENOERROR).
co_err_eos	Encountered an end-of-stream token (CO_EEOS).
co_err_invalid_arg	The stream argument is NULL (CO_ENULL_PTR) or the buffersize argument is smaller than the stream's packet

width in bytes
(CO_EINVALID_STREAM_WIDTH).

co_err_not_open The stream is not open for read
(CO_ENOT_OPEN).

Notes

The co_stream_read function must be used only on streams that have been opened with mode O_RDONLY.

The co_stream_read function blocks (waits) if the stream is empty or until the writer closes the stream.

CO_STREAM_READ_NB

```
co_error co_stream_read_nb(co_stream stream,
                          void *buffer,
                          size_t buffersize);
```

Header File

co.h

Description

This function reads a data packet from a previously opened stream but does not block if the stream is empty. This function must be called from within a process run function.

Arguments

The arguments for co_stream_read_nb are as follows:

co_stream stream A pointer to a stream as passed on the process argument list.

void *buffer A pointer to the destination variable where read data will be stored. The destination address is typically a local variable or array element.

size_t buffersize The size of the destination buffer, in bytes. This number must be at least as large as the

stream's packet width (in bytes), as specified by co_stream_create.

Return Value

Possible return values are listed, with the resulting value of co_errno shown
in parentheses:

co_err_none	Success (CO_ENOERROR).
co_err_eos	Encountered an end-of-stream token (CO_EEOS).
co_err_invalid_arg	The stream argument is NULL (CO_ENULL_PTR) or the buffersize argument is smaller than the stream's packet width in bytes (CO_EINVALID_STREAM_WIDTH).
co_err_not_open	The stream is not open for reading (CO_ENOT_OPEN).

Notes

The co_stream_read_nb function must be used only on streams that have
been opened with mode O_RDONLY.

The co_stream_read_nb function does not block if the stream is empty.
You must therefore check for a co_err_none return value to determine that a
data value has been successfully read from the stream.

CO_STREAM_WRITE

```
co_error co_stream_write(co_stream stream,
                         const void *buffer,
                         size_t buffersize);
```

Header File

co.h

Description

This function writes a data packet to a previously opened stream. This function must be called from within a process run function.

Arguments

The arguments for co_stream_write are as follows:

co_stream stream	A pointer to a stream as passed on the process argument list.
const void *buffer	A pointer to the source data. The source is typically a local variable or array element.
size_t buffersize	The size of the source buffer in bytes. This number must be at least as large as the stream's packet width (in bytes), as specified by co_stream_create.

Return Value

Possible return values are listed, with the resulting value of co_errno shown in parentheses:

co_err_none	Success (CO_ENOERROR).
co_err_eos	Encountered an end-of-stream token (CO_EEOS).
co_err_invalid_arg	The stream argument is NULL (CO_ENULL_PTR) or the buffersize argument is smaller than the stream's packet width in bytes (CO_EINVALID_STREAM_WIDTH).
co_err_not_open	The stream is not open for write (CO_ENOT_OPEN).

Notes

The co_stream_write function must be used only on streams that have been opened with mode O_WRONLY.

The co_stream_write function blocks (waits) if the output stream is already full.

COSIM_LOGWINDOW_CREATE

cosim_logwindow cosim_logwindow_create(const char *name);

Header File

cosim_log.h

Description

This function creates an Application Monitor log window. This function may be called from anyplace in an application.

Arguments

The argument for cosim_logwindow_create is as follows:

const char *name   A programmer-defined name. This name
                                may be any text string. It is used to visually
                                identify a specific log window when using
                                the Application Monitor.

Return Value

A pointer to the created log window. This return value may subsequently be used as an argument to the functions cosim_logwindow_write and cosim_logwindow_fwrite.

Returns NULL (and sets co_errno to CO_ELOGWINDOW_NOT_INITIALIZED) if a previous call to cosim_logwindow_init failed or cosim_logwindow_init was not called in the configuration function.

Also returns NULL (and sets co_errno to CO_ENULL_PTR) if a NULL name pointer ("") was given as an argument.

COSIM_LOGWINDOW_FWRITE

int cosim_logwindow_fwrite(cosim_logwindow log,
                           const char *format, ...);

Header File

cosim_log.h

Description

This function writes a formatted message to an Application Monitor log window. This function may be called from anyplace in an application.

Arguments

The arguments for cosim_logwindow_fwrite are as follows:

cosim_logwindow log	A log window pointer as returned by a preceding call to function cosim_logwindow_create.
const char *format	A null-terminated character string representing the format string for the message to be written, followed by zero or more arguments. Format strings and formatting characters follow the same rules as the standard C printf function.

Return Value

Returns 1 if successful. Returns 0 (and sets co_errno to CO_ENULL_PTR) if the log argument is NULL.

COSIM_LOGWINDOW_INIT

```
int cosim_logwindow_init();
```

Header File

cosim_log.h

Description

This function initializes the Application Monitor, which must be running before this function is called. This function must be called from an application's configuration subroutine.

Arguments

None.

Return Value

Returns 0 (and sets co_errno to CO_ECONFIG_ONLY) if called outside of the
configuration function, or if the Application Monitor is not running.

COSIM_LOGWINDOW_WRITE

```
int cosim_logwindow_write(cosim_logwindow log,
                          const char *msg);
```

Header File

cosim_log.h

Description

This function writes a message to an Application Monitor log window. This
function may be called from anyplace in an application.

Arguments

The arguments for cosim_logwindow_write are as follows:

cosim_logwindow log	A log window pointer as returned by a preceding call to function cosim_logwindow_create.
const char *msg	A null-terminated character string representing the message to be written.

Return Value

Returns 1 if successful. Returns 0 (and sets co_errno to CO_ENULL_PTR) if
the log argument is NULL.

APPENDIX D

Triple-DES
Source Listings

DES_HW.C

```
//////////////////////////////////////////////////////////
// 3DES encryption example.
//
// Copyright(c) 2003-2004 Impulse Accelerated Technologies, Inc.
//
// This implementation is based on public domain C source code by
// P. Karn and is similar to the algorithm described in Part V of
// Applied Cryptography by Bruce Schneier.
//

#ifdef WIN32
#include <windows.h>
#endif
#include <stdio.h>
#include "co.h"
#include "co_math.h"

#include "cosim_log.h"

#include "des.h"

#define BUFSIZE 16              /* buffer size for FIFO in hardware */

extern void des_producer(co_stream config_out_encrypt, co_stream config_out_decrypt,
```

```
                                  co_stream blocks_out,
                                  co_signal encrypt_c,
                                  co_parameter blocks_param);
extern void des_consumer(co_stream ic_decrypted_blocks, co_signal decrypt_finished);
extern void des_c(co_signal task_signal, co_signal next_task_signal);


// Primitive function F.
// Input is r, subkey array in keys, output is XORed into l.
// Each round consumes eight 6-bit subkeys, one for
// each of the 8 S-boxes, 2 longs for each round.
// Each long contains four 6-bit subkeys, each taking up a byte.
// The first long contains, from high to low end, the subkeys for
// S-boxes 1, 3, 5 & 7; the second contains the subkeys for S-boxes
// 2, 4, 6 & 8 (using the origin-1 S-box numbering in the standard,
// not the origin-0 numbering used elsewhere in this code)
// See comments elsewhere about the pre-rotated values of r and Spbox.
//
#define F_IC(l,r,key){\
    work = ((r >> 4) | (r << 28)) ^ key[0];\
    l ^= Spbox6[work & 0x3f];\
    l ^= Spbox4[(work >> 8) & 0x3f];\
    l ^= Spbox2[(work >> 16) & 0x3f];\
    l ^= Spbox0[(work >> 24) & 0x3f];\
    work = r ^ key[1];\
    l ^= Spbox7[work & 0x3f];\
    l ^= Spbox5[(work >> 8) & 0x3f];\
    l ^= Spbox3[(work >> 16) & 0x3f];\
    l ^= Spbox1[(work >> 24) & 0x3f];\
}


// This is the 3DES Impulse C hardware process. The inner loops will be
// automatically pipelined by the Impulse C compiler to take advantage of the
// target FPGA architecture.
//
void des_ic(co_stream config_in, co_stream blocks_in, co_stream blocks_out,
            co_parameter decrypt_param)
{
    co_error error;
    co_int8 i, k;
    co_uint8 block[8];
    co_uint32 ks[DES_KS_DEPTH][2];
    co_uint32 left, right, work;
    co_uint32 Spbox0[SPBOX_Y];
    co_uint32 Spbox1[SPBOX_Y];
    co_uint32 Spbox2[SPBOX_Y];
    co_uint32 Spbox3[SPBOX_Y];
    co_uint32 Spbox4[SPBOX_Y];
    co_uint32 Spbox5[SPBOX_Y];
```

```
        co_uint32 Spbox6[SPBOX_Y];
        co_uint32 Spbox7[SPBOX_Y];
        co_int32 decrypt = (int)decrypt_param;
        IF_SIM( cosim_logwindow log; )

        if ( decrypt == DES_DECRYPT ) {
                IF_SIM ( log = cosim_logwindow_create("des_ic decrypt"); )
        }
        else {
                IF_SIM ( log = cosim_logwindow_create("des_ic encrypt"); )
        }

        //  Get the key schedule, fill local array (note: this could also be
        //  done using a shared memory scheme).
        //
        co_stream_open(config_in, O_RDONLY, UINT_TYPE(32));

        IF_SIM( cosim_logwindow_fwrite(log, "3DES [Impulse C] reading keyschedule and SP
boxes...\n"); )
        for ( k = 0; k < 2; k++ ) {
                for ( i = 0; i < DES_KS_DEPTH; i++ ) {
                        if ( co_stream_read(config_in, &ks[i][k], sizeof(uint32)) == co_err_none ) {
                        }
                }
        }

        for ( k = 0; k < SPBOX_Y; k++ ) {
                co_stream_read(config_in, &Spbox0[k], sizeof(uint32));
        }
        for ( k = 0; k < SPBOX_Y; k++ ) {
                co_stream_read(config_in, &Spbox1[k], sizeof(uint32));
        }
        for ( k = 0; k < SPBOX_Y; k++ ) {
                co_stream_read(config_in, &Spbox2[k], sizeof(uint32));
        }
        for ( k = 0; k < SPBOX_Y; k++ ) {
                co_stream_read(config_in, &Spbox3[k], sizeof(uint32));
        }
        for ( k = 0; k < SPBOX_Y; k++ ) {
                co_stream_read(config_in, &Spbox4[k], sizeof(uint32));
        }
        for ( k = 0; k < SPBOX_Y; k++ ) {
                co_stream_read(config_in, &Spbox5[k], sizeof(uint32));
        }
        for ( k = 0; k < SPBOX_Y; k++ ) {
                co_stream_read(config_in, &Spbox6[k], sizeof(uint32));
        }
        for ( k = 0; k < SPBOX_Y; k++ ) {
                co_stream_read(config_in, &Spbox7[k], sizeof(uint32));
```

```
    }

    co_stream_close(config_in);

    /* Now read in the data, one block at a time */
    co_stream_open(blocks_in, O_RDONLY, UINT_TYPE(8));
    co_stream_open(blocks_out, O_WRONLY, UINT_TYPE(8));

    IF_SIM( cosim_logwindow_fwrite(log, "3DES [Impulse C] processing blocks...\n"); )
    while ( co_stream_read(blocks_in, &block[0], sizeof(uint8)) == co_err_none ) {
        for ( i = 1; i < DES_BLOCKSIZE; i++ ) {
            error = co_stream_read(blocks_in, &block[i], sizeof(uint8));
            if ( error != co_err_none ) break;
        }
        if ( error != co_err_none ) break;

        // Process the block...
        // Read input block and place in left/right in big-endian order
        //
        left = ((co_uint32)block[0] << 24)
            | ((co_uint32)block[1] << 16)
            | ((co_uint32)block[2] << 8)
            | (co_uint32)block[3];
        right = ((co_uint32)block[4] << 24)
            | ((co_uint32)block[5] << 16)
            | ((co_uint32)block[6] << 8)
            | (co_uint32)block[7];

        // Hoey's clever initial permutation algorithm, from Outerbridge
        // (see Schneier p 478)
        //
        // The convention here is the same as Outerbridge: rotate each
        // register left by 1 bit, i.e., so that "left" contains permuted
        // input bits 2, 3, 4, ... 1 and "right" contains 33, 34, 35, ... 32
        // (using origin-1 numbering as in the FIPS). This allows us to avoid
        // one of the two rotates that would otherwise be required in each of
        // the 16 rounds.
        //
        work = ((left >> 4) ^ right) & 0x0f0f0f0f;
        right ^= work;
        left ^= work << 4;
        work = ((left >> 16) ^ right) & 0xffff;
        right ^= work;
        left ^= work << 16;
        work = ((right >> 2) ^ left) & 0x33333333;
        left ^= work;
        right ^= (work << 2);
        work = ((right >> 8) ^ left) & 0xff00ff;
        left ^= work;
```

```
right ^= (work << 8);
right = (right << 1) | (right >> 31);
work = (left ^ right) & 0xaaaaaaaa;
left ^= work;
right ^= work;
left = (left << 1) | (left >> 31);

/* First key */
F_IC(left,right,ks[0]);
F_IC(right,left,ks[1]);
F_IC(left,right,ks[2]);
F_IC(right,left,ks[3]);
F_IC(left,right,ks[4]);
F_IC(right,left,ks[5]);
F_IC(left,right,ks[6]);
F_IC(right,left,ks[7]);
F_IC(left,right,ks[8]);
F_IC(right,left,ks[9]);
F_IC(left,right,ks[10]);
F_IC(right,left,ks[11]);
F_IC(left,right,ks[12]);
F_IC(right,left,ks[13]);
F_IC(left,right,ks[14]);
F_IC(right,left,ks[15]);

/* Second key (must be created in opposite mode to first key) */
F_IC(right,left,ks[16]);
F_IC(left,right,ks[17]);
F_IC(right,left,ks[18]);
F_IC(left,right,ks[19]);
F_IC(right,left,ks[20]);
F_IC(left,right,ks[21]);
F_IC(right,left,ks[22]);
F_IC(left,right,ks[23]);
F_IC(right,left,ks[24]);
F_IC(left,right,ks[25]);
F_IC(right,left,ks[26]);
F_IC(left,right,ks[27]);
F_IC(right,left,ks[28]);
F_IC(left,right,ks[29]);
F_IC(right,left,ks[30]);
F_IC(left,right,ks[31]);

/* Third key */
F_IC(left,right,ks[32]);
F_IC(right,left,ks[33]);
F_IC(left,right,ks[34]);
F_IC(right,left,ks[35]);
F_IC(left,right,ks[36]);
```

```
            F_IC(right,left,ks[37]);
            F_IC(left,right,ks[38]);
            F_IC(right,left,ks[39]);
            F_IC(left,right,ks[40]);
            F_IC(right,left,ks[41]);
            F_IC(left,right,ks[42]);
            F_IC(right,left,ks[43]);
            F_IC(left,right,ks[44]);
            F_IC(right,left,ks[45]);
            F_IC(left,right,ks[46]);
            F_IC(right,left,ks[47]);


            /* Inverse permutation, also from Hoey via Outerbridge and Schneier */
            right = (right << 31) | (right >> 1);
            work = (left ^ right) & 0xaaaaaaaa;
            left ^= work;
            right ^= work;
            left = (left >> 1) | (left << 31);
            work = ((left >> 8) ^ right) & 0xff00ff;
            right ^= work;
            left ^= work << 8;
            work = ((left >> 2) ^ right) & 0x33333333;
            right ^= work;
            left ^= work << 2;
            work = ((right >> 16) ^ left) & 0xffff;
            left ^= work;
            right ^= work << 16;
            work = ((right >> 4) ^ left) & 0x0f0f0f0f;
            left ^= work;
            right ^= work << 4;


            /* Put the block into the output stream with final swap */
            block[0] = (co_int8) (right >> 24);
            block[1] = (co_int8) (right >> 16);
            block[2] = (co_int8) (right >> 8);
            block[3] = (co_int8) right;
            block[4] = (co_int8) (left >> 24);
            block[5] = (co_int8) (left >> 16);
            block[6] = (co_int8) (left >> 8);
            block[7] = (co_int8) left;

            for ( i = 0; i < DES_BLOCKSIZE; i++ ) {
                  co_stream_write(blocks_out, &block[i], sizeof(uint8));
            }
      }
}
IF_SIM( cosim_logwindow_fwrite(log, "3DES [Impulse C] done processing!\n"); )

co_stream_close(blocks_out);
co_stream_close(blocks_in);
```

```
    }

    ////////////////////////////////////////////////
    // This is the configuration function.
    //
    void config_des(void *arg)
    {
        int blocks = (int)arg;

        co_stream config_encrypt_ic,
                        config_decrypt_ic,
                        blocks_plaintext_ic,
                        blocks_encrypted_ic,
                        blocks_decrypted_ic;

        co_signal encrypt_c, encrypt_finished_c, decrypt_finished_c;

        co_process producer_process;
        co_process des_encrypt_ic_process;
        co_process des_encrypt_c_process;
        co_process des_decrypt_ic_process;
        co_process des_decrypt_c_process;
        co_process consumer_process;

        IF_SIM( if ( ! cosim_logwindow_init() ) {
                printf("cosim_logwindow_init failed!\n");
        })

        encrypt_c = co_signal_create("encrypt_c");
        encrypt_finished_c = co_signal_create("encrypt_finished_c");
        decrypt_finished_c = co_signal_create("decrypt_finished_c");

        config_encrypt_ic = co_stream_create("config_encrypt_ic", UINT_TYPE(32), BUFSIZE);
        config_decrypt_ic = co_stream_create("config_decrypt_ic", UINT_TYPE(32), BUFSIZE);
        blocks_plaintext_ic = co_stream_create("blocks_plaintext_ic", UINT_TYPE(8), BUFSIZE);
        blocks_encrypted_ic = co_stream_create("blocks_encrypted_ic", UINT_TYPE(8), BUFSIZE);
        blocks_decrypted_ic = co_stream_create("blocks_decrypted_ic", UINT_TYPE(8), BUFSIZE);

        producer_process = co_process_create("des_producer", (co_function)des_producer,
                            5,
                            config_encrypt_ic,
                            config_decrypt_ic,
                            blocks_plaintext_ic,
                            encrypt_c,
                            blocks);

        des_encrypt_ic_process = co_process_create("des_encrypt_ic", (co_function)des_ic,
                                4,
                                config_encrypt_ic,
```

```
                                          blocks_plaintext_ic,
                                          blocks_encrypted_ic,
                                          DES_ENCRYPT);

    des_encrypt_c_process  = co_process_create("des_encrypt_c", (co_function)des_c,
                                          2,
                                          encrypt_c,
                                          encrypt_finished_c);

    des_decrypt_ic_process = co_process_create("des_decrypt_ic", (co_function)des_ic,
                                          4,
                                          config_decrypt_ic,
                                          blocks_encrypted_ic,
                                          blocks_decrypted_ic,
                                          DES_DECRYPT);

    des_decrypt_c_process  = co_process_create("des_decrypt_c", (co_function)des_c,
                                          2,
                                          encrypt_finished_c,
                                          decrypt_finished_c);

    consumer_process = co_process_create("des_consumer",(co_function)des_consumer,
                                          2,
                                          blocks_decrypted_ic,
                                          decrypt_finished_c);

    // Assign the encryption process to hardware
    co_process_config(des_encrypt_ic_process, co_loc, "PE0");
}

//////////////////////////////////////////////
// This is the co_initialize function, which is
// called from main(). See 3des.c
//
co_architecture co_initialize(int blocks)
{
    return(co_architecture_create("des3", "Generic_VHDL", config_des, (void *)blocks));
}
```

DES.C

```
/////////////////////////////////////////////////////////////////
// 3DES encryption example.
//
// Test bench for desktop (software) simulation.
//
// Copyright(c) 2003-2004 Impulse Accelerated Technologies, Inc.
```

```
//
// This implementation is based on public domain C source code by
// P. Karn and is similar to the algorithm described in Part V of
// Applied Cryptography by Bruce Schneier.
//
// When reading from a file, note that the C processes hold three copies
// of the file's data in memory.

#ifdef WIN32
#include <windows.h>
#include <sys/types.h>
#else  // ! WIN32
#include <unistd.h>
#endif
#include <stdio.h>
#include <sys/stat.h>
#include "co.h"
#include "co_math.h"
#include "cosim_log.h"

#include "des.h"

#define VHDL_GEN          1      /* Generate VHDL config file for test bench */

#define DEFAULT_BLOCKS 100  /* Default number of blocks to encrypt */

extern co_architecture co_initialize(int blocks); // See des_hw.c

void usage(char* ProgramName)
{
    printf("USAGE: %s <# characters> (%d is default, 0 to read all characters in file; \
            will be rounded up to nearest %i-character block)\n",
           ProgramName, DEFAULT_BLOCKS * DES_BLOCKSIZE, DES_BLOCKSIZE);
}

int parse_input_pars(int argc, char **argv)
{
    int blocks = DEFAULT_BLOCKS;
    int characters;

    if ( argc == 2 ) {
        sscanf(argv[1], "%i", &characters);
        if ( characters >= 0 ) {
            blocks = characters / DES_BLOCKSIZE;
            if ( characters % DES_BLOCKSIZE ) {
                blocks++;
            }
        }
        else {
```

```
                        usage(argv[0]);
                }
        }
        else if (argc != 1) {
                usage(argv[0]);
        }
        return blocks;
}


// Combined SP lookup table, linked in (see sp.c).
// For best results, ensure that this is aligned on a 32-bit boundary;
// Borland C++ 3.1 doesn't guarantee this!
//
extern unsigned long Spbox[SPBOX_X][SPBOX_Y];        /* Combined S and P boxes */

/* Keyschedules */
DES3_KS Ks_encrypt;
DES3_KS Ks_decrypt;

/* Block data for C processes */
unsigned char * PlaintextBlocks;
unsigned char * EncryptedBlocks;
unsigned char * DecryptedBlocks;
unsigned long NumBlocks;

void des_producer(co_stream config_out_encrypt, co_stream config_out_decrypt,
                        co_stream blocks_out,
                        co_signal encrypt_c,
                        co_parameter blocks_param)
{
        unsigned int blocks = (unsigned int)blocks_param;
        unsigned int i, k;
        unsigned char * key = (unsigned char *) "Gflk jqo40978J0dmm$%@878"; /* 24 bytes */
        co_uint8 blockElement;
        cosim_logwindow log = cosim_logwindow_create("des_producer");

        // Read plaintext from a file
        int plaintextChar,c;
        const char * plaintextFilename = "JohnDough.txt";
        FILE * plaintextFile;
        struct stat plaintextFileStat;
#ifdef VHDL_GEN          // Generate VHDL package for test bench
        int cnt;
        FILE * fout; // test bench output file
        const char * testbenchFilename = "config_des.vhd";
        fout = fopen(testbenchFilename, "w");
        if ( fout == NULL ) {
                fprintf(stderr, "Error opening test bench output file %s\n", testbenchFilename);
                printf("\nPress the Enter key to continue.\n");
```

```
            c = getc(stdin);
            exit(-1);
      }
  #endif

      plaintextFile = fopen(plaintextFilename, "r");
      if ( plaintextFile == NULL ) {
            fprintf(stderr, "Error opening plaintext file %s\n", plaintextFilename);
            printf("\nPress the Enter key to continue.\n");
            c = getc(stdin);
            exit(-1);
      }

      NumBlocks = blocks;
      if ( blocks == 0 ) {
            if ( fstat(fileno(plaintextFile), &plaintextFileStat) != 0 ) {
                  fprintf(stderr, "Error reading size of plaintext file %s\n", plaintextFilename);
                  printf("\nPress the Enter key to continue.\n");
                  c = getc(stdin);
                  exit(-1);
            }
            NumBlocks = plaintextFileStat.st_size / DES_BLOCKSIZE;
            if ( plaintextFileStat.st_size % DES_BLOCKSIZE ) {
                  NumBlocks++;
            }
      }
      // Allocate memory for the global arrays the C processes will use to get
      // at the block data.
      PlaintextBlocks = (unsigned char *) malloc(sizeof(unsigned char)
         * DES_BLOCKSIZE * NumBlocks);
      EncryptedBlocks = (unsigned char *) malloc(sizeof(unsigned char)
         * DES_BLOCKSIZE * NumBlocks);
      DecryptedBlocks = (unsigned char *) malloc(sizeof(unsigned char)
         * DES_BLOCKSIZE * NumBlocks);

      for ( i = 0; i < NumBlocks; i++ ) {
            for ( k = 0; k < DES_BLOCKSIZE; k++ ) {
                  if ( (plaintextChar = getc(plaintextFile)) != EOF ) {
                        blockElement = (co_uint8)plaintextChar;
                  }
                  else {
                        blockElement = (co_uint8)'\0';
                  }
                  PlaintextBlocks[(i * DES_BLOCKSIZE) + k] = (unsigned char)blockElement;
            }
      }

      // Generate the keyschedules for encryption and decryption
      des3key(Ks_encrypt, key, DES_ENCRYPT);
```

```
        des3key(Ks_decrypt, key, DES_DECRYPT);

        // Encrypt

        // We can start the C encryption process, since the block and keyschedule data has all
        // been written to memory.  Decryption will be triggered by the encryption process when
        // it finishes.
        co_signal_post(encrypt_c, DES_ENCRYPT);

#ifdef VHDL_GEN
        // See Tutorial #4 for information about VHDL test bench generation
        fprintf(fout,"-- Automatically generated file. See 3des.c for details.\n");
        fprintf(fout,"-- See also tutorial #4 in the Impulse C User's Guide.\n");
        fprintf(fout,"library ieee;\n");
        fprintf(fout,"use ieee.std_logic_1164.all;\n");
        fprintf(fout,"\npackage config is\n");
#endif

        // Send the keyschedule data
        co_stream_open(config_out_encrypt, O_WRONLY, UINT_TYPE(32));
#ifdef VHDL_GEN
        fprintf(fout," constant kscount : integer := 95;\n");
        fprintf(fout," type ks is array (natural range <>) of std_ulogic_vector(31 downto 0);\n");
        fprintf(fout," constant keyschedule : ks (0 to 95) := (\n");
        cnt=0;
#endif
        for ( k = 0; k < 2; k++ ) {
                for ( i = 0; i < DES_KS_DEPTH; i++ ) {
#ifdef VHDL_GEN
                        fprintf(fout,"   X\"%08x\"",Ks_encrypt[i][k]);
                        if (cnt++ == 95) fprintf(fout,"\n"); else fprintf(fout,",\n");
#endif
                        co_stream_write(config_out_encrypt, &Ks_encrypt[i][k], sizeof(unsigned long));
                }
        }
#ifdef VHDL_GEN
        fprintf(fout," );\n");
        fprintf(fout," constant spcount : integer := 512;\n");
        fprintf(fout," type sp is array (natural range <>) of std_ulogic_vector(31 downto 0);\n");
        fprintf(fout," constant spboxes : sp (0 to 511) := (\n");
        cnt=0;
#endif
        for ( i = 0; i < SPBOX_X; i++ ) {
                for ( k = 0; k < SPBOX_Y; k++ ) {
#ifdef VHDL_GEN
                        fprintf(fout,"   X\"%08x\"",Spbox[i][k]);
                        if (cnt++ == 511) fprintf(fout,"\n"); else fprintf(fout,",\n");
#endif
                        co_stream_write(config_out_encrypt, &Spbox[i][k], sizeof(unsigned long));
```

```
            }
        }
        co_stream_close(config_out_encrypt);

        // Decrypt

        // Send the keyschedule data
        co_stream_open(config_out_decrypt, O_WRONLY, UINT_TYPE(32));
        for ( k = 0; k < 2; k++ ) {
            for ( i = 0; i < DES_KS_DEPTH; i++ ) {
                co_stream_write(config_out_decrypt, &Ks_decrypt[i][k], sizeof(unsigned long));
            }
        }
        for ( i = 0; i < SPBOX_X; i++ ) {
            for ( k = 0; k < SPBOX_Y; k++ ) {
                co_stream_write(config_out_decrypt, &Spbox[i][k], sizeof(unsigned long));
            }
        }
        co_stream_close(config_out_decrypt);

        // Send the plaintext block data to the Impulse C encryption process
        co_stream_open(blocks_out, O_WRONLY, UINT_TYPE(8));
#ifdef VHDL_GEN
        fprintf(fout," );\n");
        fprintf(fout,"end package;\n");
#endif
        cosim_logwindow_write(log, "Sending plaintext...\n\n");
        for ( i = 0; i < NumBlocks; i++ ) {
            for ( k = 0; k < DES_BLOCKSIZE; k++ ) {
                blockElement = PlaintextBlocks[(i * DES_BLOCKSIZE) + k];
                co_stream_write(blocks_out, &blockElement, sizeof(blockElement));
            }
        }
        cosim_logwindow_write(log, "\n\nDone sending plaintext.\n");
        co_stream_close(blocks_out);
}

// Primitive function F.
// Input is r, subkey array in keys, output is XORed into l.
// Each round consumes eight 6-bit subkeys, one for
// each of the 8 S-boxes, 2 longs for each round.
// Each long contains four 6-bit subkeys, each taking up a byte.
// The first long contains, from high to low end, the subkeys for
// S-boxes 1, 3, 5 & 7; the second contains the subkeys for S-boxes
// 2, 4, 6 & 8 (using the origin-1 S-box numbering in the standard,
// not the origin-0 numbering used elsewhere in this code)
// See comments elsewhere about the pre-rotated values of r and Spbox.
//
#define F(l,r,key){\
```

```
    work = ((r >> 4) | (r << 28)) ^ key[0];\
    l ^= Spbox[6][work & 0x3f];\
    l ^= Spbox[4][(work >> 8) & 0x3f];\
    l ^= Spbox[2][(work >> 16) & 0x3f];\
    l ^= Spbox[0][(work >> 24) & 0x3f];\
    work = r ^ key[1];\
    l ^= Spbox[7][work & 0x3f];\
    l ^= Spbox[5][(work >> 8) & 0x3f];\
    l ^= Spbox[3][(work >> 16) & 0x3f];\
    l ^= Spbox[1][(work >> 24) & 0x3f];\
}


// This is the plain C 3DES process.  It reads the keyschedule and block
// inputs from global variables initialized in the des_producer process
// and does all processing using standard C code.
//
// An Impulse C signal is used to notify the process of which task (encrypt or
// decrypt) to start. Another signal, sent after the task is completed, notifies
// the listener of which task it should start.  cosim_logwindow functions are
// used to display input and output for comparison with the des_ic processes.
// Otherwise, this function is plain C.
//
void des_c(co_signal task_signal, co_signal next_task_signal)
{
    int i;
    unsigned int blockCount = 0;
    unsigned char block[8];
    unsigned long left,right,work;
    co_int32 task;
    DES3_KS * ks;
    unsigned char * inputBlocks;
    unsigned char * outputBlocks;
    IF_SIM( cosim_logwindow log; )

    co_signal_wait(task_signal, &task);
    if (task == DES_ENCRYPT ) {
        IF_SIM ( log = cosim_logwindow_create("des_c encrypt"); )
        ks = &Ks_encrypt;
        inputBlocks = PlaintextBlocks;
        outputBlocks = EncryptedBlocks;
    }
    else if ( task == DES_DECRYPT ) {
        IF_SIM ( log = cosim_logwindow_create("des_c decrypt"); )
        ks = &Ks_decrypt;
        inputBlocks = EncryptedBlocks;
        outputBlocks = DecryptedBlocks;
    }
    else {
        // Don't know what to do, so just exit.
```

```
            return;
    }

    IF_SIM( cosim_logwindow_write(log, "3DES [C] processing blocks...\n"); )
    while ( blockCount < NumBlocks ) {
            for ( i = 0; i < DES_BLOCKSIZE; i++ ) {
                    block[i] = inputBlocks[(blockCount * DES_BLOCKSIZE) + i];
            }

            // Process the block...
            // Read input block and place in left/right in big-endian order
            //
            left = ((unsigned long)block[0] << 24)
                    |((unsigned long)block[1] << 16)
                    | ((unsigned long)block[2] << 8)
                    | (unsigned long)block[3];
            right = ((unsigned long)block[4] << 24)
                    | ((unsigned long)block[5] << 16)
                    | ((unsigned long)block[6] << 8)
                    | (unsigned long)block[7];

            // Hoey's clever initial permutation algorithm, from Outerbridge
            // (see Schneier p 478)
            //
            // The convention here is the same as Outerbridge: rotate each
            // register left by 1 bit, i.e., so that "left" contains permuted
            // input bits 2, 3, 4, ... 1 and "right" contains 33, 34, 35, ... 32
            // (using origin-1 numbering as in the FIPS). This allows us to avoid
            // one of the two rotates that would otherwise be required in each of
            // the 16 rounds.
            //
            work = ((left >> 4) ^ right) & 0x0f0f0f0f;
            right ^= work;
            left ^= work << 4;
            work = ((left >> 16) ^ right) & 0xffff;
            right ^= work;
            left ^= work << 16;
            work = ((right >> 2) ^ left) & 0x33333333;
            left ^= work;
            right ^= (work << 2);
            work = ((right >> 8) ^ left) & 0xff00ff;
            left ^= work;
            right ^= (work << 8);
            right = (right << 1) | (right >> 31);
            work = (left ^ right) & 0xaaaaaaaa;
            left ^= work;
            right ^= work;
            left = (left << 1) | (left >> 31);
```

```
/* First key */
F(left,right,(*ks)[0]);
F(right,left,(*ks)[1]);
F(left,right,(*ks)[2]);
F(right,left,(*ks)[3]);
F(left,right,(*ks)[4]);
F(right,left,(*ks)[5]);
F(left,right,(*ks)[6]);
F(right,left,(*ks)[7]);
F(left,right,(*ks)[8]);
F(right,left,(*ks)[9]);
F(left,right,(*ks)[10]);
F(right,left,(*ks)[11]);
F(left,right,(*ks)[12]);
F(right,left,(*ks)[13]);
F(left,right,(*ks)[14]);
F(right,left,(*ks)[15]);

/* Second key (must be created in opposite mode to first key) */
F(right,left,(*ks)[16]);
F(left,right,(*ks)[17]);
F(right,left,(*ks)[18]);
F(left,right,(*ks)[19]);
F(right,left,(*ks)[20]);
F(left,right,(*ks)[21]);
F(right,left,(*ks)[22]);
F(left,right,(*ks)[23]);
F(right,left,(*ks)[24]);
F(left,right,(*ks)[25]);
F(right,left,(*ks)[26]);
F(left,right,(*ks)[27]);
F(right,left,(*ks)[28]);
F(left,right,(*ks)[29]);
F(right,left,(*ks)[30]);
F(left,right,(*ks)[31]);

/* Third key */
F(left,right,(*ks)[32]);
F(right,left,(*ks)[33]);
F(left,right,(*ks)[34]);
F(right,left,(*ks)[35]);
F(left,right,(*ks)[36]);
F(right,left,(*ks)[37]);
F(left,right,(*ks)[38]);
F(right,left,(*ks)[39]);
F(left,right,(*ks)[40]);
F(right,left,(*ks)[41]);
F(left,right,(*ks)[42]);
F(right,left,(*ks)[43]);
```

```
        F(left,right,(*ks)[44]);
        F(right,left,(*ks)[45]);
        F(left,right,(*ks)[46]);
        F(right,left,(*ks)[47]);

        /* Inverse permutation, also from Hoey via Outerbridge and Schneier */
        right = (right << 31) | (right >> 1);
        work = (left ^ right) & 0xaaaaaaaa;
        left ^= work;
        right ^= work;
        left = (left >> 1) | (left  << 31);
        work = ((left >> 8) ^ right) & 0xff00ff;
        right ^= work;
        left ^= work << 8;
        work = ((left >> 2) ^ right) & 0x33333333;
        right ^= work;
        left ^= work << 2;
        work = ((right >> 16) ^ left) & 0xffff;
        left ^= work;
        right ^= work << 16;
        work = ((right >> 4) ^ left) & 0x0f0f0f0f;
        left ^= work;
        right ^= work << 4;

        /* Put the block into the output stream with final swap */
        block[0] = (co_int8) (right >> 24);
        block[1] = (co_int8) (right >> 16);
        block[2] = (co_int8) (right >> 8);
        block[3] = (co_int8) right;
        block[4] = (co_int8) (left >> 24);
        block[5] = (co_int8) (left >> 16);
        block[6] = (co_int8) (left >> 8);
        block[7] = (co_int8) left;

        for ( i = 0; i < DES_BLOCKSIZE; i++ ) {
                outputBlocks[(blockCount * DES_BLOCKSIZE) + i] = block[i];
        }

        ++blockCount;
    }
    IF_SIM( cosim_logwindow_write(log, "3DES [C] done processing!\n"); )

    if ( task == DES_ENCRYPT ) {
        co_signal_post(next_task_signal, DES_DECRYPT);
    }
    else {
        co_signal_post(next_task_signal, DES_ENCRYPT);
    }
}
```

```
void des_consumer(co_stream ic_decrypted_blocks, co_signal decrypt_finished)
{
    uint8 icData;
    int i = 0;
    int nonMatchingBlockData = 0, nonMatchingPlaintextData = 0, failure = 0;
    cosim_logwindow log = cosim_logwindow_create("des_consumer");

    co_signal_wait(decrypt_finished, NULL);

    co_stream_open(ic_decrypted_blocks, O_RDONLY, UINT_TYPE(8));

    cosim_logwindow_write(log, "Consumer comparing decrypted data...\n");
    while ( co_stream_read(ic_decrypted_blocks, &icData, sizeof(uint8)) == co_err_none ) {
        printf("%c", (unsigned char)DecryptedBlocks[i]);
        if ( icData != DecryptedBlocks[i] ) {
            nonMatchingBlockData++;
        }
        if ( icData != PlaintextBlocks[i] ) {
            nonMatchingPlaintextData++;
        }
        if ( DecryptedBlocks[i] != PlaintextBlocks[i] ) {
            nonMatchingPlaintextData++;
        }
        i++;
    }
    co_stream_close(ic_decrypted_blocks);

    if ( nonMatchingBlockData > 0 ) {
        cosim_logwindow_fwrite(log, "\nFailure!  Decrypted output of Impulse C and C \
            implementations doesn't match: %i differences", nonMatchingBlockData);
        printf("\n\nFailure!  Decrypted outputs of Impulse C and C implementations do \
            not match: %i differences", nonMatchingBlockData);
        failure = 1;
    }
    if ( nonMatchingPlaintextData > 0 ) {
        cosim_logwindow_fwrite(log, "\nFailure!  Decrypted output doesn't match plaintext \
            output: %i differences", nonMatchingPlaintextData);
        printf("\n\nFailure!  Decrypted output doesn't match plaintext output: %i differences",
            nonMatchingPlaintextData);
        failure = 1;
    }

    if ( failure != 1 ) {
        cosim_logwindow_write(log, "\nSuccess!  Decrypted output of Impulse C and C \
            implementations is the same and matches the plaintext!");
        printf("\n\nSuccess!  Decrypted output of Impulse C and C implementations is the \
            same and matches the plaintext (%i characters)!", i);
    }
```

```
        free(DecryptedBlocks);
        free(EncryptedBlocks);
        free(PlaintextBlocks);
    }

    int main(int argc, char *argv[])
    {
        co_architecture my_arch;
        int c;
        int blocks = parse_input_pars(argc, argv);

        printf("Impulse C is Copyright 2003 Impulse Accelerated Technologies, Inc.\n");

        my_arch = co_initialize(blocks);  // See des_hw.c
        co_execute(my_arch);

        printf("\n\nApplication complete. Press the Enter key to continue.\n");
        c = getc(stdin);

        return(0);
    }
```

DES_SW.C

```
//////////////////////////////////////////////////////////
// 3DES encryption example.
//
// Copyright(c) 2003 Impulse Accelerated Technologies, Inc.
//
// This implementation is based on public domain C source code by
// P. Karn and is similar to the algorithm described in Part V of
// Applied Cryptography by Bruce Schneier.
//

#ifdef WIN32
#include <windows.h>
#else
#include "xparameters.h"
#define TIMED_TEST 1
#ifdef TIMED_TEST
#include "xtmrctr.h"
#endif
#endif
#include <stdio.h>
```

```
#include "co.h"

#include "des.h"

/* 3DES constants, don't change these */
#define BLOCKSIZE 8      /* unsigned chars per block */
#define KS_DEPTH 48      /* key pairs */

#ifdef IMPULSE_C_TARGET
#define printf xil_printf
#ifdef TIMED_TEST
XTmrCtr TimerCounter;
#endif
#endif

extern co_architecture co_initialize(void *);

/* Block data for C process */
static unsigned char Blocks[]={0x6f,0x98,0x26,0x35,0x02,0xc9,0x83,0xd7};
static unsigned long Iterations=1000;
static int Asmversion = 0;

#include "sp.c"

// Combined SP lookup table, linked in
// For best results, ensure that this is aligned on a 32-bit boundary;
// Borland C++ 3.1 doesn't guarantee this!
//
#define SPBOX_X 8
#define SPBOX_Y 64
extern unsigned long Spbox[SPBOX_X][SPBOX_Y];      /* Combined S and P boxes */

/* Keyschedule */
DES3_KS Ks;

/* Portable C code to create DES key schedules from user-provided keys
 * This doesn't have to be fast unless you're cracking keys or UNIX
 * passwords
 */

/* Key schedule-related tables from FIPS-46 */

/* permuted choice table (key) */
static unsigned char pc1[] = {
    57, 49, 41, 33, 25, 17,  9,
     1, 58, 50, 42, 34, 26, 18,
    10,  2, 59, 51, 43, 35, 27,
    19, 11,  3, 60, 52, 44, 36,
```

```
        63, 55, 47, 39, 31, 23, 15,
         7, 62, 54, 46, 38, 30, 22,
        14,  6, 61, 53, 45, 37, 29,
        21, 13,  5, 28, 20, 12,  4
    };

    /* number left rotations of pc1 */
    static unsigned char totrot[] = {
        1,2,4,6,8,10,12,14,15,17,19,21,23,25,27,28
    };

    /* permuted choice key (table) */
    static unsigned char pc2[] = {
        14, 17, 11, 24,  1,  5,
         3, 28, 15,  6, 21, 10,
        23, 19, 12,  4, 26,  8,
        16,  7, 27, 20, 13,  2,
        41, 52, 31, 37, 47, 55,
        30, 40, 51, 45, 33, 48,
        44, 49, 39, 56, 34, 53,
        46, 42, 50, 36, 29, 32
    };

    /* End of DES-defined tables */


    /* bit 0 is left-most in byte */
    static int bytebit[] = {
        0200,0100,040,020,010,04,02,01
    };


    /// Generate key schedule for encryption or decryption
    //  depending on the value of "decrypt"
    //
    void deskey(k,key,decrypt)
    unsigned long k[16][2];        /* Key schedule array */
    unsigned char *key;    /* 64 bits (will use only 56) */
    int decrypt;           /* 0 = encrypt, 1 = decrypt */
    {
        unsigned char pc1m[56];    /* place to modify pc1 into */
        unsigned char pcr[56];     /* place to rotate pc1 into */
        register int i,j,l;
        int m;
        unsigned char ks[8];

        for (j=0; j<56; j++) {     /* convert pc1 to bits of key */
            l=pc1[j]-1;    /* integer bit location */
            m = l & 07;    /* find bit     */
```

```
            pc1m[j]=(key[l>>3] &    /* find which key byte l is in */
               bytebit[m]) /* and which bit of that byte */
               ? 1 : 0;   /* and store 1-bit result */
      }
    for (i=0; i<16; i++) {     /* key chunk for each iteration */
       memset(ks,0,sizeof(ks));   /* Clear key schedule */
       for (j=0; j<56; j++)   /* rotate pc1 the right amount */
          pcr[j] = pc1m[(l=j+totrot[decrypt? 15-i : i])<(j<28? 28 : 56) ? l: l-28];
          /* rotate left and right halves independently */
       for (j=0; j<48; j++){   /* select bits individually */
          /* check bit that goes to ks[j] */
          if (pcr[pc2[j]-1]){
             /* mask it in if it's there */
             l= j % 6;
             ks[j/6] |= bytebit[l] >> 2;
          }
       }
       /* Now convert to packed odd/even interleaved form */
       k[i][0] = ((long)ks[0] << 24)
        | ((long)ks[2] << 16)
        | ((long)ks[4] << 8)
        | ((long)ks[6]);
       k[i][1] = ((long)ks[1] << 24)
        | ((long)ks[3] << 16)
        | ((long)ks[5] << 8)
        | ((long)ks[7]);
       if(Asmversion){
          /* The assembler versions pre-shift each subkey 2 bits
           * so the Spbox indexes are already computed
           */
          k[i][0] <<= 2;
          k[i][1] <<= 2;
       }
    }
}

// Generate key schedule for triple DES in E-D-E (or D-E-D) mode.
//
// The key argument is taken to be 24 bytes. The first 8 bytes are K1
// for the first stage, the second 8 bytes are K2 for the middle stage
// and the third 8 bytes are K3 for the last stage
//
void des3key(k,key,decrypt)
unsigned long k[48][2];
unsigned char *key;    /* 192 bits (will use only 168) */
int decrypt;         /* 0 = encrypt, 1 = decrypt */
{
   if(!decrypt){
      deskey(&k[0],&key[0],0);
```

```
        deskey(&k[16],&key[8],1);
        deskey(&k[32],&key[16],0);
    } else {
        deskey(&k[32],&key[0],1);
        deskey(&k[16],&key[8],0);
        deskey(&k[0],&key[16],1);
    }
}

void des_test(co_stream config_out, co_stream blocks_out, co_stream input_stream,
        co_parameter iter_param)
{
  int iterations = (int)iter_param;
  int i, k;
  unsigned char block[8];
  uint8 blockElement;
  unsigned long data,err;
#ifdef IMPULSE_C_TARGET
#ifdef TIMED_TEST
  Xuint32 counter;
#endif
#endif

  /* Send the keyschedule data */
  HW_STREAM_OPEN(des_test,config_out, O_WRONLY, UINT_TYPE(32));

  for ( k = 0; k < 2; k++ ) {
   for ( i = 0; i < KS_DEPTH; i++ ) {
    data=Ks[i][k];
    HW_STREAM_WRITE(des_test,config_out,data);
   }
  }

  for ( i = 0; i < SPBOX_X; i++ ) {
   for ( k = 0; k < SPBOX_Y; k++ ) {
    data=Spbox[i][k];
    HW_STREAM_WRITE(des_test,config_out,data);
   }
  }

  HW_STREAM_CLOSE(des_test,config_out);

  /* Send the same random block data to both processes */
  HW_STREAM_OPEN(des_test,blocks_out, O_WRONLY, UINT_TYPE(8));
  HW_STREAM_OPEN(des_test,input_stream,O_RDONLY,UINT_TYPE(8));

#ifdef IMPULSE_C_TARGET
#ifdef TIMED_TEST
  XTmrCtr_Reset(&TimerCounter,0);
```

```
#endif
#endif

  for ( i = 0; i < iterations; i++ ) {
    for ( k = 0; k < BLOCKSIZE; k++ ) {
      blockElement = Blocks[k];
      HW_STREAM_WRITE(des_test,blocks_out,blockElement);
    }

    for ( k = 0; k < BLOCKSIZE; k++ ) {
      HW_STREAM_READ(des_test,input_stream,blockElement,err);
      block[k]=blockElement;
    }
  }

#ifdef IMPULSE_C_TARGET
#ifdef TIMED_TEST
  counter=XTmrCtr_GetValue(&TimerCounter,0);
#endif
#endif

  HW_STREAM_CLOSE(des_test,blocks_out);
  HW_STREAM_CLOSE(des_test,input_stream);

#ifdef IMPULSE_C_TARGET
#ifdef TIMED_TEST
  xil_printf("FPGA processing done (%d ticks).\n\r",counter);
#else
  xil_printf("FPGA processing done.\n\r");
#endif
#endif

  printf("FPGA block out:");
  for (i=0; i<BLOCKSIZE; i++) {
    printf(" %02x",block[i]);
  }
  printf("\n\r");
}


// Primitive function F.
// Input is r, subkey array in keys, output is XORed into l.
// Each round consumes eight 6-bit subkeys, one for
// each of the 8 S-boxes, 2 longs for each round.
// Each long contains four 6-bit subkeys, each taking up a byte.
// The first long contains, from high to low end, the subkeys for
// S-boxes 1, 3, 5 & 7; the second contains the subkeys for S-boxes
// 2, 4, 6 & 8 (using the origin-1 S-box numbering in the standard,
// not the origin-0 numbering used elsewhere in this code)
// See comments elsewhere about the pre-rotated values of r and Spbox.
```

```
//
#define F(l,r,key){\
    work = ((r >> 4) | (r << 28)) ^ key[0];\
    l ^= Spbox[6][work & 0x3f];\
    l ^= Spbox[4][(work >> 8) & 0x3f];\
    l ^= Spbox[2][(work >> 16) & 0x3f];\
    l ^= Spbox[0][(work >> 24) & 0x3f];\
    work = r ^ key[1];\
    l ^= Spbox[7][work & 0x3f];\
    l ^= Spbox[5][(work >> 8) & 0x3f];\
    l ^= Spbox[3][(work >> 16) & 0x3f];\
    l ^= Spbox[1][(work >> 24) & 0x3f];\
}

// This is the plain C 3DES process.  It reads the keyschedule and block
// inputs from global variables initialized in the des_producer process
// and does all processing using standard C code.
//
//
void des_c()
{
  int i;
  unsigned int blockCount = 0;
  unsigned char block[8];
  unsigned long left,right,work;
#ifdef IMPULSE_C_TARGET
#ifdef TIMED_TEST
  Xuint32 counter;
#endif
#endif

#ifdef IMPULSE_C_TARGET
#ifdef TIMED_TEST
  XTmrCtr_Reset(&TimerCounter,0);
#endif
#endif

  for (blockCount=0; blockCount<Iterations; blockCount++) {
    for ( i = 0; i < BLOCKSIZE; i++ ) {
      block[i] = Blocks[i];
    }

    // Process the block...
    // Read input block and place in left/right in big-endian order
    //
    left = ((unsigned long)block[0] << 24)
         | ((unsigned long)block[1] << 16)
         | ((unsigned long)block[2] << 8)
         | (unsigned long)block[3];
```

```
        right = ((unsigned long)block[4] << 24)
            | ((unsigned long)block[5] << 16)
            | ((unsigned long)block[6] << 8)
            | (unsigned long)block[7];

//   Hoey's clever initial permutation algorithm, from Outerbridge
//   (see Schneier p 478)
//
//   The convention here is the same as Outerbridge: rotate each
//   register left by 1 bit, i.e., so that "left" contains permuted
//   input bits 2, 3, 4, ... 1 and "right" contains 33, 34, 35, ... 32
//   (using origin-1 numbering as in the FIPS). This allows us to avoid
//   one of the two rotates that would otherwise be required in each of
//   the 16 rounds.
//
        work = ((left >> 4) ^ right) & 0x0f0f0f0f;
        right ^= work;
        left ^= work << 4;
        work = ((left >> 16) ^ right) & 0xffff;
        right ^= work;
        left ^= work << 16;
        work = ((right >> 2) ^ left) & 0x33333333;
        left ^= work;
        right ^= (work << 2);
        work = ((right >> 8) ^ left) & 0xff00ff;
        left ^= work;
        right ^= (work << 8);
        right = (right << 1) | (right >> 31);
        work = (left ^ right) & 0xaaaaaaaa;
        left ^= work;
        right ^= work;
        left = (left << 1) | (left >> 31);

        /* First key */
        F(left,right,Ks[0]);
        F(right,left,Ks[1]);
        F(left,right,Ks[2]);
        F(right,left,Ks[3]);
        F(left,right,Ks[4]);
        F(right,left,Ks[5]);
        F(left,right,Ks[6]);
        F(right,left,Ks[7]);
        F(left,right,Ks[8]);
        F(right,left,Ks[9]);
        F(left,right,Ks[10]);
        F(right,left,Ks[11]);
        F(left,right,Ks[12]);
        F(right,left,Ks[13]);
        F(left,right,Ks[14]);
```

```
F(right,left,Ks[15]);

/* Second key (must be created in opposite mode to first key) */
F(right,left,Ks[16]);
F(left,right,Ks[17]);
F(right,left,Ks[18]);
F(left,right,Ks[19]);
F(right,left,Ks[20]);
F(left,right,Ks[21]);
F(right,left,Ks[22]);
F(left,right,Ks[23]);
F(right,left,Ks[24]);
F(left,right,Ks[25]);
F(right,left,Ks[26]);
F(left,right,Ks[27]);
F(right,left,Ks[28]);
F(left,right,Ks[29]);
F(right,left,Ks[30]);
F(left,right,Ks[31]);

/* Third key */
F(left,right,Ks[32]);
F(right,left,Ks[33]);
F(left,right,Ks[34]);
F(right,left,Ks[35]);
F(left,right,Ks[36]);
F(right,left,Ks[37]);
F(left,right,Ks[38]);
F(right,left,Ks[39]);
F(left,right,Ks[40]);
F(right,left,Ks[41]);
F(left,right,Ks[42]);
F(right,left,Ks[43]);
F(left,right,Ks[44]);
F(right,left,Ks[45]);
F(left,right,Ks[46]);
F(right,left,Ks[47]);

/* Inverse permutation, also from Hoey via Outerbridge and Schneier */
right = (right << 31) | (right >> 1);
work = (left ^ right) & 0xaaaaaaaa;
left ^= work;
right ^= work;
left = (left >> 1) | (left << 31);
work = ((left >> 8) ^ right) & 0xff00ff;
right ^= work;
left ^= work << 8;
work = ((left >> 2) ^ right) & 0x33333333;
right ^= work;
```

```
    left ^= work << 2;
    work = ((right >> 16) ^ left) & 0xffff;
    left ^= work;
    right ^= work << 16;
    work = ((right >> 4) ^ left) & 0x0f0f0f0f;
    left ^= work;
    right ^= work << 4;

    /* Put the block into the output stream with final swap */
    block[0] = (int8) (right >> 24);
    block[1] = (int8) (right >> 16);
    block[2] = (int8) (right >> 8);
    block[3] = (int8) right;
    block[4] = (int8) (left >> 24);
    block[5] = (int8) (left >> 16);
    block[6] = (int8) (left >> 8);
    block[7] = (int8) left;
  }
#ifdef IMPULSE_C_TARGET
#ifdef TIMED_TEST
  counter=XTmrCtr_GetValue(&TimerCounter,0);
  xil_printf("CPU processing done (%d ticks).\n\r",counter);
#else
  xil_printf("CPU processing done.\n\r");
#endif
#endif

  printf("CPU block out:");
  for (i=0; i<BLOCKSIZE; i++) {
    printf(" %02x",block[i]);
  }
  printf("\n\r");
}

int main(int argc, char *argv[])
{
  unsigned char * key = (unsigned char *) "Gflk jqo40978J0dmm$%@878"; /* 24 bytes */
  co_architecture my_arch;
  IF_SIM(int c;)

#ifdef IMPULSE_C_TARGET
#ifdef TIMED_TEST
  XTmrCtr_Initialize(&TimerCounter, XPAR_OPB_TIMER_0_DEVICE_ID);
  XTmrCtr_SetResetValue(&TimerCounter,0,0);
  XTmrCtr_Start(&TimerCounter,0);
#endif
#endif

  printf("Impulse C 3DES DEMO\n\r");
```

```
    des3key(Ks, key, 0);  /* Create a keyschedule for encryption */

    printf("Running encryption test on FPGA ...\n\r");

    my_arch = co_initialize((void *)Iterations);
    co_execute(my_arch);

    printf("Running encryption test on CPU ...\n\r");

    des_c();

    IF_SIM(printf("Press Enter key to continue...\n");)
    IF_SIM(c=getc(stdin);)
    return(0);
}
```

DES.H

```
/* Signal values that indicate which task to do */
#define DES_ENCRYPT   0
#define DES_DECRYPT   1

/* 3DES constants, don't change these */
#define DES_BLOCKSIZE 8             /* unsigned chars per block */
#define DES_KS_DEPTH 48             /* key pairs */
#define SPBOX_X 8
#define SPBOX_Y 64

typedef unsigned long DES_KS[16][2]; /* Single-key DES key schedule */
typedef unsigned long DES3_KS[48][2];      /* Triple-DES key schedule */

/* In deskey.c: */
void deskey(DES_KS,unsigned char *,int);
void des3key(DES3_KS,unsigned char *,int);
```

Image Filter Listings

IMG_HW.C

```
/**************************************************************************
*
* img_hw.c - Image Convolution Hardware Processes
*
* Copyright (c) 2004 by Green Mountain Computing Systems, Inc.
* All rights reserved.
*
**************************************************************************/

#include <stdio.h>
#include "co.h"
#include "cosim_log.h"
#include "img.h"

extern void call_fpga(co_memory imgmem, co_signal start, co_signal end);

void to_stream(co_signal go, co_memory imgmem, co_stream output_stream)
{
    int16 i, j;
    uint32 offset, data, d0;
    uint32 row[IMG_WIDTH / 2];
    IF_SIM(cosim_logwindow log;)
```

```
    IF_SIM(log = cosim_logwindow_create("to_stream");)

    co_signal_wait(go, &data);
    co_stream_open(output_stream, O_WRONLY, INT_TYPE(32));
    offset = 0;
    for ( i = 0; i < IMG_HEIGHT; i++ ) {
            co_memory_readblock(imgmem, offset, row, IMG_WIDTH * sizeof(int16));
            for ( j = 0; j < (IMG_WIDTH / 2); j++ ) {
#pragma CO PIPELINE
                    d0 = row[j];
                    data  = ((d0 >> 8) & 0xf8) << 16;
                    data |= ((d0 >> 3) & 0xf8) << 8;
                    data |= (d0 << 2) & 0xfc;
                    co_stream_write(output_stream, &data, sizeof(int32));
                d0 = d0 >> 16;
                    data  = ((d0 >> 8) & 0xf8) << 16;
                    data |= ((d0 >> 3) & 0xf8) << 8;
                    data |= (d0 << 2) & 0xfc;
                    co_stream_write(output_stream, &data, sizeof(int32));
            }
            offset += IMG_WIDTH * sizeof(int16);
    }
    data = 0;
    co_stream_write(output_stream, &data, sizeof(int32));
    co_stream_write(output_stream, &data, sizeof(int32));
    co_stream_close(output_stream);
}

/* The following process generates a marching three-pixel column in 3 separate
   output streams from the single input stream. This process can be used
   to feed pixels to any marching cubes image processing algorithm using a 3x3
   kernel. It can easily be extended for other kernel sizes. For example,
   given a 5x5 image with the input stream:
    0 1 2 3 4 5 6 7 8 9 10 11 12 13 14 15 16 17 18 19 20 21 22 23 24 25
   this process will generate:
    0  1  2  3  4   5  6  7  8  9   10 11 12 13 14
    5  6  7  8  9   10 11 12 13 14  15 16 17 18 19
    10 11 12 13 14  15 16 17 18 19  20 21 22 23 24

   This process generates 3-pixel columns every 2 cycles.
*/
void prep_run(co_stream input_stream, co_stream r0, co_stream r1, co_stream r2)
{
    int32 i, j;
    int32 B[IMG_WIDTH], C[IMG_WIDTH];
    int32 A01, A02, p02, p12, p22;
    IF_SIM(cosim_logwindow log;)
```

```
        IF_SIM(log = cosim_logwindow_create("prep_run");)

    co_stream_open(input_stream, O_RDONLY, INT_TYPE(32));
    co_stream_open(r0, O_WRONLY, INT_TYPE(32));
    co_stream_open(r1, O_WRONLY, INT_TYPE(32));
    co_stream_open(r2, O_WRONLY, INT_TYPE(32));

    co_stream_read(input_stream, &A01, sizeof(int32));
    co_stream_read(input_stream, &A02, sizeof(int32));

    for ( j = 0; j < IMG_WIDTH; j++ )
            co_stream_read(input_stream, &B[j], sizeof(int32));

    for ( j = 0; j < IMG_WIDTH; j++ )
        co_stream_read(input_stream, &C[j], sizeof(int32));

    co_stream_write(r0, &A01, sizeof(int32));
    co_stream_write(r1, &B[IMG_WIDTH - 2], sizeof(int32));
    co_stream_write(r2, &C[IMG_WIDTH - 2], sizeof(int32));

    co_stream_write(r0, &A02, sizeof(int32));
    co_stream_write(r1, &B[IMG_WIDTH - 1], sizeof(int32));
    co_stream_write(r2, &C[IMG_WIDTH - 1], sizeof(int32));

    for ( i = 2; i < IMG_HEIGHT; i++ ) {
            j = 0;
            do {
                    p02 = B[j];
                    p12 = C[j];
                    co_stream_read(input_stream, &p22, sizeof(int32));
                    co_stream_write(r0, &p02, sizeof(int32));
                    co_stream_write(r1, &p12, sizeof(int32));
                    co_stream_write(r2, &p22, sizeof(int32));
                    B[j] = p12;
                    C[j] = p22;
                    j++;
            } while ( j < IMG_WIDTH );
    }

    co_stream_close(input_stream);
    co_stream_close(r0);
    co_stream_close(r1);
    co_stream_close(r2);
}

#define RED(rgb) ((uint8)((rgb) >> 16))
#define GREEN(rgb) ((uint8)((rgb) >> 8))
#define BLUE(rgb) ((uint8)(rgb))
```

```
/* This process inputs a marching column of pixels from the prep_run process and
   applies a 3x3 convolution to the image to produce an output image on the output
   stream.  A pipeline is utilized to generate one output pixel every two cycles.
   This particular convolution performs edge detection on the image.
*/
void filter_run(co_stream r0, co_stream r1, co_stream r2, co_stream output_stream)
{
  uint32 data,res;
  uint32 p00, p01, p02, p10, p11, p12, p20, p21, p22;
  uint16 d0;
  IF_SIM(cosim_logwindow log;)

  IF_SIM(log = cosim_logwindow_create("filter_run");)

  co_stream_open(r0, O_RDONLY, INT_TYPE(32));
  co_stream_open(r1, O_RDONLY, INT_TYPE(32));
  co_stream_open(r2, O_RDONLY, INT_TYPE(32));
  co_stream_open(output_stream, O_WRONLY, INT_TYPE(32));

  p00 = 0; p01 = 0; p02 = 0;
  p10 = 0; p11 = 0; p12 = 0;
  p20 = 0; p21 = 0; p22 = 0;
  while ( co_stream_read(r0, &data, sizeof(int32)) == co_err_none ) {
#pragma CO PIPELINE
#pragma CO set stageDelay 256
    p00 = p01; p01 = p02;
    p10 = p11; p11 = p12;
    p20 = p21; p21 = p22;

    p02 = data;
    co_stream_read(r1, &p12, sizeof(int32));
    co_stream_read(r2, &p22, sizeof(int32));

    d0 = RED(p11) << 3;
    d0 = d0 - RED(p00);
    d0 = d0 - RED(p01);
    d0 = d0 - RED(p02);
    d0 = d0 - RED(p10);
    d0 = d0 - RED(p12);
    d0 = d0 - RED(p20);
    d0 = d0 - RED(p21);
    d0 = d0 - RED(p22);
    d0 &= (d0 >> 15) - 1;
    res = d0 & 0xff;

    d0 = GREEN(p11) << 3;
    d0 = d0 - GREEN(p00);
    d0 = d0 - GREEN(p01);
    d0 = d0 - GREEN(p02);
```

```
        d0 = d0 - GREEN(p10);
        d0 = d0 - GREEN(p12);
        d0 = d0 - GREEN(p20);
        d0 = d0 - GREEN(p21);
        d0 = d0 - GREEN(p22);
        d0 &= (d0 >> 15) - 1;
        res = (res << 8) | (d0 & 0xff);

        d0 = BLUE(p11) << 3;
        d0 = d0 - BLUE(p00);
        d0 = d0 - BLUE(p01);
        d0 = d0 - BLUE(p02);
        d0 = d0 - BLUE(p10);
        d0 = d0 - BLUE(p12);
        d0 = d0 - BLUE(p20);
        d0 = d0 - BLUE(p21);
        d0 = d0 - BLUE(p22);
        d0 &= (d0 >> 15) - 1;
        res = (res << 8) | (d0 & 0xff);

        co_stream_write(output_stream, &res, sizeof(int32));
    }
    co_stream_close(r0);
    co_stream_close(r1);
    co_stream_close(r2);
    co_stream_close(output_stream);
}

void from_stream(co_stream input_stream, co_memory imgmem, co_signal done)
{
    uint8 err;
    int16 i;
    int32 offset, low, data, d0;
    int32 rowout[IMG_WIDTH / 2];
    IF_SIM(cosim_logwindow log;)

    IF_SIM(log = cosim_logwindow_create("from_stream");)

    co_stream_open(input_stream, O_RDONLY, INT_TYPE(32));
    offset = 0;
    do {
        for ( i = 0; i < (IMG_WIDTH / 2); i++ ) {
#pragma CO PIPELINE
            err = co_stream_read(input_stream, &d0, sizeof(d0));
            if ( err != co_err_none ) break;
            low = (d0 >> 19) & 0x1f;
            low = (low << 5) | ((d0 >> 11) & 0x1f);
            low = (low << 6) | ((d0 >>  2) & 0x3f);
            err = co_stream_read(input_stream, &d0, sizeof(d0));
```

```
                    if ( err != co_err_none) break;
                    data = d0 >> 19;
                    data = (data << 5) | ((d0 >> 11) & 0x1f);
                    data = (data << 6) | ((d0 >>  2) & 0x3f);
                    rowout[i] = (data << 16) | low;
            }
            if ( err != co_err_none) break;
            co_memory_writeblock(imgmem, offset, rowout, IMG_WIDTH * sizeof(int16));
            offset += IMG_WIDTH * sizeof(int16);
        } while ( 1 );
        co_stream_close(input_stream);
        co_signal_post(done, 0);
    }

    void config_img(void *arg)
    {
        int error;
        co_signal startsig, donesig;
        co_memory shrmem;
        co_stream istream, row0, row1, row2, ostream;
        co_process reader, writer;
        co_process cpu_proc, prep_proc, filter_proc;

        startsig = co_signal_create("start");
        donesig  = co_signal_create("done");
        shrmem = co_memory_create("image", "heap0",
                                  IMG_WIDTH * IMG_HEIGHT * sizeof(uint16));
        istream = co_stream_create("istream", INT_TYPE(32), IMG_HEIGHT/2);
        row0 = co_stream_create("row0",   INT_TYPE(32), 4);
        row1= co_stream_create("row1",   INT_TYPE(32), 4);
        row2 = co_stream_create("row2",   INT_TYPE(32), 4);
        ostream = co_stream_create("ostream", INT_TYPE(32), IMG_HEIGHT/2);

        cpu_proc = co_process_create("cpu_proc", (co_function)call_fpga,   3,
                                     shrmem,   startsig, donesig);
        reader = co_process_create("reader",   (co_function)to_stream,  3,
                                   startsig, shrmem,   istream);
        prep_proc = co_process_create("prep_proc", (co_function)prep_run,   4,
                                      istream, row0,    row1,    row2);
        filter_proc = co_process_create("filter",   (co_function)filter_run, 4,
                                        row0,    row1,    row2,    ostream);
        writer = co_process_create("writer",   (co_function)from_stream, 3,
                                   ostream, shrmem,  donesig);

        co_process_config(reader, co_loc, "PE0");
        co_process_config(prep_proc, co_loc, "PE0");
        co_process_config(filter_proc, co_loc, "PE0");
        co_process_config(writer, co_loc, "PE0");
```

```
    IF_SIM(error = cosim_logwindow_init();)
}

co_architecture co_initialize()
{
    return(co_architecture_create("img_arch", "altera_nios2", config_img, NULL));
}
```

IMG_SW.C

```
/****************************************************************************
*
* img_sw.c - Image Convolution Software Processes and Testbench
*
* Copyright (c) 2004 by Green Mountain Computing Systems, Inc.
* All rights reserved.
*
****************************************************************************/

#include <stdio.h>
#include "co.h"
#include "cosim_log.h"
#include "img.h"

#ifndef USE_FPGA
#define USE_FPGA 1
#endif

extern co_architecture co_initialize(void *);

#include "test.h"

static void prn_img(int width, int height, int rowwidth, uint16 *datain, int invert)
{
    int x, y;
    unsigned int data;
    uint16 *bp;

    bp = datain;
    for ( y = 0; y < height; y++ ) {
            for ( x = 0; x < width; x++ ) {
                    data  = ((*bp) >> 8) & 0xf8;
                    data += ((*bp) >> 3) & 0xf8;
                    data += (*(bp++) << 2) & 0xfc;
                    data /= 3;
                    if (invert) data = 255 - data;
                    data = (data & 0x10) ? (data >> 5) + 1 : data >> 5;
```

```
                        switch ( data ) {
                        case 8:
                        case 7:
                                printf("#");
                                break;
                        case 6:
                                printf("&");
                                break;
                        case 5:
                                printf("G");
                                break;
                        case 4:
                                printf("O");
                                break;
                        case 3:
                                printf("o");
                                break;
                        case 2:
                                printf(",");
                                break;
                        case 1:
                                printf(".");
                                break;
                        case 0:
                                printf(" ");
                                break;
                    }
                }
                bp += rowwidth - width;
                printf("\n");
        }
}

void call_fpga(co_memory imgmem, co_signal start, co_signal end)
{
    IF_NSIM(long dur;)
    int i, j, k;
    uint16 *inp, *datain, *dataout;
    uint32 data;
    IF_SIM(cosim_logwindow log;)

    IF_SIM(log = cosim_logwindow_create("call_fpga");)

    datain = co_memory_ptr(imgmem);
    k = 0;
    for ( i = 0; i < IMG_HEIGHT; i++ ) {
            for ( j = 0; j < IMG_WIDTH; j++ ) {
                    if ( (j%32) == 0 ) inp = testimage16 + (i % 32) * 32;
                    datain[k++] = *(inp++);
```

```
          }
      }

      printf("Image in:\n");
      prn_img(32, 32, IMG_WIDTH, datain, 1);

      printf("Running...\n");
      IF_SIM(cosim_logwindow_write(log, "Running...\n");)

      co_signal_post(start, 0);

      co_signal_wait(end, &data);

      printf("Done\n");
      IF_SIM(cosim_logwindow_write(log, "Done\n");)

      dataout = co_memory_ptr(imgmem);

      printf("Image out:\n");
      prn_img(32, 32, IMG_WIDTH, dataout, 0);
}

#if !USE_FPGA
void filter()
{
      IF_NSIM(long dur;)
      int i, j, k;
      uint16 *dp, *datain, *dataout;
      uint16 p00, p01, p02, p10, p11, p12, p20, p21, p22, d0;
      uint32 data;

      dataout = malloc((IMG_HEIGHT + 1) * IMG_WIDTH * sizeof(uint16));
      datain = dataout + IMG_WIDTH;
      k = 0;
      for ( i = 0; i < IMG_HEIGHT; i++ ) {
              for ( j = 0; j < IMG_WIDTH; j++ ) {
                      if ( (j%32) == 0 )
                          dp = testimage16 + (i % 32) * 32;
                      datain[k++] = *(dp++);
              }
      }

      printf("Image in:\n");
      prn_img(32, 32, IMG_WIDTH, datain, 1);

      printf("Running...\n");

      dp = dataout;
      k = IMG_WIDTH;
```

```
for ( i = 1; i < IMG_HEIGHT - 2; i++ ) {
    for ( j = 0; j < IMG_WIDTH; j++ ) {
        p00 = datain[k - (IMG_WIDTH + 1)];
        p01 = datain[k - IMG_WIDTH];
        p02 = datain[k - (IMG_WIDTH - 1)];
        p10 = datain[k - 1];
        p11 = datain[k];
        p12 = datain[k + 1];
        p20 = datain[k + (IMG_WIDTH - 1)];
        p21 = datain[k + IMG_WIDTH];
        p22 = datain[k + (IMG_WIDTH + 1)];
        /* red */
        d0  = ((p11 >> 8) & 0xf8) << 3; /* center * 8 */
        d0 -= (p00 >> 8) & 0xf8;
        d0 -= (p01 >> 8) & 0xf8;
        d0 -= (p02 >> 8) & 0xf8;
        d0 -= (p10 >> 8) & 0xf8;
        d0 -= (p12 >> 8) & 0xf8;
        d0 -= (p20 >> 8) & 0xf8;
        d0 -= (p21 >> 8) & 0xf8;
        d0 -= (p22 >> 8) & 0xf8;
        d0 &= (d0 >> 15) - 1;
        data = (d0 & 0xf8) << 8;
        /* green */
        d0  = ((p11 >> 3) & 0xf8) << 3; /* center * 8 */
        d0 -= (p00 >> 3) & 0xf8;
        d0 -= (p01 >> 3) & 0xf8;
        d0 -= (p02 >> 3) & 0xf8;
        d0 -= (p10 >> 3) & 0xf8;
        d0 -= (p12 >> 3) & 0xf8;
        d0 -= (p20 >> 3) & 0xf8;
        d0 -= (p21 >> 3) & 0xf8;
        d0 -= (p22 >> 3) & 0xf8;
        d0 &= (d0 >> 15) - 1;
        data |= (d0 & 0xf8) << 3;
        /* blue */
        d0  = (p11 & 0x3f) << 5; /* center * 8 */
        d0 -= (p00 & 0x3f) << 2;
        d0 -= (p01 & 0x3f) << 2;
        d0 -= (p02 & 0x3f) << 2;
        d0 -= (p10 & 0x3f) << 2;
        d0 -= (p12 & 0x3f) << 2;
        d0 -= (p20 & 0x3f) << 2;
        d0 -= (p21 & 0x3f) << 2;
        d0 -= (p22 & 0x3f) << 2;
        d0 &= (d0 >> 15) - 1;
        data |= (d0 & 0xfc) >> 2;

        (*dp++) = data;
```

```
                    k++;
            }
    }

    printf("Done\n");

    printf("Image out:\n");
    prn_img(32, 32, IMG_WIDTH, dataout, 0);

}
#endif

int main(int argc, char *argv[])
{
    IF_SIM(int c;)
    co_architecture my_arch;

    printf("Edge Detect Demo\n");
    printf("---------------\n");

#if USE_FPGA
    my_arch = co_initialize(NULL);
    co_execute(my_arch);
#else
    filter();
#endif

    IF_SIM(printf("Press Enter key to continue...\n");)
    IF_SIM(c = getc(stdin);)

    return(0);
}
```

IMG.H

```
/* Defines the image height and width */
#define IMG_WIDTH 512
#define IMG_HEIGHT 512
```

APPENDIX F

Selected References

Ashenden, Peter. *The Designer's Guide to VHDL*, Second Edition. San Francisco, CA: Morgan Kaufmann, 2003.

G. Estrin, B. Bussel, R. Turn, and J. Bibb. *Parallel Processing in a Restructurable Computer System.* Washington, DC: IEEE Computer Society, 1963.

Frigo, Jan, David Palmer, Maya Gokhale, and Marc Popkin-Paine. *Gamma-Ray Pulsar Detection Using Reconfigurable Computing Hardware.* Washington, DC: IEEE Computer Society, 2003.

Gokhale, Maya B., Janice M. Stone, Jeff Arnold, Mirek Kalinowski. *Stream-Oriented FPGA Computing in the Streams-C High Level Language.* Washington, DC: IEEE Computer Society, 2000.

Hoare, C.A.R. *Communicating Sequential Processes.* Upper Saddle River, NJ: Prentice Hall, 1985.

Kernighan, Brian, and Dennis Ritchie. *C Programming Language*, Second Edition. Upper Saddle River, NJ: Prentice Hall, 1988.

Mandelbrot, Benoit B. *The Fractal Geometry of Nature.* New York, NY: W. H. Freeman.

Maxfield, Clive. *The Design Warrior's Guide to FPGAs*. Burlington, MA: Newnes, 2004.

Pellerin, David, and Michael Holley. *Practical Design Using Programmable Logic*. Upper Saddle River, NJ: Prentice Hall, 1991.

Pellerin, David, and Douglas Taylor. *VHDL Made Easy*. Upper Saddle River, NJ: Prentice Hall, 1991.

Schneier, Bruce. *Applied Cryptography: Protocols, Algorithms, and Source Code in C, Second Edition*. New York, NY: John Wiley and Sons.

Welch, P.H. *An OCCAM Approach to Transputer Engineering*. New York, NY: ACM Press, 1988.

Wolf, Wayne. *FPGA-Based System Design*. Upper Saddle River, NJ: Prentice Hall, 2004.

Index

A

ABEL, 21
   design file, 22
abstraction, 8
Actel, 3
Advanced Boolean Expression Language.
See ABEL
algorithm, 5
   acceleration, 25, 41
   concurrency, 5
   partitioning, 5
   validation, 36
Altera, 3, 23
   SOPC Builder , 178
ANSI C, 94, 129
application
   bottleneck, 14
   characteristics, 42
   domains, 41
   monitoring, 14, 98, 158
   parallelism, 14
   prototyping, 11
   structure of, 56
   supercomputing, 301
Application Monitor, 95
arithmetic operation, 129
ARM, 7, 26

array
   data, 130
   splitting, 140-144, 199-201, 227
   two-dimensional, 131
Assisted Technology, 21
Avalon bus, 26, 83

B

bandwidth considerations, 6, 14, 40
Base System Builder, 178
basic blocks, 139
behavioral simulation, 12
behavioral synthesis, 13
Bekker, Scott, 325
Birkner, John, 19
bitmap image file, 213
block, C language, 121
BMP format, 285
buffer width, 64

C

C++, 4
C language, 4, 13
   analysis, 104
   constraints for hardware, 129
   debugger, 123
   expanded source, 121

for hardware design, xv
loop unrolling, 106
optimizer, 127
pointer, 130
preprocessing, 104
programming, 11, 195
semantics, 135
Celoxica, 42
Center for Computational Biology, 326
Chameleon, 28
CHAR_TYPE, 66
Chua, H. T., 19
CISC, 25
clock
    connecting, 185
    dual, 177
    edge, 116
    generation, 109
    operating frequency, 124, 134
    rate, 109
    secondary, 185
    skew, 124
co.h, 51
co_architecture_create, 57, 342-343
co_bit_extract, 343-344
co_bit_extract_u, 344-345
co_bit_insert, 346-347
co_bit_insert_u, 347-348
co_err_already_open, 66
co_err_not_open, 66
co_execute, 52, 54, 348-349
co_initialize, 51, 104, 349-350
co_memory_blockwrite, 221
co_memory_create, 77, 350-351
co_memory_ptr, 351-352
co_memory_readblock, 78, 352-353
co_memory_writeblock, 353-354
co_parameter, 109
co_par_break, 127, 354
co_process_config, 57, 108, 355-356
co_process_create, 56, 62, 356-357
co_register_create, 357-358
co_register_get, 358
co_register_put, 359

co_register_read, 359-360
co_register_write, 360-361
co_signal_create, 361-362
co_signal_post, 362-363
co_signal_wait, 74, 115, 363
co_stream_close, 53, 113, 364
co_stream_create, 57, 64, 365-366
co_stream_eos, 68, 366
co_stream_open, 53, 367
co_stream_read, 368-369
co_stream_read_nb, 296, 369-370
co_stream_write, 53, 67, 370-371
coarse-grained, 37
column generator, 223
Common Universal tool for Programmable
Logic. *See* CUPL
communicating sequential processes.
See CSP
compiler
    Impulse C, 104-108
    processing flow, 103
complex datatypes, 130
concurrency, 5, 12, 38
configuration subroutine, 56, 62
consumer, 55
    process, 53, 96-97
cosim_logwindow_create, 97, 372
cosim_logwindow_fwrite, 96, 372-373
cosim_logwindow_init, 373-374
cosim_logwindow_write, 374
Cray, Inc., 33, 301
CSP, 39-41
    application, 40
    bandwidth considerations, 40
    idealized, 40
    programming model, 42
C-to-hardware, 27
CUPL, 21
cycle-accurate simulation, 14, 123

D

data
    dependencies, 139

bandwidth, 14, 40
Data General Eclipse, 18
Data I/O Corporation, 21
data movement, analyzing, 147
dataflow, 14, 42, 65
datatype, 59
DDR, 317
deadlocks, 69
  and pipelining, 72-73
  avoiding, 69-73
debugging
  hardware, 119-125
DES encryption, 149
design
  methods, 7
  prototyping, 9
desktop simulation, 60, 97
Digital Equipment Corporation, 33
digital signal processing. *See* DSP
diode matrix, 18
direct memory access. *See* DMA
DMA
  as alternative to streaming, 222
  input process, 222-223
  transfer, 223
DSP
  applications, 8, 13
  programmers, 11
dual clock, 185
  generating, 109
dynamically reconfigurable, 4
  computing, 25, 28, 303

E

EDA, 12
edge detection, 210
EDIF netlist, 106
embedded processor, 5, 163
  advantages of, 311-312
  as a test generator, 165-167
  disadvantages of, 312-313
  MicroBlaze, 25, 165, 178
  Nios, 48, 231

PowerPC, 25, 48, 167, 311, 317
  soft versus hard, 310
encryption, DES, 147
end-of-stream, 67
examples
  DES encryption, 147-161, 168-208
  FIR filter, 87-101
  fractal object generation, 280-299
  HelloFPGA, 50-57
  image filter, 209-255, 259-277
  SERDES interface, 326-332
  uClinux, 259-277
exporting
  from the Impulse tools, 183
  generated hardware and software, 262

F

Fast Simplex Link. *See* FSL
fetch-and-execute, 35
field upgrades, 23
field-programmable, 2
  gate array. *See* FPGA
  logic array. *See* FPLA
FIFO, 60, 64, 209, 331
FIR filter
  coefficients, 88
  consumer process, 97
  expanded source code, 121
  hardware generation, 104
  performance, 91
  producer process, 94
  source code, 89
  test bench, 90
  window size, 87
fixed-point, 42
  conversion, 285-286
fixed-width integer, 42
Fletcher, Bryan H., 309
floating point, 130
fMax, 129
FPGA, 2, 17
  as high-performance computer, 302-305
  as parallel computing machine, 27, 35

bitmap, 4, 191, 267
compiler tools, 35
computing, 133-134, 301-307
embedded processor, 310
    MicroBlaze, 25, 165, 178
    Nios, 48, 231
    PowerPC, 25, 48, 167, 311, 317
design philosophy, xvii
history of, 17
netlist, 106
operating frequency, 319
place-and-route, 10, 106
platforms, xix, 5
synthesis, 127
FPLA, 18
fractal
    accuracy of, 282
    geometry, 280
    objects, 280
FSL, 85, 160, 165, 201, 223, 259
    connections, 266
function table, 19
fusible link, 18

G

gate array, 2
gate delays, 124, 127
gcc, 45
General Electric, 18
Generate Options dialog, 262
generic, VHDL, 109
geophysics, 41
Gokhale, Maya, 46
grid computing, 326, 332

H

hand-crafted HDL, 13
Handel-C, 42
hardware, 106
    acceleration, 311, 312, 321-322
    accelerator, 163
    analysis, 159
    compiler, 13, 129

description, 14
description language. *See* HDL
development, 11
engineer, 12
generation, 125, 156-159
platform, 4
process, 41, 58, 59-63
    synchronization of, 59, 65
    communication with, 51
    constraints of, 58
prototype, 160
simulation, 91
synthesis, 159
hardware/software
    applications, 15
    interface, 14
    partitioning, 12
    solution, 13
Harris Semiconductor, 17
HDL, 4
    generation from C, 108-109
    simulation, 116
    top-level module, 108
Hello FPGA, 50
heterogeneous parallelism, 37
high-performance computing, 2
history
    of programmable platforms, 17
Hoare, Anthony, 39
Hyperterminal, 269

I

IBM, 17
IDE, 156
image data
    reading and writing, 213
image filter, 210
    partitioning, 219
image processing, 41
Impulse C, 45
    and ANSI C, 42
    datatypes, 59
    for streaming applications, 42

library, 49, 341-373
  co_architecture_create, 57, 342-343
  co_bit_extract, 343-344
  co_bit_extract_u, 344-345
  co_bit_insert, 346-347
  co_bit_insert_u, 347-348
  co_err_already_open, 66
  co_err_not_open, 66
  co_execute, 52, 54, 348-349
  co_initialize, 51, 104, 349-350
  co_memory_blockwrite, 221
  co_memory_create, 77, 350-351
  co_memory_ptr, 351-352
  co_memory_readblock, 78, 352-353
  co_memory_writeblock, 353-354
  co_parameter, 109
  co_par_break, 354
  co_process_config, 57, 108, 355-356
  co_process_create, 56, 62, 356-357
  co_register_create, 357-358
  co_register_get, 358-359
  co_register_put, 359
  co_register_read, 360
  co_register_write, 360-361
  co_signal_create, 361-362
  co_signal_post, 362-363
  co_signal_wait, 74, 115
  co_stream_close, 53, 113, 364
  co_stream_create, 57, 64, 365-366
  co_stream_eos, 68, 366
  co_stream_open, 53, 367-368
  co_stream_read, 368-369
  co_stream_read_nb, 296, 369-370
  co_stream_write, 53, 67, 370-371
  minimal program, 50-57
  motivation behind, 47
  origin, 46
  parameter, 63
  programming model, 42, 48-50
  simulation library, 60
inline function, 129
INMOS, 34
input rate, 126
input stream, 67

instruction scheduling, 47, 128, 135-136
instruction stages, 139
instruction-set simulators, 14
in-system debugging, 10
INT_TYPE, 66
integer
  datatypes, 129
  division, 129
integrated development environment. *See*
IDE
Intel, 17
  4004, 17
  8051, 25
Intersil Corporation, 18
IP core, 25
IPFlex, 28

J

Java, 4
JTAG, 108, 159

K

Kernighan, Brian, 50

L

language-based design, 12
latency, pipeline, 126
Lattice, 3
legacy
  algorithm, 36
  C code, 147
  programming, 14
Linux. *See* uClinux operating system
load balancing, 280
locality, 40
log window
  creating, 99
  initializing, 99
  writing to, 99
logic
  equation, 20
  synthesis, 23, 106

loop
    considerations, 199
    pipelining, 47, 106, 125, 204-207
    unrolling, 126, 141-142, 203-204
Los Alamos National Laboratories, 46

M

machine model, 40
    MIMD, 33
    shared memory, 34
    SIMD, 33
    SISD, 32
    von Neumann, 32
macro interface, stream, 174-175
main function, 51
Mandelbrot
    image generation, 279-299
    set, 279
Mandelbrot, Benoit, 280
Memec, 156, 169, 259, 309
    V2MB1000 board, 169, 259
memory, 47, 58. *See also* shared memory
    access, impacts of, 128
    accessing, 200
    block read and write, 139
    controller, 312-313
    for data communication, 78
    embedded, 81
    external, 81, 310, 317
    reducing accesses to, 139
    usage, 316
message passing interface. *See* MPI
MicroBlaze, 25, 165, 178-193, 201, 223, 315
    Development Kit, 156
    stream performance, 85
MIMD, 33
MinGW library, 94
mixed processor design, 7
model
    machine, 40
    programming, 31, 32, 47
        communicating processes, 41
        Impulse C, 41-43

streams-oriented, 42
monitoring, 99
    functions, 63
Monolithic Memories, 19
Montana State University, 326
Moore's Law, 25
Morphics, 28
MPI, 38
multicomputer, 33
multitasking, 32
multithreaded, 32

N

National Institutes of Health, 326
National Science Foundation, 326
National Semiconductor
    DS92LV16, 326
neural data, 326
Nios, 48
    stream performance, 83
Nios II, 25, 231

O

O_RDONLY, 66, 115
O_WRONLY, 67, 114
obsolescence mitigation, 311
Occam, 34, 36
on-chip interface, 26
On-chip Peripheral Bus. *See* OPB
OPB, 26, 84, 317
    timer, 187, 188
operating frequency, maximum, 124
operating system
    embedded, 257-277
    uClinux, 257-277
optimization, 137-139
    C code, 195
    expression-level, 137-139
    level, 315
    through experimentation, 323
    within basic blocks, 139
optimizer
    operation, 135-139

Stage Master, 135
output streams, 66

P

PACT, 28
PAL, 19
    Assembler, 19
PALASM, 19
parallel
    computing, 27
    processes, 31
    programming, xvii, 15
parallelism
    coarse-grained, 36
    extreme levels of, 302
    programming for, 36, 38
    spatial, 303
    statement-level, 87, 133-145, 290
    system-level, 87, 219, 290
partitioning, 14
    system-level, 219, 290
PCI, 307
Pentium, 7
peripheral
    integration, 25
Photoshop, 307
PIC processor, 25
picoChip, 28
pipeline
    generation, 136-137, 206
    goal, 205-208
    hardware size, 207
    loop, 204
    performance, 145, 207
    rate, 144-145
    system-level, 113, 219-231
PIPELINE pragma, 106
pipelining, 106, 142-145
pixel stream, 210
place-and-route, 10, 106
platform, 4
    FPGA-based, xix, 4-5
    selection, 169-170

Platform Studio, 26, 178, 264
Platform Support Package, 157, 261
PLB, 84
Posix, 60
potassium channels, 326
PowerPC, 25, 48, 167, 311, 317
pragma
    PIPELINE, 106, 125, 136, 214
    StageDelay, 126, 139
    UNROLL, 125, 137, 203, 214
process, 41, 58-63
    synchronization of, 12, 59, 65
    understanding, 59-63
    run function, 53
processing
    elements (PEs), 38
    machine, 38
    model, 32
processor
    as test generator, 37
    benchmarks, 313
    considerations, 167
    core, 25
    embedded, 40, 309-323
    hard core, 167
    MicroBlaze, 25, 165, 178
    performance, 313
    peripherals, 312-313
    soft, 37, 310
Processor Local Bus. *See* PLB
producer, 51
    process, 51, 94-96
programmable
    array logic. *See* PAL
    hardware platform, 4, 26
    logic, 2, 17
        origins of, 18-23
programming
    abstraction, 12-16
    model, 31, 32, 47
        communicating processes, 41
        Impulse C, 41-43
        streams-oriented, 42
prototyping, 7, 9

of hardware, 16
rapid, 27

Q

Queensland, University of, 258
Quicklogic, 3
Quicksilver, 28

R

Radiation, Inc., 18
rapid prototyping, 27
rate
  introduction, 126
  pipeline, 126
reconfigurable computing, 25, 28, 303
recursive function call, 129
register
  overview of, 74-76
Register-Transfer-Logic. *See* RTL
reprogrammability, 23
reset, 109
RISC, 25, 26
Ritchie, Dennis, 50
RTL, 12, 49
  design, 9
  simulation, 168

S

SDRAM, 317
secondary clock, 185
SERDES
  for data streaming, 327-329
  handshaking, 328
  initializing, 328
  synchronization pattern, 328
  transceiver, 329
serial interface, 326-327
serializer/deserializer. *See* SERDES
shared memory, 49, 76
  performance considerations, 81-86
  using, 76-78
shift operand, 130

signal
  creating, 73-74
  interface, 113
  overview of, 73-74
  posting, 74
  value, 74
  wait mode, 115
  waiting for, 74
signed type, 59
Signetics Corporation, 18
Silicon Graphics, 301
SIMD, 33
simulation
  consumer process, 53
  cycle-accurate, 123
  DES encryption, 155-156
  desktop, 60
  hardware, 116
  library, 60
  producer process, 51
  software, 155-156
  source-level debugging, 121
  test bench, 50
  tools, 10
  VHDL, 116
simulator
  hardware, 168
SISD, 32
Snider, Dr. Ross, 326
soft processor, 37, 48
software
  acceleration, xix
  process, 106
  simulation, 155-156
  test bench, 50
software-based methods, 9
solution space, 12
SOPC Builder, 26, 178
SP box, 149, 200
Spartan-3, 317
spatial parallelism, 303
SRAM, 317
stage, 121
  delay, 126

Stage Master, 126, 127
StageDelay pragma, 126, 139, 142
standard processor, 8
state machine, 18
    generated from C, 120
    SERDES interface, 330
stdio.h, 51
stream, 63-66
    closing, 66
    custom interface, 325-332
    datatype, 67
    deadlocks, 69
    for input, 67
    for output, 66
    hardware, 64
    I/O, 65-68
    interface, 43, 109, 112
    macro interfaces, 174-175
    mode
        read, 115
        write, 114
    nonblocking, 71
    opening, 54
    overview of, 63-66
    parameters, 109
    performance, 201-202
        considerations, 81-86
    protocol, 113
    read mode, 115
    reading, 68
    write mode, 114
Streams-C, 46
struct, 130
structured ASIC, 7
SUIF, 46
supercomputing, 301
synchronization
    pattern, 328
    process, 12, 59, 65
synthesis, 58
system
    architect, 5
    integration, 5, 306
    on a programmable chip, 25-27, 37

SystemC, 15, 94
system-level pipeline, 113, 219-231

T

target platform
    limitations of, 103
technology mapping, 106
test bench, 16, 97, 155
    embedded, 163-194
test fixture, 97
test generator, 165
test vector, 20
Texas Instruments, 17
TFTP, 272
Thinking Machines, 33
thread programming, 60, 63
threads, 38
TIFF format, 259
timed C, 61
tools, role of, 8-9
Transputer, 33, 34
TTL device, 19

U

UART, 25
uClinux operating system
    FTP client, 259, 274
    kernel image, 269
    overview, 257-259
    RAM disk, 259
UINT_TYPE, 66
union, 130
unit test, 165
UNROLL pragma, 106, 203
unsigned type, 59
untimed C, 61

V

V2MB1000 board, 169
value engineering, 7
VAX 11/750, 18
Verilog, xviii, 4, 15

VHDL, xviii, 4, 15, 49
video stream, 210
Virtex-4, 167
Virtex-II, 156, 326
Virtex-II Pro, 167
Visual Studio, 45, 149, 157
von Neumann, John, 32

W

Williams, John, 258
Windows
    bitmap format file, 282
    GDI, 213
wireless communications, 41

X

xil_printf, 315
Xilinx, 23
    EDK tools, 178, 264
    MicroBlaze, 25, 48, 165, 178-193, 201, 223
    Platform Studio, 262
    Spartan-3, 317
    Virtex-4, 167
    Virtex-II, 156, 326
    Virtex-II Pro, 167
    Xygwin shell, 271

Z

Zilog Z-80, 25